Sustainable Energy and the Environment

Sustainable Energy and the Environment

Editor: Ted Weyland

www.callistoreference.com

Callisto Reference,
118-35 Queens Blvd., Suite 400,
Forest Hills, NY 11375, USA

Visit us on the World Wide Web at:
www.callistoreference.com

© Callisto Reference, 2017

ISBN: 978-1-63239-866-6 (Hardback)

The publisher's policy is to use permanent paper from mills that operate a sustainable forestry policy. Furthermore, the publisher ensures that the text paper and cover boards used have met acceptable environmental accreditation standards.

Trademark Notice: Registered trademark of products or corporate names are used only for explanation and identification without intent to infringe.

Printed in the United States of America.

Cataloging-in-publication Data

Sustainable energy and the environment / edited by Ted Weyland.
    p. cm.
Includes bibliographical references and index.
ISBN 978-1-63239-866-6
1. Renewable energy sources. 2. Environmental protection. 3. Sustainable development. I. Weyland, Ted.
TJ808 .S66 2017
621.042--dc23

Table of Contents

Preface

Energy is required to do the most basic to the most complex actions in today's world. Exploitation of resources can cause severe energy crisis in the coming years. Therefore, sustainable use of energy is the need of the hour. The different alternatives to fossil-fuel based energy are hydroelectricity, wind energy, solar power, bioenergy, tidal power, geothermal power, etc. This book will provide knowledgeable insights into the field of sustainable energy and its positive impacts on environment. Some of the diverse topics covered in it will address the varied branches that fall under this category. As this field is emerging at a rapid pace, the contents of this text will help the readers understand the modern concepts and applications of the subject. Students, researchers, experts and all associated with sustainable energy and environmental conservation will benefit alike from the book.

This book is a result of research of several months to collate the most relevant data in the field.

When I was approached with the idea of this book and the proposal to edit it, I was overwhelmed. It gave me an opportunity to reach out to all those who share a common interest with me in this field. I had 3 main parameters for editing this text:

1. Accuracy – The data and information provided in this book should be up-to-date and valuable to the readers.

2. Structure – The data must be presented in a structured format for easy understanding and better grasping of the readers.

3. Universal Approach – This book not only targets students but also experts and innovators in the field, thus my aim was to present topics which are of use to all.

Thus, it took me a couple of months to finish the editing of this book.

I would like to make a special mention of my publisher who considered me worthy of this opportunity and also supported me throughout the editing process. I would also like to thank the editing team at the back-end who extended their help whenever required.

Editor

Energy in Perspective of Sustainable Development in Nigeria

Sunday Olayinka Oyedepo[*]

Mechanical Engineering Department, Covenant University, Ota
*Corresponding author: Sunday.oyedepo@covenantuniversity.edu.ng

Abstract Sustainable energy systems are necessary to save the natural resources avoiding environmental impacts which would compromise the development of future generations. Delivering sustainable energy will require an increased efficiency of the generation process including the demand side. This paper reviews the pattern of energy-use in Nigeria and makes a case for the implementation of an energy efficiency policy as a possible strategy to address the nation's energy crisis. The study as well explores the role of industrial energy use in sustainable development in Nigeria and the potential sources to increase energy efficiency in industrial sector. The study showed that the pattern of electrical energy consumption in the industries reviewed was majorly from generating set while power supply from national grid compliment generating set if available; this is due to either low voltage or epileptic power supply from national grid. Direct and indirect sources that lead to electrical energy waste and in-efficient energy utilization in the industries were identified such as energy loss as a result of aging electric motor, worn out or slack / misaligned machine parts, excessive heating and cooling, use of low efficient lightings etc. The review paper shows that industrial energy efficiency in Nigeria is a readily achievable, cost effective and has potential of reduction in industrial consumption using good energy management practices and energy efficient equipment. This study will be of help to government, industrialists and industrial policy makers.

Keywords: energy, energy efficiency, sustainable development, energy management, industry

1. Introduction

Energy has a major impact on every aspect of our socio-economic life. It plays a vital role in the economic, social and political development of our nation. Inadequate supply of energy restricts socio-economic activities, limits economic growth and adversely affects the quality of life. Improvements in standards of living are manifested in increased food production, increased industrial output, the provision of efficient transportation, adequate shelter, healthcare and other human services. These will require increased energy consumption. Thus, our future energy requirements will continue to grow with increase in living standards, industrialization and a host of other socio-economic factors [1].

As the very basis of development, energy use is closely related to the level of productivity in the industry, commerce, agriculture and even in office activities. Energy consumption per capita is one of the indicators or benchmarks for measuring the standard of living of a people or nation. The unprecedented use of energy which began with the industrial revolution certainly brought about massive increase in productivity and change in lifestyle. Since then energy demand has been in the increase- to produce more products, travel further and faster or to be more comfortable. Physically, energy is

defined as the capacity for doing work. The capacities of energy to do work are inherent properties of energy carriers. Although energy cannot be created nor destroyed according to classical thermodynamics, its capacity for doing work can be degraded and destroyed due to system irreversibility in line with the logic of the second law of thermodynamics [2].

Some of the common energy carriers or sources are coal, petroleum, natural gas, nuclear fuels, biomass etc. Of all these, the most widely used energy sources are the hydrocarbon compounds or fossil fuels which account for more than 80% of global primary energy consumption [3]. For instance, fossil energies provide about 67% of the energy needed to produce electricity - a veritable and the most terminal form of energy for transmission and distribution for industrial production processes [4]. Energy usage has become an important concern in the past years and there has been growth awareness and an increase in taking personal responsibilities in preventing environmental pollution by minimizing energy waste. Energy has been the key to economic development worldwide, but in the way it is sourced, produced and used, two major drawbacks have emerged. First, the overall energy system has been very inefficient. And second, major environmental and social problems, both local and global, have been associated with the energy system [5]. Climate change and environmental externalities associated with energy consumption have become a major international issue. It has been observed that among the various sectors contributing to green house gas (GHG) emissions, industrial sector contribution was significant; thus mitigating GHG emissions from the sector offers one of the best ways of confronting the climate change problem. Energy efficiency is a major key in this regard. An estimated 10-30% reduction can be achieved at little or no cost by improving efficiency of energy use in the industry [6].

Although Nigeria is relatively endowed with abundant fossil fuels and other renewable energy sources, the energy situation in the country is yet to be structured and managed in such a way as to ensure sustainable energy development, most especially in the industrial sector. Nigeria as a nation is passing through a serious energy crisis and it has been even more affected not by a lack of energy resources, but largely due to poor resource and financial management, a crippling dependence on imports particularly second-hand goods built with out-dated, inefficient technology etc [7]. As a nation that has limited technological capacity but sees industrialization as constituting a crucial leverage and pre-condition for meaningful development, Nigeria should be wise enough to manage her scarce energy resources judiciously.

The use of energy pervades every aspect of modern society but it is not efficiently used in many industries. In view of the fact that there is an incessant increase in fuel costs, energy efficiency studies are thus rapidly becoming more important. Several millions of dollar can be saved in accumulated energy cost when energy is properly managed. Based on this fact, several researchers have reported on the energy consumption, conservation potential and environmental impact of energy use of different industrial process operations both within and outside Nigeria. Nagesha [8] presented the energy consumption pattern in a textile dyeing industrial cluster and environmental implications in terms of emission of GHGs due to energy use. The study identified substantial scope for energy efficiency and analysed energy consumption in the cluster from an economic perspective. All the economic performance indicators adopted in the study seemed to have significant association with energy efficiency in the cluster. Also, it was observed that the small scale industries which are energy efficient performed better on the economic front and experienced 'higher returns to scale'. The study concluded that the firms in the energy intensive product clusters must aim at enhancing their energy efficiency as it leads to multiple benefits and ensures sustainable development in the long run. Fawkes [9] investigated energy efficiency in South African Industry. This study showed that strong incentives exist for energy efficiency improvement in South African industry, in particular, the potential for increasing profit, the need to reduce greenhouse gas (GHG) emissions, the need to maintain economic competitiveness, and the need to delay the cost of new peak-load electricity generation facilities. In their study, Lung et al [10] investigated the impacts that several emerging technologies have had in the U.S food processing industry. This paper assessed the energy efficiency potential for four of these technologies in the U.S. food processing industry. Based on the assessments of these four emerging and newly commercialized technologies, the potential for energy savings in the U.S. food industry is quite strong. In

addition, these technologies have yielded important productivity and other benefits. Depending on the available market portions in which these technologies can be implemented, sector-wide energy savings could range from 1572 GJ and 134 million kWh to 2342 GJ and 186 million kWh. In addition, non-energy benefits such as improved product quality, better production and reduced greenhouse gas emissions are likely. Aiyedun et al [11] assessed the energy efficiency in Nigerian Eagle Flour Mills Limited, Ibadan. The study which is limited based on the available years of data collected (1996-2000) analyzed the energy consumption, productivity and efficiency of the company. The results of the study showed that energy is not quite efficiently utilized in this industry because the energy productivity increased substantially from 0.369 $MJkg^{-1}$ in 1996 to 0.716 $MJkg^{-1}$ in the year 2000. An average of 47,810.59 GJ of energy was consumed annually within this period with 44.68%, 0.23%, 42.16% and 12.93% of this energy accruing from electricity, lubricants, diesel and petrol, respectively. The average energy productivity, the average intensity of energy and the average cost of energy input per unit kg are 0.527 $MJkg^{-1}$, 1.084 GJm^{-2} and 28 kobo/kg, respectively. The average value of the normalized performance indicator (NPI) obtained is 0.199 GJm^{-2} which indicates substantial energy consumption for the building type. The areas where the industry uses and wastes energy, and where actions for energy conservation can be implemented were identified.

Aderemi, et al [12] examined the pattern of energy consumption in selected food companies in South-western Nigeria; identified the sources of electrical energy waste and assessed the effectiveness of the strategies for electrical energy savings in the industry. Four sub-sectors of food and drinks industry in the category of Small and Medium Enterprises were examined. They include; beverage, bakery and confectionery, grain mills and storage of cold food products. The study revealed that the pattern of electrical energy consumption in the food companies was mainly from generating set; this was due to either low voltage or epileptic power supply from national grid. Also, the study identified 12 direct sources that lead to electrical energy waste and inefficient energy utilization in the food industry. One of these, among others was the energy loss as a result of worn out or slack / misaligned belts that needed timely replacement or tensioning. Other indirect sources identified include lack of training and retraining of staff, power factor of electrical equipment, and equipment age, among others. In their study, Noah et al [13] carried out a comprehensive energy audit of Vitamalt Nigeria Plc, Agbara using portable thermal and electrical instruments with the objective of studying the pattern of energy consumption and identifying the possibilities of saving energy in the plant. A five year (2000-2004) data on energy consumption of Vitamalt Nig. Plc was collected and analysed. The study showed that the Normalized performance indicator (NPI) calculated over the span of five years gave an average of 1.2 GJ/m^2 indicating a fair range in energy performance level classification (1.0 - 1.2) while significant savings and improvement in energy usage is achievable. The authors concluded that maximizing efficiency of existing system, optimizing energy input requirement and significant capital investment in procuring new energy conserving equipment must be made for the energy performance level to fall into a good range classification (less than 0.8).

The increasing role of energy efficiency as a catalyst for sustainable industrial development is realism in the industrialized countries of the world. In Nigeria the story is different at the moment as the huge benefits derivable from adoption of energy efficiency and conservation measures by industries remain largely untapped due largely to lack of awareness of the economic and social benefits of energy efficiency measures. This, in addition to high incidence of power outages resulting to large scale use of own power generation and lack of investment capital have given rise to high specific energy content of goods produced by industries in Nigeria. The cumulative effect is loss of competitive edge in the global market by these industries and low after- tax returns. This constitutes a major disincentive to investment and sustainable industrial growth. As a matter of utmost importance, industries in Nigeria should take advantage of opportunities in low level, low risk but high worth energy efficient measures that reduces the bottom line of any business enterprise. In so doing, a lead time will be created to pursue high-tech driven production processes that will find support at maturity in an already established energy efficient culture.

As earlier presented of the previous works on energy efficiency in Nigerian industries, none of these researchers discussed possible ways of achieving sustainable development in Nigeria in perspective of effective utilization of energy in manufacturing industries in the country. Therefore, the prime objectives of this paper are (i) to review energy consumption pattern in Nigeria (ii) to explore the potential sources to increase energy efficiency in industrial sector in Nigeria and (iii) to explore the role of industrial energy use in achieving sustainable development in Nigeria.

2. Energy Situation in Nigeria

Nigeria has an abundant supply of natural energy sources, both fossil and renewable. Energy plays a double role in Nigeria's economy: as an input into all economic activities and as the mainstay of Nigeria's foreign exchange earnings through the export of crude oil and, more recently, from increasing natural gas exports. Nigeria's economy is heavily dependent on the oil sector and now on gas too, since both together account for 90-95% of export revenues, over 90% of foreign exchange earnings and nearly 80% of government revenues. The majority of Nigeria's exports of crude are destined for markets in the United States and Western Europe with Asia becoming an increasingly important market of late.

The National energy is at present almost entirely dependent on fossil fuels and firewood which are depleting fast. According to Chendo [14], recent estimates indicated that the reserve for crude oil stood at about 23 billion barrels in 1998, natural gas 4293 billion m^3 at the beginning of 1999, made up of 53% associated gas and 47% non associated gas. Coal and lignite stood at 2.7 billion tones, Tar sands at 31 billion barrels of oil equivalent and large-scale hydropower at 10,000 MW. Table 1 and Table 2 show various conventional and non-conventional energy sources and their estimated reserves in Nigeria [15].

Table 1. Nigeria's conventional energy resources

Resources	Reserve	Resources in Energy units (billion tonnes)	%Total conventional energy
Crude oil	23 billion barrels	3.128	21.0
Natural gas	4293 billion m^3	3.679	24.8
Coal and lignite	2.7 billion tonnes	1.882	12.7
Tar sands	31 billion barrels of Oil equivalent	4.216	28.4
Hydropower	10, 000MW	1.954(100yrs)	13.1
Total	Conventional/Commercial Energy resources	14.859	100%

Source: Ref [15]

Table 2. Nigeria's non conventional energy resources

Resources	Reserves	Reserves (billion tonnes)
Fuel wood	43.3 million tonnes	1.6645 (over 100 years)
Animal wastes		
And crop residue	144 million tonnes	3.024 (over 100 years)
Small scale hydropower	734.2 MW	0.143 (over 100 years)
Solar radiation	1.0 kWm^{-2} Land area (peak)	-
Wind	2.0-4.0 ms^{-1}	-

Source: Ref. [15]

The current energy supply mix in the country is characterised by Figure 1. This further confirms the fact that presently, renewable-energy use in the country is split essentially between hydroelectricity and traditional fuel wood [16].

From the energy point of view, the Nigeria economy can be disaggregated into industry, transport, commercial, household and agricultural sectors. However, the household sector presently dominated energy consumption in Nigeria. This makes it the most important energy sector of the Nigeria economy [17]. Figure 2 shows sectoral distribution of National Final Energy Consumption.

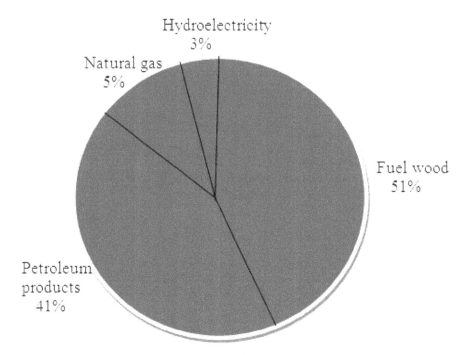

Figure 1. Typical energy supply mix in Nigeria (*Source: Ref. [16]*)

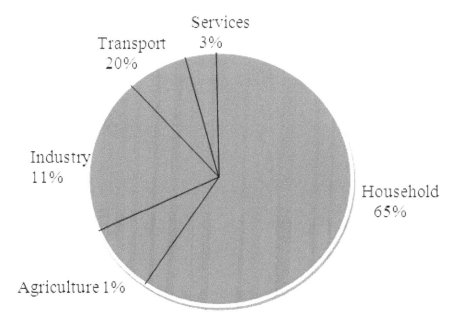

Figure 2. Sectoral distribution of national final energy consumption (PJ) (*Source: Ref. [17]*)

2.1. Energy Consumption Pattern in Nigeria from the Pre Colonial Till Date

The forms of energy consumed in Nigeria have increasingly diversified with innovations in science and technology. In the earliest times, like everywhere else in the primitive world, energy was consumed in the primary renewable form, essentially as biomass in the form of wood fuel and solar energy. Later, at the turn of the present century, primary non-renewable energy forms were introduced from fossil fuels. Only coal was used initially, but later, petroleum products and natural gas were included. These primary forms, which were used mainly for transportation, dominated the energy scene for several years until the principal secondary form, electricity, was introduced.

2.1.1. Coal Production and Consumption

Coal was the first fossil fuel to be discovered and used in Nigeria. Oil was discovered about 47 years after discovery of coal [18]. Coal production in Nigeria started in 1916 with an output of 24,500 tonnes for that year. Production rose to a peak of 905,000 tonnes in the 1958/59 year with a contribution of over 70% to commercial quantities in 1958 and the conversion of railway engines from coal to diesel, production of coal fell from the beginning of the sixties of only 52,700 tonnes in 1983 [19].

During the civil war years of 1967-1969, production stood at only 20,400-35,000 t. This rose to 323,000 t in 1972 and progressively declined to 118,000 t in 1980. The estimated cumulative production between 1916 and 1980 is about 25.3 million metric tonnes. In 1980 coal contributed less than 1% to commercial energy consumption in the country as compared to 70% for oil, 25% for natural gas and about 5% for electricity.

Over 95% of the Nigerian coal production has been consumed locally, chiefly for railway transportation, electricity production, and industrial heating in cement production lines. Between 1952 and 1958, coal consumption by the Nigerian Railway Corporation accounted for about 60% of the overall consumption. Due to its diesel conversion programme, commenced in 1966, its share of coal consumption fell progressively to <30% in 1966, and to an insignificant level in 1986. The other major consumer, the National Electric Power Authority, NEPA (now the Power Holding Company of Nigeria Plc, PHCN), had about a 16-30% share of the total consumption between 1952 and 1966. As a result of the civil war, its sole coal fired electricity generating station was put into disuse. Consequently, its coal consumption has been insignificant since 1970. The cumulative effect of the decline in demand by the two major consumers, together with the ease and cost-effectiveness associated with the use of other energy sources, resulted in the edging out of coal in the national energy scheme [20].

2.1.2. Petroleum Production and Consumption

Petroleum exploration in Nigeria witnessed steady growth over the past few years. Proven recoverable reserves of crude oil amount to about 1.48×10^6 billion tonnes, with commercial production commencing in 1958. The production then was only 3.1 million metric tonnes, but rapidly increasing market demand forced this to rise to 20.3 million tonnes in 1960, 54.2 million tonnes in 1970, and 104.1 million tonnes in 1980. The observed increases in production have been determined by external market forces, rather than increased local demand. The latter rose from only 1.3 million tonnes in 1970 to barely 6.5 million tonnes / year in 1980. The average domestic consumption to the total production standing at an average value of 3% within the period. Thus, some 97% of the total production is exported, usually as crude oil. The local consumption of petroleum products is supplied from three refineries with a total combined capacity of about 13.5 million tonnes/year since 1980. Supply for the petroleum products up till 1980 was supplemented by imports, while 30% of the output of residual oil was exported. Based on various oil prospects already identified especially in the deepwater terrain and the current development efforts, it is projected that proven reserves will reach about 40 billion barrels by year 2020 and potentially 68 billion barrels by year 2030. Oil production in the country also increased steadily over the years; however, the rate of increase is dependent on economic and geopolitics in both producing and consuming countries. Nigeria's current production capacity is about 2.4 million barrels per day even though actual production is averaging around 2.4 million barrels per day partly due to the problems in the Niger Delta and

OPEC production restriction. Average daily production is projected to increase to 4.0 million barrels per day by 2010 and potentially to over 5.0 million per day in year 2030 [21]. The consumption of petroleum products stood between 80 and 90% of the total commercial energy consumption over the 13 yr from 1971 to 1984. The growth rate over the period averaged about 18%, with gasoline 22%, kerosene 17% and diesel 16%. Gasoline and gas oils are mainly used for transport (77%), household uses (12%) and industrial/commercial operations (11%). Half of the household consumption was used for operating standby electricity generators.

2.1.3. Production and Consumption of Natural Gas

The Nigerian reserves of natural gas are estimated at 4.67×10^{12} m^3 at a mean specific volume of 1.56×10^{-3} m^3/kg, a mean gauge pressure of about 12 bar and a calorific value of 35 MJ/m^3 [22]. The current production rate stands at about 1.8×10^{10} m^3/yr, usually as associated gas. About 8.5 m^3 of gas can be expected per barrel of crude petroleum. Since the infrastructure for gas utilization is underdeveloped in Nigeria, as high as 75% of the gas produced was being flared in the past. However, gas flaring was reduced to about 36% as a result of strident efforts by the Government to monetize natural gas. Domestic utilization of Natural gas is mainly for power generation which accounted for over 80% while the remaining are in the industrial sector and very negligible in the household sector. Given the current reserves and rate of exploitation, the expected life-span of Nigerian crude oil is about 44 years, based on about 2mb/d production, while that for natural gas is about 88 years, based on the 2005 production rate of 5.84 bscf/day [20].

2.1.4. Production and Consumption of Wood Fuel

Wood fuel contributed about 5% of the fuel consumption of the Electricity Corporation of Nigeria (Now Power Holding Company Nigeria (PHCN)) in 1952/1953, but this decreased gradually to zero by 1960 [23]. Small quantities were also used for rail transport during the same period. The most significant use of fuel wood, however, is for domestic cooking and baking and heating in small-scale industries such as bakeries and brickworks. Presently, the largest sources of fuel wood are communal bushes and private farmlands, from whence fire wood is fetched freely or at a small fee. Between 1961 and 1973, the fuel wood consumption increased by 37% from 41.5 million m^3 in round wood equivalent to 56.8 million m^3 [20]. The 1973 consumption amounted to about 9.1×10^{11} MJ, or an equivalent of 12.1 million tonnes of oil [14]. The projected consumption for 1985 and 2000 are, respectively, 17.8 and 23.6 million tonnes of oil equivalent.

2.1.5. Electricity Production and Consumption

Electricity generation, distribution and consumption became noticeable in Nigeria in the 1930s. By 1961, the total installed capacity was about 185 MW, which increased to 805 MW by 1970 and 2800 MW by 1983 [24]. Small generating units of < 16 MW accounted for about 82% of the total installed capacity in 1965, but production is now heavily dominated by large hydroelectric schemes and gas or oil fired stations. The lone coal fired steam turbine at Oji has a capacity of 30 MW, but has not been reactivated after the 1967-1970 civil war. In spite of the high installed capacity, the average power is only 1400 MW, while average daily production is 1200 MW with typical availability of about 40% of the installed capacity between 1974 and 1984.

Figure 3 shows total electricity consumption in MWh and the various sectoral consumptions from 1979 to 2007. Public sector electricity utilization by the industrial sector has been fairly static because of the unreliability nature of the public electric supply system in the country. Thus,

many companies had resolved in providing their own power generating sets for more reliable self generation of electricity leading to high costs of their products and services [25]. Okafor [14] observed that power distribution to the industrial sector in Nigeria also remains abysmally irregular. The effect of irregular power on the cost of production by manufacturing industries was assessed by Adebayo and Alake [26]. The study observed that cost of operating on self power generating sets is 50 times cost of operating on power supply from national grid by PHCN.

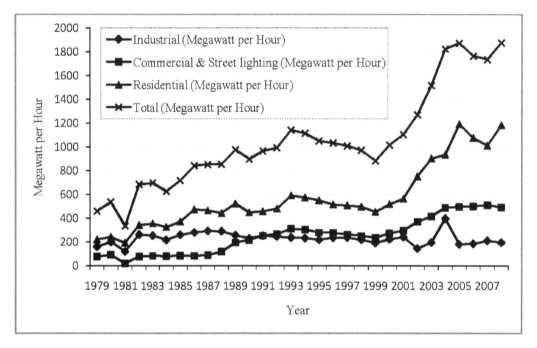

Figure 3. Electricity consumption pattern in Nigeria (Source: Ref. 14)

3. Overview of Industrial Sector Trends

The industrial sector represents more than one third of both global primary energy use and energy-related carbon dioxide emissions [27]. In developing countries, the portion of the energy supply consumed by the industrial sector is frequently in excess of 50% and can create tension between economic development goals and a constrained energy supply.

The industrial sector uses 160 exajoules (EJ) of global primary energy, which is about 37% of total global energy use. Primary energy includes upstream energy loses from electricity, heat, petroleum and coal products production. The industrial sector is extremely diverse and includes a wide range of activities. This sector is particularly energy intensive, as it requires energy to extract natural resources, convert them into raw materials, and manufacture finished products. The industrial sector can be broadly defined as consisting of energy-intensive industries (e.g., iron and steel, chemicals, petroleum refining, cement, aluminium, pulp and paper) and light industries (e.g., food processing, textiles, wood products, printing and publishing, metal processing) [28].

The aggregate energy use depends on technology and resource availability, but also on the structure of the industrial sector. The share of energy-intensive industry in

the total output is a key determinant of the level of energy use.

3.1. Energy Use Efficiency in Industry

Energy efficiency is a term used in different ways, depending on the context and possibly on the person using it. But it is more commonly understood to mean the utilisation of energy in the most cost effective manner to carry out a manufacturing process or provide a service whereby energy waste is minimised and the overall consumption of primary energy resources is reduced. In other words, energy efficient practices or systems will seek to use less energy while conducting any energy-dependent activity; and at the same time, the corresponding (negative) environmental impacts of energy consumption are minimised. According to The Aspen Institute Centre for Business Education [29], energy efficiency is defined as the ability to generate the same economic output with less energy input.

Industrial energy efficiency or conversely, energy intensity, defined as the amount of energy used to produce one unit of a commodity is determined by the type of processes used to produce the commodity, the vintage of the equipment used, and the efficiency of production, including operating conditions [30]. Energy intensity varies between products, industrial facilities, and countries depending upon these factors. The objective of an energy efficient industrial system is analogous to "just in time" manufacturing—to provide the appropriate level of service needed to support the production process, to have a backup plan to address emergencies, and to keep the entire system well-maintained and well-matched to production needs over time.

Energy efficiency is rising toward the top of many national agendas for a number of compelling reasons that are economic, environmental and intergovernmental in nature. As many industries are energy-intensive, this is resulting in new impetus to industrial energy efficiency policies. The economic reasons are quite clear. Most important has been the rise in energy prices from 2005-2006 and their likely continuation at a high level. Increasing concerns over energy security (reliability of supply) are a second factor. Energy supply in many countries increasingly depends on imported oil and gas, and supply is being constrained by geopolitical events while global economic growth is resulting in greater energy demand. Additionally, in many developing countries energy efficiency is also a way to alleviate the investment costs for expanding energy supply infrastructure in the face of tight fiscal constraints [31].

Despite the potential, policy makers frequently overlook the opportunities presented by industrial energy efficiency to have a significant impact on climate change mitigation, security of energy supply, and sustainability. The common perception holds that energy efficiency of the industrial sector is too complex to be addressed through public policy and, further, that industrial facilities will achieve energy efficiency through the competitive pressures of the marketplace alone. Neither premise is supported by the evidence from countries that have implemented industrial energy efficiency programs. The opportunities for improving the efficiency of industrial facilities are substantial, on the order of 20-30% [32], even in markets with mature industries that are relatively open to competition.

Industrial energy efficiency is dependent on operational practices, which change in response to variations in production volumes and product types. Due to this dependence, industrial energy efficiency cannot be fully realized through policies and programs that focus solely on equipment components or specific technologies. Companies that actively manage their energy use seek out opportunities to upgrade the efficiency of equipment and processes because they have an organizational context that supports doing this wherever cost effective, while companies without energy management policies do not. Providing technology-based financial incentives in the absence of energy management will not result in significant market shifts *because there is no organizational context* to respond to and integrate the opportunity into ongoing business practice [30].

There are many benefits of increased energy efficiency. These can broadly be categorised into financial/economic, environmental and social benefits. The relative importance of each of these benefits depends on the actual situation in a given country or area, including for example the prices of different types of energy, the cost of energy efficiency measures and equipment, the tax regime and the current levels of energy efficiency already being achieved [33]. For private companies, the most important benefits of higher energy efficiency will be linked to the financial benefits of lower costs for running the business. This applies to typical manufacturing companies as well as to energy suppliers such as electricity generating plants and oil refineries. The drivers for improving industrial energy efficiency include the desire to reduce overall costs of production in order to maintain competitiveness, reducing vulnerability to rapidly increasing energy prices and price spikes, responding to regulatory requirements for cleaner production (including air quality, solid waste, and greenhouse gas emissions), and meeting consumer demand for greener, more environmentally-friendly products [30].

Notably, while energy-use is increasing in many developing countries, especially, Nigeria the imperatives to enhance energy efficiency in industries have received little attention. This gives rise to the question: if energy efficiency pays, why is it not happening in developing countries? Many studies have been carried out on energy utilization in Nigerian industries. The results have shown opportunities for energy efficiency in industrial sector.

3.2. Opportunities for Industrial Energy Efficiency in Nigeria

In Nigeria, energy savings opportunities in the industrial sub-sector of the economy have remained a matter for speculation over the years due to uncoordinated efforts at addressing issues relating to energy efficiency and management. It is in the bid to create necessary awareness on the huge potentials for energy savings in the sector that Energy Commission of Nigeria in collaboration with United Nations Industrial Development Organization (UNIDO) and other stakeholders have for some time now engaged themselves in industrial energy efficiency programs in Nigeria.

Strictly speaking, two forms of energy carriers are commonly used in the industry: electricity and heat. However, among all the energy forms, electricity is the most widely deployed in the industry for the transformation of raw materials into the desired end products. Electricity consumption in the industry is usually for lighting and motor power-drives of various kinds of equipment, such as pumps, fans, compressors, blowers, conveyors, air conditioners and various machine tools. It is also used in electric furnaces and electrolysis. Improvement in the efficiency of electric motors in particular can result into large energy and cost savings. On the other hand, thermal energy is mostly used in boilers for process steam generation and in kilns such as in cement production [34].

Energy Audits and Surveys are the roadmaps to energy efficiency and conservation measures in the industry. They provide information required to make decisions on which are the most cost-effective measures of energy efficiency program to implement. Answers such as the types of energy use in an industry, how much is being used, the cost, where it is being used, factors affecting consumption, savings potentials and economic assessments are equally met by energy efficiency programs. Investigation carried out in some industries in Nigeria reveals areas of energy conservation (savings) in Nigerian industries.

Below are highlights of walk-through energy audits of some industries in Nigeria.

3.2.1. A Foundry Industry

Walk-through energy audit carried out in this industry shows that there are opportunities for energy savings in the industry. For instance the total current measured is above the rated value for the mains breaker, leading to unacceptable overheating and frequent tripping of the breaker which is now superficially overcome by the installation of a big standing fan to dissipate the generated heat. This is a source of energy waste and can be avoided replacement of the mains breaker especially due to the fact that the contacts may melt completely and results into shutdown and loss of production man-hours. Furthermore based on the measured TDS value of the high frequency induction furnace cooling water system, it is inferred that the ion exchange resin has expired and are therefore due for recharging or replacement to avoid scale formation and rust along the piping network.

3.2.2. Plastic Industry:

In this industry it was observed that energy savings potentials in the plastic company include: Repair of badly damaged insulation, Elimination of fuel, gas, oil and water leaks, Reduction in excessive heating and cooling, Cleaning of dirty surfaces of heat exchangers, motors and lamps, Prompt replacement of worn out belts, Greater use of diffused light, Replacement of the large number of incandescent light bulbs with energy efficient CFLs.

3.2.3. Bottling Company

In this industry, it was observed that the electricity supply is 100% from 3No 800kVA diesel generators while thermal energy for the boiler is from low pour fuel oil (LPFO). Two, out of the three diesel generators are run at a time (24hrs/day) and the other stays on standby. To say the least, this scenario is replicate of most industries in Nigeria which is due to poor electricity supply situation in the country. In terms of energy efficiency, the compounding wastes along the energy supply line are better imagined.

The scenario in the bottling company in relation to energy efficiency is that, a 10 bar, 2 metric ton per hour capacity, low pour fuel oil fired steam boiler, produces steam at a pressure of 4-5 bar (about 140°C-150°C) use in bottle washing that requires hot water at temperature of about 80°C-90°C. It was observed that the steam produced at a high temperature of about 140°C has to be throttled to reduce the temperature to the required level for bottle washing. Ironically the runoff water from the final washing stage comes out at a temperature of 60-70°C and is emptied into the drain. While this practice is considered proper from point of view of avoidance of contamination, it is suggested that a low pressure steam boiler operated at 2 bars can meet the steam requirement and thus save thermal energy.

Furthermore in the compressed air unit, the water-cooled single stage compressor delivers at a temperature of about 80°C while the cooling water comes off at 60°C and is again let off the drain. Opportunity for energy efficiency here is that the heat of the air compression can be recovered to heat the boiler feed water and this may result to about 5% energy savings.

3.2.4. Beer Manufacturing Company:

The investigation reveals the following areas of energy savings opportunities: In the De-aerator – copious amount of steam loss from the deaerator by deliberate action of operators; Steam line leakages from loose joints and holes along the piping network; The Wort Kettle- Loss of latent heat in the evaporation of water from the kettle; Exposed steam lines- Radiation loss from un-insulated parts; Boiler fuel-not sufficiently atomized for efficient combustion; Boiler oversized and operates on part load most of the time; Cooling Tower- treated water allowed to over flow and thick ice formation along NH_3 pipeline; Brine motor pump-Use of constant speed motor drive which run continuously even at no load; Large quantity of water waste at the bottle cleaning section; Boiler TDS not monitored, feed water make up not measured and condensate not recovered; Generator frequency low at 47Hz and power factor low at times.

3.2.5. Chemical Industry

In this industry the source of electricity used in this factory includes: a 1,250KVA gas generator that runs for 24hrs except when it is under maintenance, a 1000KVA diesel generator is used to support the main generator during repairs, also a 153KVA diesel generator used to run the factory when there are no activities. A 500KVA transformer connected to PHCN to generate electricity. Walk-through energy audit carried out in this industry reveals the following areas of energy savings opportunities: Replacement of high capacity generators with smaller capacity generator for load shedding as this will minimize energy wastage. A lot of energy is wasted as the high capacity generators are not fully utilized; Replacement of the large number of high pressure sodium bulbs with energy efficient CFLs; Replacement (especially large) standard electric motors with high efficiency types (especially in the mill-hopper section); Installation heat-reclamation equipment – economizers and air heaters for flue gas and heat exchangers / heat pumps for boiler blow down. Replacement of high capacity generators with smaller capacity generator for load shedding as this will minimize energy wastage.

3.2.6. Food Industry

Study on pattern of electrical energy consumption from 210 selected micro and small-scale food and beverage companies in Nigeria was carried out by Aderemi et al [12]. The study showed that the pattern of electrical energy consumption in the food companies was mainly from generating set; this was due to either low voltage or epileptic power supply from national grid. Direct and indirect sources that lead to electrical energy waste and in-efficient energy utilization in the industry were identified such as energy loss as a result of worn out or slack / misaligned belts that need timely replacement or tensioning, training and retraining of staff, power factor of electrical equipment among others. Three out of eleven strategies were effective in reducing the companies' electricity bill by 3% for the same quantity of production. These include: switching off most lighting during day time; instant replacement / tensioning of worn out / slack belts or chains and; disconnection of all faulty equipment. This finding shows that 72.8% of all the acclaimed strategies to reduce energy consumption were not effective. The study concluded that the factors that constituted electrical energy waste and energy use inefficiency in the food companies in the study area were very identical and recommendations for effective energy use efficiency in the firms were proposed.

3.3. Industrial Energy Efficiency Programme and its Influence on Economy

Industrial energy efficiency has emerged as one of the key issues in ensuring that the per capita income of citizens of a country is improved upon [35]. For example, in 2002, India's total primary energy consumption amounted to 538 million tonnes of oil equivalent. The industrial sector used about 40% of total energy in the country. Six energy-intensive industrial sub-sectors—Aluminum, Fertilizer, Iron and Steel, Cement, Pulp and Paper and Chloral–alkali—consumed 60% of industrial energy. The expenditure in energy among the total production costs in India's industrial sector is generally very high, accounting for 15% in textiles, 25% in pulp and paper, and 40% in glass, ceramics and cement. The Indian Government is therefore interested in developing better policies and measures to supply energy more efficiently and save energy effectively in the industrial sector. A number of energy efficiency policies and measures have been developed over the past 25 years. These include (i) disclosing companies' particulars on energy efficiency; (ii) accelerated depreciation for energy efficiency and pollution control equipment; (iii) setting up the Energy Management Centre under the Ministry of Energy; (iv) the deregulation to promote industrial competitiveness; (v) energy price reforms to guide energy efficiency initiatives and to encourage international competitiveness; and (vi) enforcement of the Energy Conservation Act and the Electricity Act. From the experience of India, Nigeria stands a lot to gain from energy efficient programmes which would make cost of production to reduce. This programme is very low and new to many manufacturing industries in Nigeria as a result of the following:

- Over 40% of energy used is wasted in old, ageing and obsolete industrial equipment which are most of the times very much inefficient.
- 25% saving potential from good house-keeping measures.
- Retrofitting in industries would be able to save about 35% of energy currently in use.

Since this sector spends close to 30% of total cost of production then such an amount can be spent on retrofitting and employment of energy saving programme such as lighting system; using compact fluorescent lamps (cfls), energy-efficient motors, improved steam boilers, etc [36].

3.4. Energy Use Efficiency and Sustainable Development

Sustainable energy can be defined as energy which provides affordable, accessible and reliable energy services that meet economic, social and environmental needs within the overall developmental context of society, while recognising equitable distribution in meeting those needs [5]. In practice, sustainable energy has meant different things to different people. Some think of it as the energy related to renewable energy and energy efficiency. Some include natural gas under the heading of sustainable energy because of its more favourable environmental quality. Whatever approach is used, sustainable energy always implies a broad context which covers resource endowment, existing energy infrastructure, and development needs.

Sustainable energy will, however, require new approaches in the mobilisation of energy resources for development. This would involve: shifts to renewable energy sources; development and wide dissemination of sustainable and renewable energy technologies; energy efficiency and conservation; and technological developments that allow the use of fossil fuels in a cleaner way [37].

Sustainable development is defined as development that meets the present needs and goals of the population without compromising the ability of future generations to meet theirs.

Energy is related to the multidimensional aspects of sustainable development: the economic social and environmental perspectives. Adequate and affordable energy supplies have been the key to economic development and the transition from subsistence agricultural economies to modern industrial and service-oriented societies. Energy is central to improved social and economic well-being, and is indispensable to most industrial and commercial wealth generation. It is the key to relieving poverty, improving human welfare and raising living standards. But no matter how essential it may be for development, energy is only a means to an end. The end is good health, high living standards, a sustainable economy and a clean environment. No form of energy — coal, solar, nuclear, wind or any other — is good or bad in itself, and each is only valuable in as far as it can deliver this end [38].

The requisite of sustainable development is that the production and use of energy should not endanger the quality of life of current and future generations and should not exceed the carrying capacity of ecosystems. Of all the measures that will contribute to meeting this requisite and/or of challenge of sustainable development and limiting climate change, one obvious solution is to use energy more efficiently. That means consuming less energy to produce goods and services,

Environmental-friendly new behaviours and working methods, coupled with the use of new technologies that offer better energy performance [39]. Energy-use efficiency is the fastest, cheapest, cleanest way to address these challenges. The efficient use of energy and supplies that are reliable, affordable and less-polluting are widely acknowledged as important and even indispensable components of sustainable development [40,41].

3.5. Highlights of Policy Implications from the Study

From the study a number of important policy implications are implicit. These include:

• Electricity from the national grid is heavily subsidised and does not give incentive for energy efficiency investment;

• High energy cost has adversely affected employment situation in the industry. Retrenchment of workers/reduction in number of shifts is always an easier way of reducing cost than other options such as energy efficiency;

• Energy reduction is another substitute for job reduction and both companies and government as well as development partners should be actively engaged in industrial energy efficiency options in developing countries, like Nigeria;

• In some of the companies' plants, many machines are very old and thus do not meet the highest energy efficiency standard;

• In some companies, lots of heat energy are generated as by-products but are not being reused at other parts of the plants but simply thrown away;

• National benchmarks for energy consumption in the various industrial processes are not available;

• Some of the companies expressed the need for external consulting, auditing and advice on energy efficiency opportunities;

• Possibility of joint production of power by companies is there but such independent power generation depends on reliable gas supply and competitive energy prices;

• Need for case studies and concrete measures which can be followed by companies to become more energy efficient. In other words, technological advice on energy efficient options will be very useful;

• The need to help in developing cases for small and medium scale power plants and providing information about industrial energy-use;

• Finance for investment in energy efficiency not readily available either from retained earnings or bank loans due mainly to the financial crisis; and

• Need for incentives or subsidies on investments in energy efficiency. Since companies pay fine for polluting the environment with generators, they should be rewarded for greening the environment with energy efficient machines/processes.

3.6. Benefits of Industrial Energy Efficiency Measures

Benefits generally derivable from industrial energy efficiency measures are highlighted below:

3.6.1. Cost Savings

Energy represents cost; therefore saving energy through efficient use saves production cost. In addition to reducing costs it releases funds for further investment for other purposes. Depending on the type of industry, an estimated 10-25% energy savings are achievable. Replacing old and rewound motors for example by energy efficient motors would result to a considerable savings in terms of energy and running cost which is 8 to 10 times the cost of investment for the same motor. For electric motors the kW savings margin is given by the following relation:

$\%kW \ savings$

$= \left(New \ Efficiency - Old \ Efficiency \right) / New \ Efficency \times 100$

3.6.2. Environmental Savings

Improving energy efficiency is one of the most effective means of improving environmental performance. Energy efficiency programs in the industrial sector in Nigeria will provide a source of reducing greenhouse gas emissions under a clean development mechanism scheme as laid out in article 12 of the Kyoto protocol. This is true especially when viewed against the backdrop of the fact that the production and use of energy account for between 50 to 60 % of the greenhouse emissions into the atmosphere1. Energy not used cannot pollute the environment.

3.6.3. Resource Savings

Energy efficiency and conservation measures act as a quicker and cheaper way to save scarce energy and material resources. For instance, in boiler operations, a 3mm diameter hole on a pipe line carrying $7kg/cm^2$ steam would waste 32,650 litres of fuel oil per year. A simple housekeeping measure that fixes the hole saves that amount of resources.

3.6.4. Enhances Competitive Edge

A company that produces the same quantity of products with reduced energy input or produces higher quantities of products with the same amount of energy will maintain a competitive edge over a similar company with high energy bills. This simply indicates that more energy-efficient practices can effectively reduce operating cost and enhance competitive edge for a relatively small investment.

3.6.5. Promotes Sustainable Industrial Development

There is evidently no gainsaying that energy efficiency and conservation measures will be a very effective pathway of promoting sustainable development which has been described as "meeting the needs of the present without compromising the ability of future generations to meet their own needs". In a rule of thumb sustainable industrial development can similarly be described as "keeping the industries working today, tomorrow and in the future through a systematic approach to energy efficiency measures that meets the needs of the present without compromising the needs of upcoming generations." Therefore increasing energy availability through rational use is one way to ensure sustainable industrial development in Nigeria.

3.6.6. Promotes Corporate Social Responsibility

Corporate social responsibility is a concept focusing on the business contribution to sustainability. It is a process by which companies manage their relationships with a variety of stakeholders who can have real influence on their license to operate. Energy efficiency measures can therefore provide industries with instruments to deal with new challenges and requirements to meet with global expectations particularly curtailing CO_2 emissions and other pollutants.

3.6.7. Promotes Increased Productivity

Energy efficiency measures leads to increased productivity and reduces industrial hazards and risks to worker health.

4. Conclusion

Energy is an important production factor and therefore should be managed in parallel with land, labour and capital. Energy efficient production process should be seen as a quick and cheaper source of new energy supply as the cost of providing energy can be several times the cost of saving it. Increasingly energy efficiency is considered to include not only the physical efficiency of the technical equipment and facilities but also the overall economic efficiency of the energy system. Hence the adoption of energy efficiency measures in the industrial subsector in Nigeria will enhance profitability, reduce greenhouse gas emissions, promote sustainable development, and improve corporate social responsibility. The time to begin aggressive campaigns for energy efficiency measures in the Nigeria industrial sub-sector in particular and the whole economy chain in general is long overdue.

Pertinent outcomes from this study are (i) the general level of information in Nigeria on industrial energy efficiency is low; (ii) very few companies have adequate awareness and knowledge about implementing energy efficiency projects; (iii) most companies have never carried out an external energy audit to determine areas where efficiency can be enhanced; (iv) most companies need active policy on identifying and repairing leakages such as air, heat and steam, through a combination of internal and external energy audit; (v) the relative low price of fuel in Nigeria, combined with the high investment costs for machines result in long payback period for investments in energy efficiency; (vi) despite the major problem of energy supply facing the companies, a number of them have no clear information on energy efficiency options; and (vii) finance for investment in energy efficiency not readily available either from retained earnings or bank loans due mainly to the financial crisis.

Energy demands by industries in Nigeria will continue to grow. Presently most of these industries are financially and environmentally unstable. With increasing pressure on available resources due to large population, very low GDP, loss of competitive edge in the global market of goods produced in Nigeria, and a drive to catch up with the rest of the world in improving the standard of living of her citizens by at least 2020, Nigeria cannot afford to waste her energy resources through inefficient industrial production processes. There is therefore an urgent need to promote energy efficiency and management measures for sustainable industrial development in Nigeria.

The key policy challenge is the need to address the subsisting paradox where companies pay fine for polluting the environment with generators but are not rewarded for greening the environment with energy efficient machines/processes. This paper therefore recommends the need for incentives or subsidies on investments in energy efficiency.

References

[1] ECN (Energy Commission of Nigeria)(2003), National Energy Policy.

[2] Nag, P.K (2004), Power Plant Engineering (2nd Ed.); Tata McGraw Hill, India.

[3] Awwad, A. A and Mohammed A. A (2007), 'World Energy Road Map-A Perspective; WEC- 2007 Energy Future in an interdependent world.

[4] Jean, P. H and Marc D.F (2007), 'From a forced dependency to positive cooperation in the field of Energy', WEC-2007 Energy Future in an Interdependent World.

[5] Davidson, O (2006), 'Energy for Sustainable Development: An Introduction'. In: Winkler, H (Ed.), 'Energy Policies for Sustainable Development in South Africa, Option for the Future'.

[6] David Y. C., Kuang-Hax, C. H and Min-Hsien, G. H (2007), 'Current Situation of energy conservation in high-energy consuming industries in Taiwan', Energy Policy 35,202-209.

[7] Eleri, E.O (1995), 'Nigeria: Energy for Sustainable Development', Volume 19, Number 1, Journal for Energy and Development, 104.

[8] Nagesha, N (2008), 'Role of energy efficiency in sustainable development of small-scale industry clusters: an empirical study', Energy for Sustainable Development, Vol. 12, No. 3: 34-39.

[9] Fawkes, H (2005), 'Energy efficiency in South African industry, Journal of Energy in Southern Africa', Vol. 16 No 4: 18-25.

[10] Lung, R.B, Masanet, E and McKane, A (2006), ' The Role of Emerging Technologies in Improving Energy Efficiency: Examples from the Food Processing Industry', Proceedings of the Twenty-Eighth Industrial Energy Technology Conference, New Orleans, LA, May 9-12, 2006.

[11] Aiyedun1,P.O, Adeyemi, O.A and Bolaji, B.O(2008), 'Energy Efficiency of a Manufacturing Industry: A Case Study of Nigeria Eagle Flour Mills Limited, Ibadan', ASSET An International Journal, 7 (2):91-103.

[12] Aderemi, A.O, Ilori, M.O, Aderemi, H.O and Akinbami, J.F.K (2009), 'Assessment of electrical energy use efficiency in Nigeria food industry', African Journal of Food Science, Vol. 3(8): 206-216.

[13] Noah, O.O, Obanor, A.I and Audu, M.L (2012), 'Energy Audit of a Brewery-A Case Study of Vitamalt Nig. Plc, Agbara', Energy and Power Engineering, 2012, 4, 137-143.

[14] Chendo, M.A.C., 2001, Non-conventional energy source: development, diffusion and impact on human development index in Nigeria, N. J. Renewable Energy, 9: 91-102.

[15] Enete, C.I and Alabi, M.O (2011), 'Potential Impacts of Global Climate Change on Power and Energy Generation', Journal of Knowledge Management, Economics and Information Technology (www.scientificpapers.org), Issue 6, pp 1-14.

[16] Akinbami, J.F.K (2001), 'Renewable energy resources and technologies in Nigeria - Present situation, future prospects and policy framework', Mitigation Adaptation Strategies Global Change, 6: 155-181.

[17] Oladosu, G.A. and A.O. Adegbulugbe (1994), 'Nigeria's Household Energy Sector: Issues and Supply/Demand Fronties', Energy Policy, 22: 538-549.

[18] Ogunsola, O.I. (1990), 'History of Energy Sources and Their Utilization in Nigeria', Energy Sources 12, pp. 181-198.

[19] Centre for People and Environment (CPE)(2009), Pre-Feasibility Study Of Electricity Generation From Nigerian Coalmine Methane, Woodbridge, USA.

[20] Enibe, S.O and Odukwe, A.O (1990), 'Patterns of Energy Consumption in Nigeria', Energy Conservation and Management Vol. 30, No. 2, pp. 69-73.

[21] Ajao, K.R, Ajimotokan, H.A, Popoola, O.T and Akande, H.F(2009), 'Electric Energy Supply In Nigeria, Decentralized Energy Approach', New York Science Journal, 2(5):84-92.

[22] Oyedepo, S.O (2012), 'Efficient energy utilization as a tool for sustainable development in Nigeria' International Journal of Energy and Environmental Engineering, 3:11, pp 1-12.

[23] Schatzle, L (1969), 'The Nigerian coal industry', Nigerian Institute of Social and Economic Research, Ibadan

[24] Bajpai, S.C and Suleiman, A.T (1985), 'Cost effectiveness of photovoltaic plants in Nigeria by the Year 2000', Sokoto Energy Research Centre, University of Sokoto, Nigeria.

[25] Enebeli E. E (2010), 'Causality Analysis of Nigerian Electricity Consumption and Economic Growth', Journal of Economics and Engineering, №4, Pp 80-85.

[26] Adebayo, A. A. and Alake, T. J (2012), 'The Impact of Irregular Power Supply on the Cost of Production of Selected Manufacturing Industries in Ilorin, Kwara State', Paper presented at National Engineering Conference on 'Strategies for Promotion of Sustainable Technology –Based and Production Driven Economy in Nigeria', Held at Kwara Hotel, Ilorin, Kwara State, Nigeria, December 3rd-7th.

[27] Price, L., de la Rue du Can, S., Sinton, J and Worrell, E (2006), 'Sectoral Trends in Global Energy Use and Greenhouse Gas Emissions', Berkeley, CA: Lawrence Berkeley National Laboratory (LBNL-56144).

[28] Oyedepo, S. O and Aremu T. O (2013), 'Energy Audit of Manufacturing and Processing Industries in Nigeria: A Case Study

of Food Processing Industry and Distillation & Bottling Company', *American Journal of Energy Research, Vol. 1, No. 3, 36-44.*

[29] The Aspen Institute Centre for Business Education (2009), 'A Closer Look at Business Education: Energy Efficiency', October Edition.

[30] McKane , A., Price, L and de la Rue du Can, S (2006), Policies for Promoting Industrial Energy Efficiency in Developing Countries and Transition Economies. Background Paper for the UNIDO Side Event on Sustainable Industrial Development on 8 May 2007 at the Commission for Sustainable Development (CSD-15). https://www.unido.org/doc/65592.

[31] Peck, M and Chipman, R (2007), 'Industrial energy and material efficiency: What role for policies?', In: Industrial Development for the 21st Century: Sustainable Development Perspectives, UN, New York.

[32] IEA (International Energy Agency) (2004), Key World Energy Statistics. Energy Statistics.

[33] Fidelis O.O and Aregbeyen, O (2011), 'Energy Use and Sustainable Development: Evidence from the Industrial Sector in Nigeria', Sixth African Economic Conference jointly organised by the United Nations Economic Commission for Africa (UNECA), the African Development Bank (AfDB), United Nations Development Programme (UNDP) and the Development Bank of Southern Africa (DBSA) to be held from 25 to 28 October 2011 in Addis Ababa, Ethiopia.

[34] Unachukwu G. O(2003), 'Energy Efficiency Measures Investigation in Cement Company: BCC Case study', Nigerian Journal of Renewable Energy Vol. 10, Issues 1&2: 85-92.

[35] Oluseyi, P.O; Akinbulire, T.O and Awosope, C.A.O (2007), "energy efficiency in the third world: the demand-side management (DSM) option", CIER 2007, Hammamet,Tunisia, November 4-6, 2007.

[36] Al-Shakarchi, M.R.G and Abu-Zei, N.S (2002), "A Study of Load management By Direct Control for Jordan's Electrical Power system", Journal of Science and Technology,vol.7,No.2.

[37] Tsighe, Z(2001), 'Opportunities and Constraints for Sustainable Energy in Eritrea', In: Habtetsion, S; Tsighe, Z and Anebrhan, A (Eds.) Sustainable Energy in Eritrea, Proceedings of a National Policy Seminar Held on October 30-31, Asmara, Eritrea.

[38] International Atomic Energy Agency (IAEA), (2005), Energy Indicators for Sustainable Development: Guidelines and Methodologies. United Nations Department of Economic and Social Affairs, International Energy Agency, Eurostat and European Environment Agency.

[39] Total (2007), Energy Efficiency, www.total.com.

[40] World Commission on Environment and Development (WCED) (1987), Our Common Future, Oxford University Press, Oxford.

[41] Goldemberg, J. and Johansson, T.B.: 1995, 'Energy as an instrument for socio-economic development', in T.B. Johansson and J. Goldemberg (eds.), *Energy for Sustainable Development: A Policy Agenda*, New York, United Nations Development Programme, pp.9-17.

Electric Power Generation from Waste Heat

Sana Ullah Khan[1,*], Irfan Khan[2], Engr. Hashmat Khan[3], Engr. Qazi Waqar Ali[3]

[1]Sarhad University of Science & IT, PAKISTAN
[2]Electrical Engineering, Sarhad University of Science & IT, PAKISTAN
[3]SARHAD University of Science & IT, PAKISTAN, MSc Electrical Power Engineering, BSc Electrical Engineering
*Corresponding author: Sanaullah_51@yahoo.com

Abstract Waste heat is developed as a by-product in power generation, commercial procedures and electric machines, among others. Huge of spend warmed are designed by industry. Thermoelectric developer is one of the alternate sources for the growth of power. Thermoelectric developer is a device which transforms heat straight into electrical power, using a phenomenon called the "Seebeck effect". In this paper we will recommended a thermoelectric developer which will use invest warmed exhausted by the places for development of electric power. The suggested program will be depending on thermoelectric content known as bismuth telluride (Bi2Te₃). This developer having no energy price because spend warmed is the opinions for this developer. In our suggested system we used voltmeters, ammeter and warmed wide range receptors to find the regards between the opinions warmed and the designed outcome power.

Keywords: *Seebeck effect, alternate, Electrical power, thermoelectric, invest heat*

1. Introduction

A thermoelectric developer is an incredible system scenario that provides immediate power modification from warmed energy(heat) due to a warmed wide range mountain into electric power based on "Seebeck effect". [1].

In 1821 Thomas Johann Seebeck, German physicist, first found the Seebeck impact. Seebeck first identified that a compass weblink deflected when placed in the position of a shut design identified of two different elements with a heated comprehensive extensive wide range distinction between the junctions. This statement provides immediate evidence that a current goes through the closed schedule inspired by the warmed wide range difference. A heat range distinction causes cost providers (electrons or holes) in the content to dissipate from the hot part to the cool part. Mobile cost providers move to the cool part and keep behind their oppositely billed and motionless nuclei at the hot part thus providing increase to a thermoelectric volts. The make up of cost providers on the amazing aspect gradually stops when an relative amount of cost providers move back to the hot aspect as a result of the power handled area designed by the cost separation. At this point, the content gets to stable state.

Only a rise in the heat range distinction can continue a accumulation of more charge providers on the cold side and thus lead to a rise in the thermoelectric volts. The voltage known as the thermoelectric emf, is created by a warm variety difference between two different elements (A and B) such as elements or semiconductors.

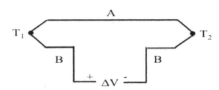

Seebeck effect: A temperature difference create a potential difference for the junction between materials A and B.

$$\Delta V = (\alpha A - \alpha B)\,\Delta T \quad [11]$$

where, αA and αB are the Seebeck coefficients of components A and B, respectively.

Now the regards between the heated wide range and voltage as shown in Graph 1.

The features of Bi2Te3 are as under, [8].

Properties	
Molecular formula	Bi_2Te_3
Molar mass	800.761 g/mol
Appearance	grey powder
Density	7.7 g/cm^3
Melting point	585 °C

Performance of a thermoelectric content can be indicated with regards to the dimensionless figure out of benefit ZT.

$$ZT = \frac{\sigma S^2 T}{\lambda} \quad [10]$$

Where λ is thermal conductivity, T is the temperature, S is the seebeck coefficient, σ is the conductivity, thermoelectric content targeted on improving the see beck coefficient (s) and decreasing the heat conductivity λ to improve the electric conductivity σ.

Temperature vs Voltage:
x=[25 30 35 40 45 50 55 60 65 70 75]
y=[0.2 2.9 5.2 5.89 6.07 7.1 8.2 8.46 8.46 8.46 8.46]
plot(x,y)
xlabel('Temprature in Degree C')
ylabel('Voltage in Volts')

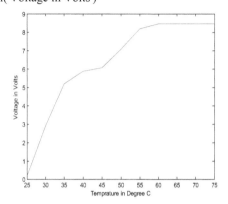

Graph 1. show relation between temperature and voltage

Temperature vs Current:
x=[25 30 35 40 45 50 55 60 65 70 75]
y=[1.20 4.37 4.40 4.54 5.06 5.70 5.93 6.50 6.80 7.00 7.20]
plot(x,y)
xlabel('Temprature in Degree C')
ylabel('Current in mA')

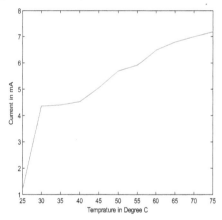

Graph 2. show relation between temperature and voltage

The value of ZT is if one then it will consider good. The value from 3 to 4 wide range for thermoelectric opponents with specific devices in performance such as mechanical generator. Up to 2012 the value ZT is reported 2.2.

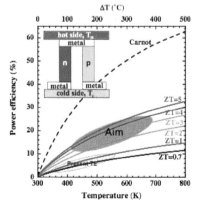

2. Power Factor

The Seebeck coefficient is not the only broad variety that selects the strength of a content in a thermoelectric developer. Under a given warmed broad range difference, the ability of a content to produce useful power is quantified by its power factor,

$$POWER\ FACTOR = \sigma S^2$$

Components with excellent energy factor are able to produce more energy in a space-constrained program, but they are not actually effective.

3. Device Efficiency

The efficiency of a thermoelectric device for electricity generation is given by η, defined as

$$\eta = \frac{energy\ provided\ to\ the\ load}{heat\ energy\ absorbed\ at\ hot\ junction}$$

Which depends on the Seebeck coefficient S and electrical conductivity σ, in an actual thermoelectric device, two materials are used. The maximum efficiency η_{max} is then given by

$$\eta_{max} = \frac{T_H - T_C}{T_H} \frac{\sqrt{1+ZT}-1}{\sqrt{1+Z\bar{T}}+\frac{T_C}{T_H}}$$

Where T_H is the temperature at the hot junction and T_C is the temperature at the surface being cooled. ZT⁻ Is the modified dimensionless figure of merit, [9]

Semiconductor materials are the most efficient, and are combined in pairs of "p type" and "n type". The electrons flow from hot to cold in the "n type," While the electron holes flow from hot to cold in the "p type." This allows them to be combined electrically in series and thermally in parallel.

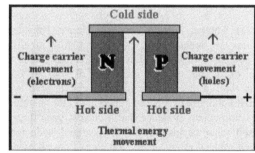

The electrons will move from the hot side to the cold side.

Because electrons on the hot side of a material are more energized than on the cold side.

If a complete circuit can be provided then electricity will flow.

By combining these elements in series so we enhance existing and power output of thermoelectric developer.

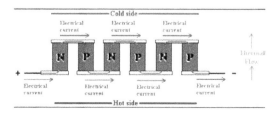

The relation between the temperature and power output is shown in Graph 3.

Temperature vs power:

x=[25 30 35 40 45 50 55 60 65 70 75]

y=[0.24 12.67 24.01 26.74 30.71 40.47 49.60 55 57.52 59.22 60.9]

plot(x,y)

xlabel('Temprature in Degree C')

ylabel('Power in mW')

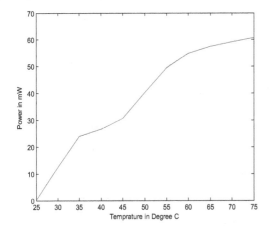

Graph 3. show relation between temperature and power

Thermoelectric generators can be applied in a variety of applications.

1. Many space probes, including the Mars Curiosity rover, generate electricity using a thermoelectric generator whose heat source is a radioactive element.

2. Human body is a heat source for TEG.

3. Cars and other automobiles produce waste heat (in the exhaust and in the cooling agents). Harvesting that heat energy, using a thermoelectric generator, can increase the fuel efficiency of the car.

4. In addition to in automobiles, waste heat is also generated many other places, such as in industrial processes and in cooking. Again, the waste heat can be reused to generate electricity. In fact, several companies have begun projects in installing large quantities of these thermoelectric devices.

5. Solar cells use only the high frequency part of the radiation, while the low frequency heat energy is wasted. Several patents about the use of thermoelectric devices in tandem with solar cells have been filed. The idea is to increase the efficiency of the combined solar/thermoelectric system to convert the solar radiation into useful electricity.

6. The power loss in the transmission line can be recovered by using TEG.

Thermoelectric developer provides several unique benefits over other technologies:

1. They have very small size.

2. They are easy, lightweight and safe;

3. They are really effective (typically exceed 100,000 hours of steady-state operation) and quiet in operation. Since they have no particular moving parts and need significantly less maintenance;

- Silent operation
- No moving parts

4. They are able of working at raised temperatures

5. They are environmentally friendly;

6. They are appropriate for small-scale and far away programs common of non-urban source of energy, where there is limited or no power [2,3,4,5].

4. Problem Statement

Development can be evaluate by a country's power consumption in creating nations industry is the biggest customer of power. In developing countries about 22% of people having no access to electric power. As the globe has being affected by the power crises, so we have to discover out the possible remedy for this problem. One of the possible alternatives as to discover out the bottlenecks in the existing energy program to create it effective enough to fulfill up with the energy need of these days. As traditional fuels are depleting, so we have to modify to different resource for power production. To apply the substitute resources of energy, to compensate the gap between supply and demand of electrical power. The most substitute resource of power in universe is the sun, so we implement the warming effect of sun as a mean of growth of power by design known as thermoelectric developer. Moreover, low-grade warmed (heat resources approximately under 100°C) [6] is also available from natural sources such as geothermal tanks and solar panel technology. Restoring this warm into useful power would preserve a lot of money through enhancing performance and reducing power expenses as well as being useful to the surroundings [7]. For any heated engine, the recommendations of thermodynamics place essential limitations on the quality of useful energy which can be created. Due to the need for any creation process to eliminate warm, the portion of warm which may be transformed relies on the consumption and fatigue temperature, with lower temperature ranges being less efficient. However, since such a lot of warmed is easily available, now the problem is that to select one of the business economics technology and settings which generates the best using of warm. Now thermoelectric developer is the most economic technology for the utilization of this warm. Thermoelectric generator requires two types of cost that is device cost and operating cost. The operating cost is negligible because the input for this generator is waste heat which is available freely. Power loss in transmission line in the form of heat and we can recover this power loss by using thermoelectric generator placed on transmission line because it is the portable one.

5. Proposed Methodology

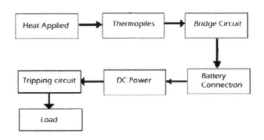

The opinions for this developer is spend warmed, which is available abundantly in nature. This warmed is used to the thermopiles (Bi2Te3) by using converging representation to focus the warmed on one factor of thermopiles to create the warmed wide variety difference which are responsible for the development of electric power. In DC generation the proper polarity connection is necessary so for this purpose we use bridge circuit to correct polarity instantly. Battery is used to store the extra generated power when we need then we use it. Tripping circuit which contains (solid state relay, operational amplifier and potentiometer) is used to provide protection to the generation section in case of fault on load side and also stops the battery from the release.

6. Conclusion

Thermoelectric developer is used to provide electric power which is designed from spend warmed produce by different equipments in industry, transmission line etc. In this paper we provided our suggested analysis in which we use a converging reflection for concentrating warmed on thermopiles. We took the Bi2Te3 thermopiles with dimensionless figure out of advantages of 1.5 and the results have confirmed that with the use of this converging representation we get an efficient result. when the load is connected to the generation section so we calculate the current with help of ammeter circuit. Graph 3, show the relation between the current and temperature.

From the graphs we observed that at room temperature the value of current and voltage is small, so that's why there is no significance power output. The number of thermopiles in our project is three(3) which provide two junctions, which are connected electrically in series and thermally in parallel. The voltage at the junction is (3.9 to 4.2) volts. At 75°C our system give maximum output , and the calculated efficiency of our system is 4.2%.

References

[1] Thermoelectric power generation using waste-heat energy as an alternative green technology by Basel i. Ismail*, wael h. Ahmed**.

[2] Riffat SB, Ma X. Thermoelectric: A review of present and potential applications. Appl Therm Eng 2003; 23: 913-935.

[3] Omer SA, Infield DG. Design and thermal analysis of two stage solar concentrator for combined heat and thermoelectric power generation. Energy Conversion & Management 2000; 41: 737-756.

[4] Yadav A, Pipe KP, Shtein M. Fiber-based flexible thermoelectric Power generator. J Power Sources 2008; 175: 909-913.

[5] Jinushi T, Okahara M, Ishijima Z, Shikata H, Kambe M. Development of the high performance thermoelectric modules for High temperature heat sources. Mater Sci Forum 2007; 534-536.

[6] A. W. Crook (ed). Pro_tingFrom Low-Grade Waste Heat. London: Institute of Electrical Engineers, 1994.

[7] D. M. Rowe (ed). CRC Handbook of Thermoelectric. Danvers, MA: CRC Press, 1995. (Modeling and application).

[8] Satterthwaite, C. B.; Ure, R. (1957). "Electrical and Thermal Properties of Bi2Te3". Phys. Rev. 108 (5).

[9] D. M. Rowe (ed). CRC Handbook of Thermoelectrics. Danvers, MA: CRC Press, 1995.

[10] Terry M. Tritt, "Part III-Semiconductors and Semimetals" Recent Trend in Thermoelectric Materials Research: (Volume 71): (Acedamic press New York 2000) p.6.

[11] Kittel, C. Introduction to Solid State Physics Wiley, 2005.

3

Simulation of a ZEB Electrical Balance with aHybrid Small Wind/PV

Mohammad Sameti, Alibakhsh Kasaeian[*], Fatemeh Razi Astaraie

Department of Renewable Energies, Faculty of New Sciences and Technologies, University of Tehran, Tehran, Iran
*Corresponding author: akasa@ut.ac.ir

Abstract Electricity production from modern renewable technologies (wind energy, solar energy, water power in small scale, and geothermal energy) is growing rapidly worldwide. In addition to the large-scale power generation, applications in the residential sector is also of interest which are classified into stand-alone systems (without connecting to the grid) and grid-connected systems. Zero energy building (ZEB) is a concept based on minimized energy demand and maximized harvest of local renewable energy resources. In this paper, the electrical energy consumption of a typical residential building is modeled. In addition to the electrical grid, the house is connected to a hybrid wind turbine and photovoltaic array together with a battery storage system. The cost of electricity purchased from the electrical grid was optimized to its minimum level. The results showed that, considering a load profile with 21.675 kWh of daily consumption, a 3 kW PV array with a 2 kW wind turbine and a 5 kWh battery could save 96% of the monthly electricity bill; from 2'377 to 66'960dollars. Excluding battery storage, this saving was reduced to 74%.

Keywords: *Zero Energy Building (ZEB), energy storage, Energy Cost Optimization*

1. Introduction

Today, it is obvious that modeling and simulation has an important role in the area of system engineering. Description of the experimental behavior of the system is not always possible. The main reasons may be the unavailability of the inputs and outputs, dangerous testing condition, high cost of testing, noncompliance of the system time constants with the human dimensions and lack of clarity of the experimental behavior of the system due to the disturbance. This will become more important in the field of renewable energies, so that almost no renewable system is being studied under real or laboratory condition before the computer simulation is performed. On the other hand, the use of the new renewables in both small and large electricity generation is growing rapidly. Because of the high cost of building construction, any innovative idea should be already modeled to assure that the system is really beneficial.

A *Net Zero Energy Building*(NZEB) is simply defined as a building in which the electricity or heat from renewable energy resources can supply the same highly reduced energy demands. Such buildings' annual energy need can be supplied by one or more conventional or modern distribution energy systems such as electricity grid, district thermal system, gas pipe network, biomass and biofuels distribution networks [1]. The electrical components of a NZEB may be summarized as following [2]:

- *Smart Meter:* It is a gateway between the building and the smart electrical grid.
- *Heat Pump:* It reduces the energy required for interior heating.
- *Battery Storage:* It serves as the backup energy during power failure or on-peak hours.
- *Solar Photovoltaic and Small Wind Turbine:* they serve as the supplementary power generation to meet the house energy demands.
- *High Efficiency Lighting:*Modern CFL, LED, OLED require less energy.
- *Home Energy Manager:* It controls and optimizes the energy flow in the house.

Figure 1 shows a schematic of the prescribed system and the *building boundary system* which energy compares to flows and canflow in and out of the system [3]. Building boundary comprises of *physical* and *balance* boundaries. Depending on renewables generated *on-site* or *off-site*, the physical boundary may vary. Also, it includes the size ranging from a single and small house to a group of buildings. Heating, cooling, ventilation, lighting and appliances can be mentioned as different energy uses and balance boundary associated with these types of energies used for writing the energy balance for our system. Here, we confine our study to balance boundary which encompasses small wind, photovoltaic array, battery and electricity grid. The model presented here exactly shows what a home energy manager does in Figure 1 to optimize the cost of electricity purchased from the grid.

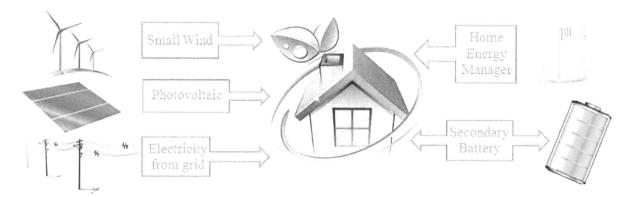

Figure 1. Schematic of the energy flow exchange between the energy-efficient building and different electricity components

1.1. Battery Storage

Electrical storage battery is a needful part of renewable power systems particularly, small wind turbines and photovoltaic and is defined as a device which allows electricity to flow in both directions: inside and outside.

In the ZEB approach, battery storage is an essential component of the standby, emergencybackup services and electric vehicles which is the concept known as vehicle-to-building (V2B) and vehicle-to-home (V2H) technology. They are able to provide beneficial storage capacity to both vehicle and the building owners by lowering cost of the electric vehicles, reducing building'spurchased energy and enabling reliable emergency power systems.

The *State of Charge* (*SOC*) of a rechargeable battery (also, known as secondary battery) at any time is defined as

$$SOC = \frac{C}{C_{rated}} \qquad (1)$$

where C is the remaining capacity and C_{rated} is the rated capacity both measured in Ampere-hours (Ah) which means delivering C/k Amperes in k hours [4].

1.2. Photovoltaic & Small Wind

The efficient solar to electrical energy conversion using photovoltaic (PV) cells is one of the promising solutions for the global future energy demand. *Solar cells* are capable of converting the solar radiation into the electrical direct current (DC) through the so-called effect called photovoltaic. *Photovoltaic modules* or *photovoltaic panels* are made of the multiple interconnected solar cells. Owing to its low output power to meet the requirement of the house demands, *photovoltaic arrays* are usually utilized which are the linked collection of photovoltaic modules. To be applicable for household devices, the array output DC power should be converted to an alternative current (AC) via a device called *inverter*. Solar arrays electrical power are generally measured in watts (*W*), kilowatts (*kW*) and sometimes megawatts (*MW*) [5]. As illustrated in Figure 2, the array power generation varies from zero to its maximum value throughout the day.

Arotational flow driven machine which converts the kinetic energy from the moving air (wind) into the mechanical energy is called a *wind turbine*. Using a generator, the output mechanical energy is converted into the useful electricity. The relationship between the extracted mechanical power and the wind speed can be shown by a plot called a *power curve*. Figure 3 illustrates an idealized wind turbine power curve. The operating limits of a wind turbine is characterized by *cut-in* and *cut-out* speeds.

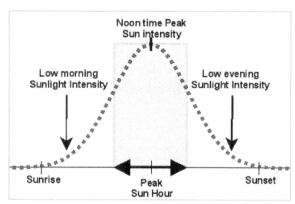

Figure 2. Typical electricity PV generation profile [6]

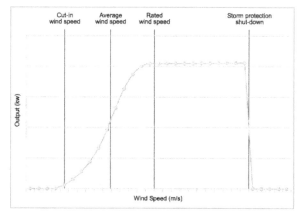

Figure 3. Idealized wind turbine power curve [7]

It can be seen from Figure 3 that the power curve is divided into four distinct zones:

- *Before cut-in wind speed* consists of low wind speeds which there is not sufficient torque exerted by the wind on the turbine blades to make them rotate so no power can be extracted.
- *Between cut-in and rated wind speed* is a transition region mainly concerned with keeping rotor torque and noise low and can be expressed as

$$P = \frac{V - V_{in}}{V_{out} - V_{in}} \times P_{rated} \qquad (2)$$

where P is the extracted power (W), P_{rated} is the rated turbine power (W), V is the wind speed (m/s), V_{in} is the cut-in speed (m/s) and V_{out} is the cut-out speed (m/s).

- *Between rated and cut-out wind speed* which is relevant to higher wind speeds, the turbine design limit the output power to a maximum level (rated output power) and no further rise will happen in the extracted power.
- *After cut-out speed (storm protection shut-down)* there is a risk of damage to turbine structure and the control system stops the rotor.

1.3. Smart Meter

During the day, the *smart meter* has the option of charging various rates which are split into three separate periods according to demand for electricity: *off-peak* (low demand), *mid-peak* (moderate demand) and *on-peak* (high demand). The lowest prices are usually at night, on weekends and on holidays and are different throughout summertime and wintertime. Figure 4 shows a typical time-of-use price periods and is used in the model of the current study. Table 1 shows the assumed electricity rates in the current study.

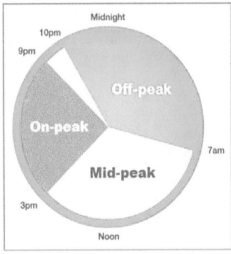

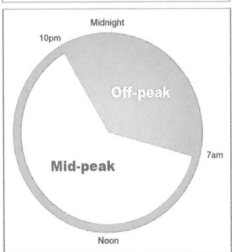

Figure 4. Hourly separate periods of electricity demand for weekdays (up) and weekends (down) [8]

Table 1. Rates of electricity in different periods (values of B_i) [9]

Period	Off-peak	Mid-peak	On-peak
Price (*cent/kWh*)	7.2	10.9	12.9

2. Model Equations

At each point of the time t_i (here, we consider each point as an hour) where $0 \le i \le 24$, we can write the energy balance on the building system boundary as

$$E_w(t_i) + E_{pv}(t_i) + E_g(t_i) = E_b(t_i) + E_c(t_i) \qquad (3)$$

where E is the energy exchanged during 1 hour and the subscripts w, pv, g, b and c stand for wind turbine, photovoltaic panel, electrical grid, battery and the home energy demand, respectively. For the case of the battery, the exchanged energy is identical to its charging when $E_c < 0$ and discharging when $E_c > 0$, in other words

$$E_b(t_i) = E_{bc}(t_i) - E_{bd}(t_i) \qquad (4)$$

where c and d denotes charging and discharging, respectively.

Taking the battery capacity into account, we can associate the state of the charge (SOC) and the exchange energy as following

$$SOC(t_i)C_{rated} = E_b(t_i) + SOC(t_{i-1})C_{rated} \qquad (5)$$

where the value of SOC is always between its minimum and maximum levels.

$$SOC_{min} \le SOC \le SOC_{max} \qquad (6)$$

The battery fault can be a reason for not assigning the values 0 and 1 to the minimum and maximum values of the SOC. As the second reason, the system owner probably like to ensure that there is always a charge remnant $SOC_{min}C_{rated}$ left in the battery. Eq. (2) is the simple case in which the battery efficiency considered to be 100% during charging and discharging. The more typical case can be written as (5).

$$E_b(t_i) = E_{bc}(t_i)/\eta_c - \eta_d E_{bd}(t_i) \qquad (7)$$

where η_c and η_d are the battery efficiencies in charging and discharging modes, respectively.

The daily price of purchased electricity $B(cent)$ can be calculated using

$$B = \sum_{i=1}^{24} B_i \qquad (8)$$

where $B_i(cent/hour)$ is the rate of electricity at the time t_i. Values of B_i can be found in Table 1. Assuming a constant daily trend throughout a month, the monthly electricity bill is calculated. The aim of the current simulation is to find the minimum value of B denoted by B_{min}.

Assumed value for the two first terms in Eq. (2) is depicted in Figure 5 and Figure 6. The PV array has a nominal power output of $3kW$ and the small wind turbine specifications can be found in Table 2.

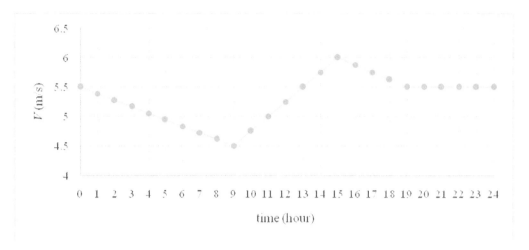

Figure 5. Distribution of daily wind speed at turbine height used for the turbine power output [11]

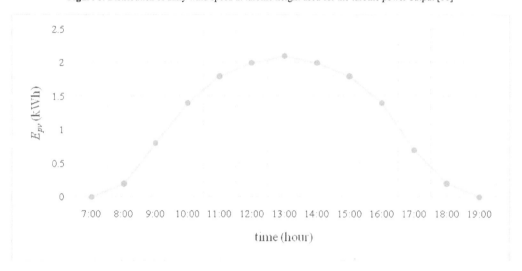

Figure 6. Distribution of hourly photovoltaic energy generation [12]

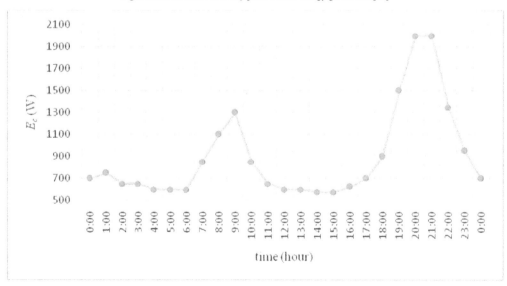

Figure 7. Distribution of hourly energy consumption for a typical family [13]

Table 2. Small wind turbine characteristics [10]

Tanfon FD-2000			
$V_{in}(m/s)$	$V_{out}(m/s)$	$P_{rated}(kW)$	Rotor Diameter (m)
3.5	9.0	2.0	3.2

Table 3. Battery specifications

η_c	η_d	C_{rated}	SOC_{min}	SOC_{max}
0.9	0.9	$5\ kWh$	0	1.0

In this study we considered a typical family which consumes electricity similar to what can be seen in Figure

8. more information about load profile patterns can be found in ref. (14).The values of constants in Eqs. (5) and (6) in the model can be found in Table 3.

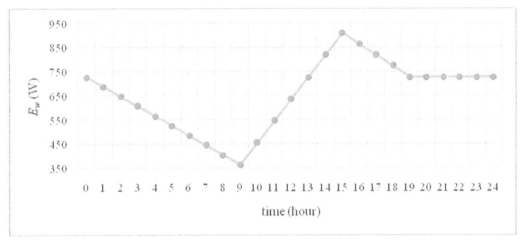

Figure 8. Distribution of daily wind turbine power used in the model

3. General Algebraic Modeling System

Equations (3) to (8) comprise a linear system of equations which can be solved using the *General Algebraic Modeling System* (GAMS) which is a high level modeling system [15] and is very similar to fourth generation programming systems. The procedure is carried out according to the defined algorithm by the software. It is capable to globally solve linear programs as well as finding local optima of nonlinear programs. In this software, the programmer can build models which are expressed in simple and concise algebraic statements to find the optimum value of a predefined function. Here, the statements are very similar to the model equations (3) to (8) except for its initial condition is required by the software to solve the problem more efficiently. We put the value of $E_w(t_i)$ at $t_i = 0$ o be $500kWh$ as can be seen in Figure 11. The available input data for wind, PV and the energy consumption along with the rate periods are imported to the system defining tables which pertains to each energy supplies/demands.

4. Results

The hourly wind turbine output power $E_w(t_i)$ is calculated using Eq. (2) and Figure 5. The output power is shown in Figure 8 and is used in the model equations. Modeling the equation systems in GAMS, the monthly electricity bill B is calculated in three cases:

- *Case I*: In presence of RET (*renewable energy technology*) with battery storage. In this case, there is a minimum value of B.
- *Case II*: In presence of RET without battery storage. Here, no minimum value of B exist.
- *Case III*: In absence of RET.

It is assumed that the daily trend of energy consumption depicted in Figure 7 continues for all the other days throughout a month. Figure 9, Figure 10, Figure 11 illustrates the simulation results. As Table 4 shows, using all elements can save 96% in electricity cost while 22% of

this reduction owes to the battery storage. The more the battery storage capacity was used the more the improvement is seen in the energy saving through shifting high amount of energy from off-peak to the on-peak hours.

Table 4. Simulation results for daily prices for different cases

Case	I	II	`III
Elements			
Daily Price ($)	79.24	608.26	2'232
Monthly Price ($)	2'377	18'240	66'960

5. Conclusion

Zero Energy Building(ZEB) is a modern concept which deals with providing the electrical or thermal energy from renewable energy resources to supply the buildings' annual energy demands. However ZEBs have different aspects, current research investigated them with electrical approach as the part of building balance boundary.

A simple model was developed to encompass the balance boundary aspect of a nearly zero energy building and the electricity cost was optimized to its minimum level using the GAMS software. Three different modes (RET with storage, RET without storage, conventional non-ZEB) were simulated to show how each element affect the monthly electricity bill. Electricity bill were calculated to be 2'377, 18'240 and 66'960 dollars per month for each of the previous cases, respectively. Hourly variations of each energy flow were shown in different graphs. Besides free resources of renewable energies, the electrical energy storage which shifts the high-cost hours to the low-cost ones, is the main reason for this reduction.

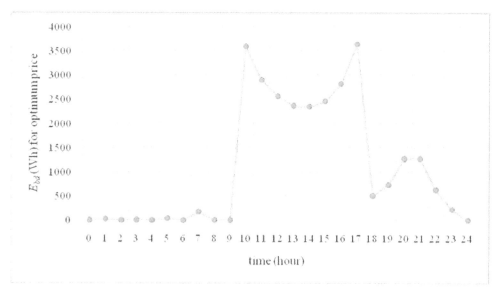

Figure 9. Simulation results for hourly variations of power flow into the balance boundary of the building through battery storage in case I

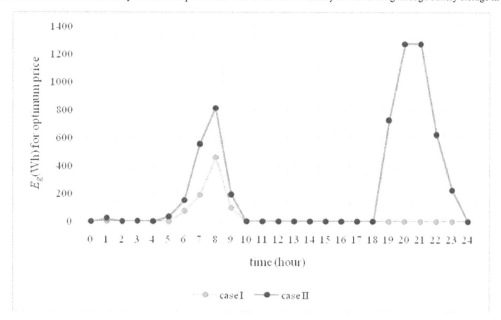

Figure 10. Simulation results for hourly variations of power flow into the balance boundary of the building through electricity grid in cases I and II

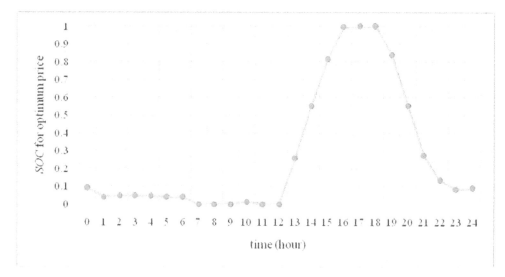

Figure 11. Simulation results for hourly variations of battery state of charge (*SOC*) in case I

References

[1] www.zeb.aau.dk/digitalAssets/29/29578_nzeb-working-definition.pdf(accessed 2013/14/11).

[2] www.sustainableconstructionservices.com.au/products/future-products (accessed 2013/14/11).

[3] Scognamiglio, A., &Røstvik, H. N. (2012). Photovoltaics and zero energy buildings: a new opportunity and challenge for design. *Progress in Photovoltaics: Research and Applications.*

[4] Patel, M. R. (2005). *Wind and solar power systems: design, analysis, and operation.* CRC press.

[5] Tiwari, G. N., &Dubey, S. (2010). Fundamentals of photovoltaic modules and their applications (No. 2). Royal Society of Chemistry.

[6] solarexpert.com/solar-electric/performance-factors (accessed 2013/21/11).

[7] www.pfr.co.uk/cloich/15/Wind-Power/119/Capacity-Factor (accessed 2013/22/11).

[8] www.solarchoice.net.au/blog/victorian-government-applauded-for-energy-efficiency-project (accessed 2013/22/11).

[9] www.ontarioenergyboard.ca/OEB/Consumers/Electricity/Electricity+Prices (accessed 2013/27/11).

[10] tanfon.en.alibaba.com/product/577125632-209438303/3kw_to_5kw_wind_turbine_for_home_and_small_office_small_factory.html (accessed 2013/27/11).

[11] www.engr.colostate.edu/ALP/ALP_91_Byers_East.html (accessed 2013/27/11).

[12] http://blogs.scientificamerican.com/solar-at-home/2010/07/30/a-solar-detective-story-explaining-how-power-output-varies-hour-by-hour (accessed 2013/30/11).

[13] Arif, M. T., Oo, A. M., & Ali, A. S. (2013). Estimation of Energy Storage and Its Feasibility Analysis.

[14] The impact of commercial and residential sectors' EEIs on electricity demand. EMET Consultants Pty Limited. 2004. Online resource: www.ret.gov.au/Documents/mce/energy-eff/nfee/_documents/consreport_07_.pdf.

[15] www.gams.com (accessed 2013/20/12).

4

Hydrogen Energy as Advance Renewable Resource

Krishna Kumar, Omprakash Sahu*

Department of Chemical Engineering, NIT Raipur, India
*Corresponding author: Lops0121@gmail.com

Abstract Reducing the demand on fossil resources remains a significant concern for many nations. Renewable-based processes like solar- or wind-driven electrolysis and photo biological water splitting hold great promise for clean hydrogen production; however, advances must still be made before these technologies can be economically competitive. Approximately 95% of the hydrogen produced today comes from carbonaceous raw material, primarily fossil in origin. Only a fraction of this hydrogen is currently used for energy purposes; the bulk serves as a chemical feedstock for petrochemical, food, electronics and metallurgical processing industries. However, hydrogen's share in the energy market is increasing with the implementation of fuel cell systems and the growing demand for zero-emission fuels. Hydrogen production will need to keep pace with this growing market. In this regard's an effort has been made to study of hydrogen as new renewable energy resources.

Keywords: *biomass, carbon dioxide, Methane, process, utilization*

1. Introduction

Energy is an essential factor in development since it stimulates, and supports economic growth, and development. Fossil fuels, especially oil and natural gas, are finite in extent, and should be regards as depleting assets, and efforts are oriented to search for new sources of energy. The clamour all over the world for the need to conserve energy and the environment has intensified as traditional energy resources continue to dwindle whilst the environment becomes increasingly degraded. The basic form of biomass comes mainly from firewood, charcoal and crop residues. Out of the total fuel wood and charcoal supplies 92% was consumed in the household sector with most of firewood consumption in rural areas.

The term biomass is generally applied to plant materials grown for non-food use, including that grown as a source of fuel. However, the economics of production are such that purpose-grown crops are not competitive with fossil-fuel alternatives under many circumstances in industrial countries, unless subsidies and/or tax concessions are applied. For this reason, much of the plant materials used as a source of energy at present is in the form of crop and forest residues, animal manure, and the organic fraction of municipal solid waste and agro-industrial processing by-products, such as bagasse, oil-palm residues, sawdust and wood off-cuts. The economics of use of such materials are improved since they are collected in one place and often have associated disposal costs [1] The technology is shrinking day-by-day in toady's world and the prefix "nano" implies one of the dimension sizes in this cutting-edge era. Nano is of the order of $10^{(-9)}$. The demand for energy is increasing at a high rate these days. Each and every thing requires energy to carry out its functions. So conservation of energy becomes an important issue. First

of all, let me define what is Energy? The rate of doing work is termed as energy. Energy can be in various forms (light, heat, work, etc). There are two major sources of energy being used in today's world, Renewable and Non-renewable sources. Let me define the non-renewable source first [2]. These are the sources which cannot be renewed after a period of time and becomes exhausted e.g., coal, fossil fuels, natural gas, etc. Such sources release harmful gases to the environment thereby polluting the atmosphere. The alternative to such sources is the renewable source of energy. These sources are clean or say, non-polluting or it reduces the effect of harmful gases to a considerable amount. Such sources can never be exhausted and hence called non-conventional sources of energy [3]. Renewable sources include solar energy, bio-mass, wind, etc. In today's economy, reliable, efficient, pollution free, abundant energy requirement is the major challenge. Our major economy needs, in terms of energy comprises of transportation sector, residential and commercial sectors. We are heavily dependent on the non-renewable sources for our energy needs. Not only these resources will deplete over time, they are also the major source of pollution, which is another key issue in front of the economy. To face these challenges there's needed to come up with the new technology that helps in reducing the problems and also improves our economy [4].

2. History of Renewable Sources of Energy

Several developments have been made and are in progress to harness the renewable source of energy. The increasing popularity of the use of solar energy, wind energy and bio-mass fuels provide the evidence that the work has been in progress to accomplish the task and improves the economy. The Energy Efficiency and Renewable Energy branch of the US Department of Energy Office [6] heads the research, development, and deployment efforts in renewable sources of energy. It develops energy efficiency technologies to provide reliable and affordable supply of energy using the solar, biomass and wind. Due to their efforts, tremendous progress has been made in bringing renewable energy technologies to the marketplace. While the efforts of DOE have started giving results but a lot more has to be done to meet current energy challenges [7].

2.1. Solar Energy

Solar Energy is the energy obtained from the sun. It's the most efficient and clean source of energy to drive the latest trends in the market. Solar energy in the form of photovoltaic cells has been extensively used in electricity and the related areas. It is the permanent and reliable source of non-conventional form of energy. Solar energy is a non-exhaustible, non-pollutant, readily available source of energy [8]. The sun is being used for many purposes in our daily life routine. It is used for several

household activities like cooking, drying clothes, generating electricity and so on and so forth. The solar energy can be used through photovoltaic cells to generate the electricity that can either be stored in the form of battery or used for many applications such as [9]:

- Desalination of salty water,
- Railway signals,
- Electrification,
- Telecommunication, etc.

The modern technology is full of electronic gadgets that utilize solar energy like solar cookers, solar cells, solar heating and cooling systems, solar timber kilns, and power towers.

2.2. Biomass

It is defined as the conversion of biodegradable waste obtained from the organic and inorganic substances into fuel or power. It is an important source of energy used in domestic as well in industrial applications. All such kind of energy sources are used to produce the pollution free atmosphere and healthy and clean surroundings. Several researches show the new trends in the use of biomass productions [10].

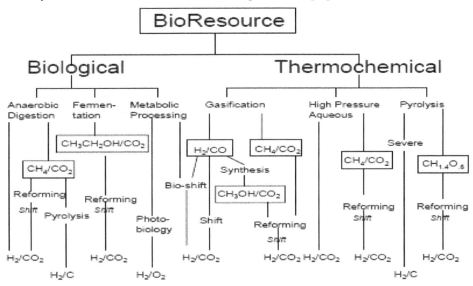

Figure 1. Pathway from biomass for production of hydrogen

2.3. Wind Energy

The air in our environment moves in many directions. The movement is caused by the temperature difference. Hot air rises while cool air comes down. The air from hot tropical region moves towards the cold polar region [11]. The wind energy can be converted into electricity by using a windmill. The wind rotates the fan on the mill which is connected to a dynamo that generates electricity. This wind electricity can also be utilized to produce hydrogen which is the most important element in Hydrogen economy [12].

3. Emerging Technology (Hydrogen)

To conserve and establish the new renewable sources, many countries are trying hard to develop new projects and harness the new renewable forms of energy. These

countries are trying to tap the energy from relatively unexplored sectors. Nanomaterials and Hydrogen fuel cell have the advantage of being smaller and portable. Therefore they have many more applications. The world has already begun the transition to cleaner fossil fuels containing less carbon and more hydrogen. As the world's supply of fossil fuels decreases, the shift to renewable energy sources will continue with a move to resources such as hydrogen, which human beings previously were unable to harness. There are five key policy reasons why this shift is necessary [13]:

(1)The environment. Emissions from vehicles are the largest source of air pollution.

(2)Human health. More than 50,000 people per year may die prematurely from exposure to fine particulates emitted by trucks and buses, power plants and factories.

(3)Economics. The costs of producing oil continue to increase, as deeper wells are drilled farther and farther from markets in harsh climates.

(4)Energy Security. Military and political costs of maintaining energy security internationally are becoming untenable.

(5)Supply. World oil supplies are finite, and are expected to reach their peak as early as 2010. (Cannon, 1998)

To say that hydrogen is an energy "source" is actually a misnomer. That is, hydrogen is not a primary energy like natural gas or oil, existing freely in nature. Instead, hydrogen is an energy "carrier," which means it is a secondary form of energy that has to be manufactured. Although hydrogen is the most abundant element in the universe, practically all of it is found in combination with other elements, for example, water (H_2O), or fossil fuels such as natural gas (CH_4) [14].

Hydrogen can be generated from many primary sources. Today, hydrogen is mainly extracted from fossil fuels through a process known as "steam reforming" (Thomas, 2001). However, most supporters of fuel cells and renewable energy are uncomfortable with the idea of making hydrogen through steam reforming because carbon dioxide (CO_2) is a byproduct of the process. To truly reap the benefits of the environmentally friendly characteristics, environmentalists argue, hydrogen should be made from clean water and clean solar energy, as well as "cleaner" nuclear energy, including fusion. Others disagree, citing that hydrogen produced from natural gas would nonetheless cut greenhouse emissions by up to 40 percent. This, natural gas supporters argue, proves that society does not have to wait for purely renewable hydrogen energy to make significant cuts in greenhouse emissions [15]. Furthermore, proponents of natural gas argue that it may serve as the bridge to a hydrogen and renewable energy society

3.1. Hydrogen Fuel Cell

Hydrogen can be used in a fuel cell which basically operates like a battery. The fuel cell consists of two electrodes and an electrolyte. Hydrogen and Oxygen are passed over the electrodes to generate electricity and Water. Hydrogen cells are used in Auto industry. Compressed hydrogen tanks are used to supply the Hydrogen and Oxygen is used from the air directly. There is no pollution caused by hydrogen fuel cell autos and the only emission is water [16]. If the hydrogen fuel cell autos become main stream instead of exception, we can eliminate autos from the global pollution problem. By the middle of the 21st century, the global community will be dependent on alternative fuels as energy sources. Alternative fuels, those that are not derived from oil, will have taken the place of fossil fuels in powering everything from automobiles, office buildings, and power plants to everyday household items such as vacuum cleaners and flashlights. Driven by environmental, health, economic and political concerns, the global community has been forced to begin developing technology and infrastructure to support the revolution fossil fuels to alternative fuels such as hydrogen [17]. In particular, the world's leaders have targeted the automotive fleet and the internal combustion engine. By replacing the internal combustion engine in automobiles with the hydrogen fuel cell, we could achieve zero emissions of pollutants into the environment [18]. The transformation of the existing transportation system is key to solving many of the world's environmental problems and significantly improving the quality of the air that we breathe. This paper will focus on the role that the Polymer Electrolyte Membrane (PEM) Fuel Cell, widely considered the most practical fuel cell, will play in the switch to alternative fuels [19].

3.2. Hydrogen Economy

The hydrogen economy is an energy system of the coming generations in the near future. The hydrogen can be generated using the renewable energy sources which are readily available. One of such sources is the wind energy that is playing the major role in the generation of hydrogen. The hydrogen economy is capable of fulfilling the human needs of the coming generations [20]. The hydrogen being in the most demand needs the technologies for their production, storage, distribution, and utilization.

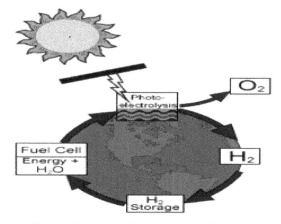

Figure 2. Generation of Hydrogen-clean, cyclic process

There's always the need for clean, efficient, convenient forms of energy which the user can easily access. Hydrogen is one of the many other convenient forms of energy which forms an energy system and satisfies the human energy needs [21].

Hydrogen in chemistry has the following properties:
- Available in huge quantity.
- Can be stored in solid, liquid or gas form.
- Can be converted into other forms of energy efficiently.
- Renewable source as made from the product of water or water vapor.
- Easily transportable.
- Hydrogen as an energy carrier is environmentally compatible.

3.3. Hydrogen Production

Several technologies have been developed to produce hydrogen. Some of the ways have been attempted to describe regarding the hydrogen production. Hydrogen is mainly being produced from fossil fuels in refineries or in industries. The fossil fuels which are used for hydrogen production are in the form of coal, crude oil or natural gas. These fuels produce carbon-dioxide gas during their production process. The processes involved are hydro-treating and hydro- cracking. To avoid the emission of carbon-dioxide gas many other technologies are coming up to produce cost effective hydrogen. Water electrolysis is one of the efficient methods to produce hydrogen but it needs electricity which is expensive [22]. If the method of water electrolysis is being used with photovoltaic (PVs) then that would be more suitable as well an effective method. But photovoltaic cells are costly to produce and install so even though highly efficient but not a good alternative. Wind energy is the other way to produce hydrogen at a low cost but this energy can be utilized in the areas where the wind energy is easily available. The energy required to produce hydrogen is more than what it releases during its utilization [23].

3.4. Hydrogen Storage

After production storage becomes and important issue which needs to be taken care of. Hydrogen can be stored as solid, liquid or gas in the form of glass micro-spheres, chemical hydrides, metal hydrides or cryo-adsorbers. Hydrogen storage in caverns, aquifers are costly and cause loss of gas and pressurized gas storage systems are similar as conventional gas storage systems. Liquid hydrogen storage is being used only in the condition of high need of hydrogen. Metal hydride storage system has an advantage of storing hydrogen in terms of safety aspect. This process requires system set up and the release of heat during the process is another important factor to make this storage system more popular [24].

3.5. Hydrogen Transport and Distribution

Hydrogen transportation by pipeline is up to 200 km from production to utilization sites but for effective transportation high capacity reciprocating compressors are used. The pipelines used for hydrogen transportation requires large diameter and more compression power. Due to low volume of hydrogen and lower pressure losses, less recompression stations are required and that too placed far apart. It has been estimated that transportation of hydrogen is cheaper comparative to electricity transport [25].

3.6. Hydrogen Utilization

The use of hydrogen as a fuel in the internal combustion engines has been found to a great extent. The hydrogen is more efficiently use by 20% in the internal combustion engines. The greater advantage is its more clean that is the use of hydrogen causes less amount of pollution compare to other gasoline engines. Hydrogen use in jet engines and turbines produces the only pollutant nitrogen oxides. Use of hydrogen in biomedical technology is becoming popular in the form of micro steam generator. Catalytic burners in household appliances are coming up with the use of combustion of hydrogen only [26].

3.7. Hydrogen Safety

Every process has its own risks and benefits. Similarly hydrogen can be a risk-full factor if the proper care is not done starting from the process of production until the process of utilization. Hydrogen has the smallest molecule so high tendency to leak through the smaller openings. Also due to low ignition energy of hydrogen the flame

becomes nearly invisible and that could be a dangerous issue as it becomes hard to detect if there is a fire [27]. Liquid hydrogen also causes the risk of cold burns. In spite of all the safety hazards hydrogen is still has a very good safety record and is actually a safer fuel than any other gas. Although the transition to a hydrogen society must happen at a relatively rapid pace in order to answer to the demands of the global community, safety considerations should not be overlooked. Although hydrogen is considered to be a safe form of energy, like gasoline or any form of fuel, it does have the potential to be dangerous under certain circumstances [28]. Regardless of whether hydrogen has the potential to be dangerous, government and industry leaders have to be proactive in shaping public perception when attempting to market the new fuel. Unfortunately, many people associate hydrogen with the hydrogen bomb of World War II and the Hindenburg disaster of the 1930s, even though neither incident relates to hydrogen fuel cell technology.

3.8. Benefit

There is an unmistakable link between energy and sustainable human development. Energy is not an end in itself, but an essential tool to facilitate social and economic activities. Thus, the lack of available energy services correlates closely with many challenges of sustainable development, such as poverty alleviation, the advancement of women, protection of the environment, and jobs creation. Emphasis on institution-building and enhanced policy dialogue is necessary to create the social, economic, and politically enabling conditions for a transition to a more sustainable future. On the other hand, biomass energy technologies are a promising option, with a potentially large impact for Sudan as with other developing countries, where the current levels of energy services are low. Hydrogen accounts for about one third of all energy in developing countries as a whole, and nearly 96% in some of least developed countries [29,30].

Climate change is a growing concern around the world, and stakeholders are aggressively seeking energy sources and technologies that can mitigate the impact of global warming. This global concern is manifest in the 1997 Kyoto Protocol, which imposes an imperative on developed nations to identify feasible options by the next Conference of the Parties to the Convention (COP) meeting later in 2001. Possible actions range from basic increases in energy efficiency and conservation, to sophisticated methods of carbon sequestration to capture the most common greenhouse gases (GHGs) emission (CO_2). On the other hand, renewable energies have always been identified as a prime source of clean energies that emit little or no net GHGs into the atmosphere. Forest ecosystems cause effects on the balance of carbon mainly by the assimilation of CO_2 by the aboveground biomass of the forest vegetation. The annual emissions of greenhouse gases from fossil fuel combustion and land use change are approximately 33×10^5 and 38×10^5 tons respectively [31]. Vegetation and in particular forests, can be managed to sequester carbon. Management options have been identified to conserve and sequester up to 90 Pg C in the forest sector in the next century, through global afforestation [32,33]. This option may become a necessity (as recommended at the Framework Convention on

Climate Change meeting held in Kyoto), but a preventative approach could be taken, reducing total GHGs emissions by substituting biomass for fossil fuels in electricity production.

Simply sequestering carbon in new forests is problematic because trees cease sequestering once they reach maturity, and as available land is used up the cost of further afforestation will grow. Indeed the cost of reducing the build-up of GHGs in the atmosphere is already lower for fossil fuel substitution than for sequestration, since fast growing energy crops are more efficient at carbon removal, and because revenue is generated by the scale of electricity. Some biomass fuel cycles can also provide the additional benefits of enhanced carbon storage [34]. The relative merits of sequestration versus fossil fuel substitution are still debated. The flow of carbon during the life cycle of the biomass should determine whether it is better left standing, used as fuel or used as long-lived timber products [35]. Where there are existing forests in good condition there is general agreement that they should not be cut for fuel and replanted [36]. This principle also concurs with the guidelines for nature protection, i.e., energy crops should never displace land uses of high ecological value. Where afforestation is undertaken, however, fossil fuel substitution, both by using wood fuel and using timber as a renewable raw material, should be more sustainable and less costly approach than sequestration could also be used to displace the harvest of more ecologically valuable forests. For efficient use of Bioenergy resources, it is essential to take account of the intrinsic energy potential. Despite the availability of basic statistics, many differences have been observed between the previous assessments of Bioenergy potential [37]. These were probably due to different assumptions or incomplete estimations of the availability, accessibility and use of by products

4. Conclusion

Hydrogen technology can not only provide fuel, but is also important for comprehensive utilization of biomass forestry, animal husbandry, fishery, evaluating the agricultural economy, protecting the environment, realizing agricultural recycling, as well as improving the sanitary conditions, in rural areas. The hydrogen energy, one of the important options, which might gradually replace the oil in facing the increased demand for oil and may be an advanced period in this century. Any county can depend on the biomass energy to satisfy part of local consumption. Development of Hydrogen technology is a vital component of alternative rural energy programme, whose potential is yet to be exploited. A concerted effect is required by all if this is to be realised. The technology will find ready use in domestic, farming, and small-scale industrial applications. Support hydrogen research and exchange experiences with countries that are advanced in this field. In the meantime, the biomass energy can help to save exhausting the oil wealth. The diminishing agricultural land may hamper Hydrogen energy development but appropriate technological and resource management techniques will offset the effects. The conclusion obtained from the above topic is that we should increase the use of renewable sources of energy and decrease the use of non renewable resources. Existing renewable resources are well established and proven. It has been seen through the various articles that available renewable energy resources are helping in the production of the other forms of energy which makes our energy system more strong and economical. Likewise the production of hydrogen, from the available wind energy, and its usage is more clean, safe and efficient. They are commercially available and are being utilized. The new upcoming technologies in renewable resources are very promising but a lot more research and infrastructure is required before it can be adapted.

References

[1] Sherif, S.A., Barbir, F., Veziroglu, T. N., Wind energy and the hydrogen economy-review of the technology, Retrieved April 19, 2006.

[2] Advanced Materials & Composites News (2001) January 2, v23 i507.

[3] Borgwardt, Robert H., "Platinum, fuel cells, and future US road transport," Transportation Research Part D 6 (2001) 199-207.

[4] Cannon, James, "Hydrogen: America's Road to Sustainable Transportation" (1988).

[5] Clureanu, Mariana and Roberge, Raymond, "Electrochemical Impedance Study of PEM Fuel Cells, Experimental Diagnostics and Modeling of Air Cathodes" (2001) The Journal of Physical Chemistry. B, Materials, surfaces, interfaces and biophysical, 105. 3531-3539.

[6] Gottesfield, Shimshon, "The Polymer Electrolyte Fuel Cell: Material Issues in a Hydrogen Fueled Power Source," (2000), http://education.lanl.gov/RESOURCES/h2/gottesfeld/education.html.

[7] Hackney, Jeremy and de Neufville, Richard, "Life cycle model of alternative fuel vehicles: emissions, energy and cost trade-offs" (2001) Transportation Research. Part A, General.

[8] Robinson, G. "Changes in construction waste management", Waste Management World, pp. 43-49, May-June 2007.

[9] Sims, R. H. "Not too late: IPCC identifies renewable energy as a key measure to limit climate change", Renewable Energy World 10 (4): 31-39, 2007.

[10] Omer, A.M., et al. "Biogas energy technology in Sudan", Renewable Energy, 28 (3): 499-507, 2003.

[11] Omer, A.M. "Review: Organic waste treatment for power production and energy supply", Cells and Animal Biology, 1 (2): 34-47, 2007.

[12] Omer, A. M. "Renewable energy resources for electricity generation", Renewable and Sustainable Energy Reviews, Vol. 11, No. 7, pp. 1481-1497, United Kingdom, September 2007.

[13] Bacaoui, A., Yaacoubi, A., Dahbi, C., Bennouna, J., and Mazet, A. "Activated carbon production from Moroccan olive wastes-influence of some factors", Environmental Technology, 19: 1203-1212. 1998.

[14] Rossi, S., Arnone, S., Lai, A., Lapenta, E., and Sonnino, A. "ENEA's activities for developing new crops for energy and industry. In: Biomass for Energy and Industry (G. Grassi, G. Gosse, G. dos Santos Eds.), Vol. 1, p. 107-113, Elsevier Applied Science, London and New York, 1990.

[15] Omer, A.M. "Renewable energy potential and future prospects in Sudan", Agriculture Development in Arab World, 3: 4-13. 1996.

[16] FAO. "State of the world's forest", Rome: FAO, 1999.

[17] Haripriye G. "Estimation of biomass in India forests", Biomass and Bioenergy, 19: 245-58, 2000.

[18] Hall O. and Scrase J. "Will biomass be the environmentally friendly fuel of the future?", Biomass and Bioenergy. 15: 357-67, 1998.

[19] Omer, A.M. "Biomass energy potential and future prospect in Sudan", Renewable & Sustainable Energy Review, 9: 1-27, 2005.

[20] Singh, et al, "Biomass conversion to energy in India: a critique", Renewable and Sustainable Energy Review, 14: 1367-1378.

[21] Duku, et al, "Comprehensive review of biomass resources and biofuels potential in Ghana", Renewable and Sustainable Energy Review, 15: 404-415.

[22] Cheng, et al, 'Advanced biofuel technologies: status and barriers', World Bank Report, WPS5411, 2010.

[23] Bessou, et al, 'Biofuels, greenhouse gases and climate change'', Agronomy for Sustainable Development.

[24] Mirodatos, C.; Pinaeva, L.; Schuurman, Y. (2001). The Carbon Routes in Dry Reforming of Methane, American Chemical Society Fuel Chemistry Division Preprints 46:pp. 88-91.

[25] Hydrogen Program (2000). Sorption Enhanced Reaction Process for Production of Hydrogen (H2-SER). U.S. DOE Hydrogen Program Annual Operating Plan FY2000 and FY2001: pp. 3-18 to 3-19.

[26] Hydrogen Program (2001). Thermal Dissociation of Methane Using a Solar-Coupled Aerosol Flow Reactor. *U.S. DOE Hydrogen Program Annual Operating Plan FY2001*: pp. 3-5 to 3-6.

[27] Abedi, J.; Yeboah, Y. D.; Bota, K. B. (2001). Development of a Catalytic Fluid Bed Steam Reformer for Production of Hydrogen from Biomass. *Fifth International Biomass Conference of the Americas, Orlando, Florida (Cancelled).* Abstracts to be published as a CD/ROM.

[28] Chornet, E.; Czernik, S.; Wang, D.; Gregoire, C., and Mann, M. (1994). Biomass to Hydrogen via Pyrolysis and Reforming. *Proceedings of the 1994 U.S. DOE Hydrogen Program Review, April 18-21, 1994, Livermore California,* NREL/CP-470-6431; CONF-9404194: pp. 407-432.

[29] Zhu, H.; Suzuki, T.; Tsygankov, A. A.; Asade, Y., and Miyake, J. (1999) (Japan). Hydrogen Production for Tofu Wastewater by Rhodobacter Sphaeroides Immobilized in Agar Gels. Int. J. Hydrogen Energy; 24: pp. 305-310.

[30] Weaver, P.; Maness, P.-C.; Markov, S., and Martin, S. (1996) (USA). Biological H2 from Syngas and from H2O. Proceedings of the 1996 U.S. DOE Hydrogen Program Review, Miami, FL, May 1-2; NREL/CP-430-21968: pp. 325-330.

[31] Sparling, R.; Risbey, D., and Poggi-Varaldo, H. M. (1997) (Canada and Mexico). Hydrogen Production from Inhibited Anaerobic Composters. Int. J. of Hydrogen Energy; 22: pp 563-566.

[32] Tanisho, D. and Ishiwata, Y. (1995) (Japan). Continuous Hydrogen Production from Molasses by Fermentation Using Urethane Foam as a Support of Flocks. Int. J. Hydrogen Energy; 20(7):pp. 541-545.

[33] Andrian, S. and Meusinger, J. (2000). Process Analysis of a Liquid-feed Direct Methanol Fuel Cell System. Journal of Power Sources 91: pp. 193-201.

[34] Karim, G. A. and Zhou, G. (1993). A Kinetic Investigation of the Partial Oxidation of Methane for the Production of Hydrogen. Int. J. Hydrogen Energy, 18(2):pp. 125-129.

[35] Qiyuan, Li. (1992 a). A New Kind of Hydrogen-Producing Catalyst for Hydrocarbon Steam Reforming. Int. J. Hydrogen Energy, 17(2):pp. 97-100.

[36] Bjorklund, A. Melaina, M. Keoleian, G. (2001) (Sweden, USA) Hydrogen as a transportation fuel produced from thermal gasification of municipal solid waste an examination of two integrated technologies Int. J. Hydrogen Energy; 26 pp. 1209-1221.

[37] Finkenwirth, O. and Pehnt, M. (2000) (Germany). Life Cycle Assessment of Innovative Hydrogen Production Path. The International Hydrogen Energy Forum 2000, September 11-15, 2000: pp. 161-179.

5

A Review of Direct and Indirect Solar Cookers

Mohammadreza Sedighi[1,*], Mostafa Zakariapour[2]

[1]Department of Mechanical Engineering, Islamic Azad University Nour Branch, Nour, Iran
[2]Department of Mechanical Engineering, K.N.Toosi University of Technology, Tehran Iran
*Corresponding author: mrsedighi67@gmail.com

Abstract The sun's free, zero-emissions energy produces no household air pollution, preserving the environment as people cook food and pasteurize drinking water. In recent years, much experience has been acquired with the solar cooking systems described. In present work a review has been made to study conducted researches in the field of solar cookers. Experimental, theoretical, numerical analyses are included to compare operation and efficiency of solar cookers. Also the article reviews and summarizes findings of conducted researches on factors influence solar cooker use rates.

Keywords: solar cookers, Developing impacts, efficiency, exergy

1. Introduction

The continuous increase in the level of greenhouse gas (GHG) emissions and the increase in fuel prices are the main driving force to utilise various source of renewable energy [1]. Among the clean energy technologies, solar energy is recognized as one of the most promising choice since it is free and provides clean and environmentally friendly energy [2-8]. The Earth receives 3.85 million EJ of solar energy each year [9]. Solar energy offers a wide variety of applications in order to harness this available energy resource. Among the thermal applications of solar energy, solar cooking is considered as one of the simplest, the most viable and attractive options in terms of the utilization of solar energy [10].

Solar cookers suggest clean and free cooking which is attraction for either modern urban life as alternative free and clean energy and rural living in developing countries that are grappled with lack of the energy. Firewood is used as fuel in family cooking in rural. In India, 47% of the energy for home cooking comes from wood, and in many Africa countries, this value is higher than 75%, such as in Mali or Burkina Faso, where it reaches 95% [11]. Over 50% of the population in Nicaragua use wood as fuel for cooking, and over 53% of the country's overall energy consumption comes from wood (GHA, 2003) [12]. Similar situation has been reported in other countries, such as, Ethiopia, Peru, and Indonesia. Wood cut for cooking purposes contributes to the 16 million hectares of forest destroyed annually.

In near future, the large-scale introduction of solar energy systems, directly converting solar radiation into heat, can be looked forward to. The continuous increase in the level of greenhouse gas emissions and the increase in fuel prices are the main driving forces behind efforts to more effectively utilize various sources of renewable energy [13].

Energy consumption for cooking in developing countries is a major component of the total energy consumption, including commercial and non-commercial energy sources. Energy requirement for cooking accounts for 36% of total primary energy consumption in India. Hence, there is a critical need for the development of alternative, appropriate, affordable mode of cooking for use in developing countries [14]. Most of the thickly populated countries are blessed with abundant solar radiation with a mean daily solar radiation in the range of 5–7 kWh/m2 and have more than 275 sunny days in a year [15].

Cooking with the energy of the sun is not a new or novel idea. According to Halacy and Halacy [16] the first scientist to experiment with solar cooking was a German physicist named Tschirnhausen (1651–1708). He used a large lens to focus the sun's rays and boil water in a clay pot. His experiments were published in 1767 by a Swiss scientist Horace de Sausure who also discovered that wooden "hotboxes" he built produced enough heat to cook fruit. French Scientist Ducurla improved on the hotbox design by adding mirrors to reflect more sunlight and insulating the box. The first book on the subject "Solar Energy and its Industrial Applications" was published by August Mouchot. In 1877, Mouchot designed and built solar cookers for French soldiers in Africa and in 1878 exhibited a solar concentrator at the Parisexhibit. The first recorded solar cooker to be used on South African soil was probably by Sir John Herschel during a scientific expedition to the Cape of Good Hope in 1885. The stove was made out of mahogany, painted black, buried into sand for better insulation and covered by a double glazing to reduce heat losses [17]. Increased public interest in solar stoves emerged in the 1950s and 1960s.

Interest in renewable energy during this period was fuelled by the aftermath of the Second World War with its fuel shortages and rationing, an increased desire to use solar energy "to help people" and as a potential area of investment [18]. Independence gained by former colonial states brought a focus on development and the need to address the "underdeveloped" state of these countries. Lastly, the oil crisis of the early 1970s also contributed to efforts to become less dependent on nonrenewable sources of energy. Growing fuel wood and other energy shortages, coupled with expanding populations in China and India, encouraged governmental research on alternatives in the 1970s with China holding its first seminar on solar cooking in 1973 [19]. Activities the 1980s and 1990s built on earlier efforts at first. China began distributing subsidized cookers in 1981. The ULOG group in Switzerland, EG Solar in Germany and Solar Cookers International were all founded during the 1980s. The work of Barbara Kerr and Shery Cole resulted in a solar cooker kit that was easy to build by the user and served as foundation for the development of a solar panel cooker by Solar Cookers International, which is still used today [19].

2. Cooking

First, The cooking is based on heating a given of food to the boiling temperature of water and in the second part the food is kept at the boiling temperature for a certain period of time depending on the nature of the food. The obviations indicate that the mass flow rate of the gas in the first part is 2-3 times greater than the second part [20].

Lof [21] has described the principles of cooking. As per his principle, the energy requirement is at maximum during the sensible heating period. Heat required for physical and chemical changes involved in cooking is less. The energy required for a specific cooking operation is not always well defined and can vary widely with the cooking methods used. During cooking, 20% of heat is spent in bringing food to boiling temperature, 35% of heat is spent in vaporization of water and 45% of heat is spent in convection losses from cooking utensils. Insulating the sides of the vessel and keeping the vessel covered with a lid can considerably reduce the heat losses.

3. Solar Cookers

A solar cooker is a device which uses the energy of direct sun rays (which is the heat from the sun) to heat, cook or pasteurize food or drink. The vast majority of solar cookers presently in use are relatively cheap, low-tech devices. Because they use no fuel and cost nothing to operate, many nonprofit organizations are promoting their use worldwide in order to help reduce fuel costs (for low-income people) and air pollution, and to slow down the deforestation and desertification caused by gathering firewood for cooking. Solar cookers are classified into direct and indirect solar cookers depending upon the heat transfer mechanism to the cooking pot. Direct type solar cookers use solar radiation directly in the cooking process while the indirect cookers use a heat transfer fluid to transfer the heat from the collector to the cooking unit.

3.1. Direct Solar Cookers

Direct solar cookers may be considered the most common type available due to their ease of construction and low-cost material [22]. Commercially successful direct type cookers are box type and concentrating type cookers. Box type solar cooker is an insulated container with a multiple or single glass cover [23]. This kind of cooker depends on the green house effect in which the transparent glazing permits the passage of shorter wavelength solar radiation, but is opaque to most of the longer wavelength solar radiation coming from relatively low temperature heated objects [24].

The inner part of the box is painted black in order to maximize the sunlight absorption. Maximum four cooking vessels are placed inside the box [25,26]. The cover of the box usually comprises a two-pane "window" that lets solar radiation enter the box but keeps the heat from escaping. This in addition to a lid with a mirror on the inside that can be adjusted to intensify the incident radiation when it is open and improve the box's insulation when it is closed [27]. The speed of the cooking depends on the cooker design and thermal efficiency. The schematic of box type cookers with single reflectors shown in Figure 1, Figure 2.

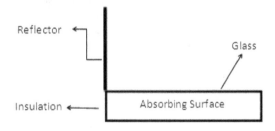

Figure 1. Schematic of box type cookers with single reflectors

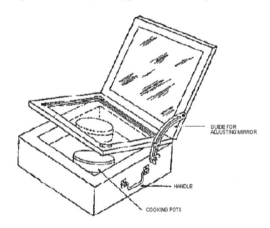

Figure 2. Schematic of box type cookers with single reflectors

Harmim et al. [28] experimentally investigated a box-type solar cooker with a finned absorber. The results indicated that solar box cooker equipped with fins was about 7% more efficient than the conventional box-type solar cooker. The time required for heating water up to the boiling temperature was reduced about 12% when a finned absorber plate was used.

Experimental studies were conducted to see the effect of sand and granular carbon used as the heat absorbing material on the surface of absorber plate in solar box cooker by A. Saxena and et al [1]. An annual performance of solar cooker provided with a mixture of material spread on the absorber tray has been estimated for the different months by considering the actual values on the day of cooking trials (Table 1). The main advantages of box-type solar cookers are: They make use of both direct and diffuse solar radiation; Several vessels can be heated at once; They can double as an oven (not for crispy baked goods); They are light and portable; They are easy to handle and operate; They needn't track the sun; The moderate temperatures make stirring unnecessary; The food can be kept warm until evening; The boxes are easy to make and repair using locally or regionally available materials; They are relatively inexpensive (compared to other types of solar cookers).

Table 1. Year around performance of box type solar cooker from April 2008 to March 2009

S. No	Month	TAmbient (°C)	TSolar box cooker (°C)	Twater (°C)	Radiation (w/m2)	Wind Speed (m/s)	Time
1	April	42	145	99	950	5.38	12.00- 14.00
2	May	40	120	97	900	5.8	12.00- 14.00
3	June	38	127	98	850	5.6	12.00- 14.00
4	July	36	105	96	830	4.12	12.00- 14.00
5	August	35	100	95	800	3.97	12.00- 14.00
6	September	31	98	93	820	5.33	12.00- 14.00
7	October	31	96	89	800	5.33	12.00- 14.00
8	November	24	92	84	780	3.42	12.00- 14.00
9	December	21.5	91	82	750	2.67	12.00- 14.00
10	January	21	90	80	720	4.22	12.00- 14.00
11	February	25	94	85	750	4.01	12.00- 14.00
12	March	32	115	98	850	4.89	12.00- 14.00

Disadvantages of solar box cookers include: slow cooking process due to low temperatures Cooking must be limited to the daylight hours; The glass cover causes considerable heat losses; Such cookers cannot be used for frying or grilling [27].

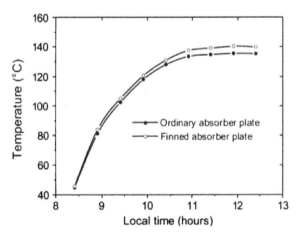

Figure 3. Finned and ordinary absorber plate temperatures

The most elementary kind of reflector cooker is one that consists of (more or less) parabolic reflectors and a holder for the cooking pot situated at the cooker's focal spot [27].

A solar parabolic cooker simply consists of a parabolic reflector with a cooking pot which is located on the focus point of the cooker and a stand to support the cooking system [29]. Concentrating type solar cooker is working on one or two axis tracking with a concentration ratio up to 50 and temperature up to 300 C, which is suitable for cooking. Cookers that concentrate light from below and cookers that concentrate light from above are the two major types of concentrating solar cookers.

Within few hours of sunshine, the cooker makes tasty meals for 4–5 persons at gentle temperatures, cooking food and preserving nutrients without burning and drying out. Figure 4 shows the commercial parabolic cooker.

The advantages of reflector cookers include: the ability to achieve high temperatures; and accordingly short cooking times; relatively inexpensive versions are possible; some of them can also be used for baking. Disadvantages are their size, cost, the risk of fires and burns and the inconvenience to adjust the cooker as it requires frequent directional adjustment to track the sun.

Figure 4. commercial parabolic cooker

3.2. Indirect Solar cookers

In indirect type solar cookers, the pot is physically displaced from the collector and a heat-transferring medium is required to convey the heat to the cooking pot. Solar cooker with flat plate collector, evacuated tube collector and concentrating type collector are commercially available cookers under this category.

Schwarzer and Silva [30]. Developed flat-plate solar cooker which can be incorporated into the construction of kitchen as shown in Figure 5. The two basic system components are the solar collectors with reflectors and a cooking unit. Peanut or sunflower oil is used as heat transfer medium and the cooker is designed with two non-removing pots. Disadvantages of this cooker are nonremovable pots, which makes cleaning and dishing food difficult.

Figure 5. Outdoors cooker with heat storage developed by Schwarzer and Silva [30]

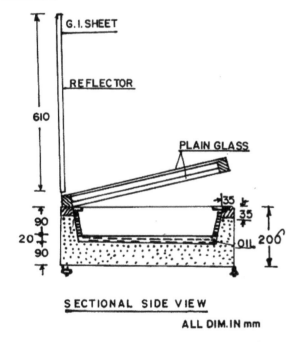

SECTIONAL SIDE VIEW

ALL DIM.IN mm

Figure 6. The schematics of a hot box storage solar cooker [15]

Balzar [31] developed vacuum tube collector-based solar cooker. It consists of a vacuum tube collector with integrated long heat pipes directly leading to the oven plate. Solar cookers using vacuum tube collectors have several advantages. They do not need tracking. They can reach high temperatures and cooking can take place in the shade or inside a building because of the spatial separation of collecting part and oven unit.

Thermal energy storage [32] is essential whenever there is a mismatch between the supply and consumption of energy. The solar cookers must contain a heat storage material to store thermal energy in order to solve the problem of cooking outdoors and impossibility of cooking food due to frequent clouds in the day or during off-sunshine hours. Thermal energy can be stored as a change in internal energy of a material as sensible heat, latent heat and thermo-chemical or combination of these. In this section, the different types of solar cookers which use sensible or latent heat storage materials are summarized.

In sensible heat storage, thermal energy is stored by raising the temperature of a solid or liquid. A hot box

solar cooker with used engine oil (Figure 6) as a storage material has been designed, fabricated and tested by Nahar [15] so that cooking can be performed even in the late evening.The maximum stagnation temperature inside the cooking chambers of the hot box solar cooker with storage material was the same as that of the hot box solar cooker without storage during the day time, but it was 23°C more in the storage solar cooker from 1700 to 2400 h. The efficiency of the hot box storage solar cooker has been found to be 27.5%.

The oil is heated up in the collectors and moves by natural flow to the cooking unit, where it transfers part of its sensible energy to the double-walled cooking pots. Manually controlled valves guide the oil flow rate either to the pots or to the storage tank. The major advantages are the possibility of indoor cooking, the use of a thermal storage tank to keep the food warm for longer periods of time or night cooking and the reach of high temperatures of the working fluid in a short period of time.

Latent heat storage [33,34] makes use of the energy stored when a substance changes from one phase to another. The use of PCMs for storing heat in the form of latent heat has been recognized as one of the areas to provide a compact and efficient storage system due to their high storage density and constant operating temperature. PCM (Phase Change Material) take advantage of latent heat that can be stored or released from a material over a narrow temperature range. PCM possesses the ability to change their state with a certain temperature range. These materials absorb energy during the heating process as phase change takes place and release energy to the environment in the phase change range during a reverse cooling process. Basically, there are three methods of storing thermal energy: sensible, latent and thermo-chemical heat or cold storage. Thermal energy storage in solid-to-liquid phase change employing phase change materials.

(PCMs) has attracted much interest in solar systems due to the follow advantages: (i) It involves PCMs that have high latent heat storage capacity; (ii) The PCMs melt and solidify at a nearly constant temperature; (iii) A small volume is required for a latent heat storage system, thereby the heat losses from the system maintains in a reasonable level during the charging and discharging of heat.

A solar cooker with latent heat storage for cooking food in the late evening was designed and tested [35]. In this design, the phase change material (PCM) was filled below the absorbing plate. Commercial grade stearic acid (melting point 55 8C, latent heat of fusion 161 kJ/kg) is used as a latent heat storage material. In such type of design, the rate of heat transfer from the PCM to the cooking pot during the discharging mode of the PCM is slow and more time is required for cooking food in the evening.

Hussein et al. [36] designed a novel indirect solar cooker with outdoor elliptical cross section, wickless heat pipes, flat-plate solar collector and integrated indoor PCM thermal storage and cooking unit as shown in Figure 7. They constructed and tested under actual meteorological conditions of Giza, Egypt. Two plane reflectors are used to enhance the insolation falling on the cooker's collector, while magnesium nitrate hexahydrate (T=89_C, latent heat of fusion 134 kJ/kg) is used as the PCM inside the indoor cooking unit of the cooker. It is found that the average daily enhancement in the solar radiation incident on the collector surface by the south and north facing reflectors is about 24%. Different experiments have been performed on the solar cooker without load and with different loads at different loading times to study the possibility of benefit from the virtues of the elliptical cross section wickless heat pipes and PCMs in indirect solar cookers to cook food at noon and evening and to keep food warm at night and in early morning. The results indicate that the present solar cooker can be used successfully for cooking different kinds of meals at noon, afternoon and evening times, while it can be used for heating or keeping meals hot at night and early morning.

Cross sectional views of the indirect solar cooker under investigation are shown in Figure 7, Figure 8.

The condenser section of the closed loop wickless heat pipes network was made of a copper tube of 9.5 mm nominal diameter and about 7 m length in the form of a helical coil as shown in Figure 3. The condensing helical coil (i.e. condenser section) was then flame heated, and its inner surface was cleaned and rinsed by the procedures performed on the evaporator assembly [37,38]. Then, it was incorporated into an indoor cooking unit that has an inner galvanized iron box of 0.56 m length, 0.28 m width and 0.165 m height as shown in Figure 7 and Figure 8.

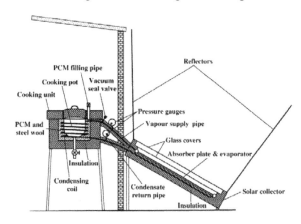

Figure 7. Cross sectional side view of the present solar cooker shows its main components

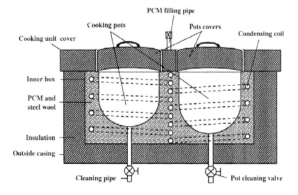

Figure 8. Cross sectional front view of the indoor PCM cooking unit shows its main components

During sunshine hours, heated water transfers its heat to the PCM and is stored in the form of latent heat through a stainless steel tubing heat exchanger. This stored heat is utilized to cook the food in the evening time or when sun intensity is not sufficient to cook the food. They

concluded that system was able to cook successfully twice (noon and evening) in a single day during Japanese summer months. Noon cooking did not affect evening cooking, and the evening cooking using the heat through PCM storage unit was found to be faster than noon cooking.

4. Numerical Analysis

El-Sebaii [39] numerically analyzed a box-type solar cooker with outer-inner reflectors. Numerical calculations were carried out for different tilt angles of the outer reflector on a typical winter day (20 January) in Tanta, Egypt. The optimum tilt angle of the outer reflector was 60_. For this specific value, it was observed that the specific and characteristic boiling times were decreased by 50% and 35%, respectively, compared to the case without the outer reflector. The overall utilization efficiency of the cooker was determined to be 31%.

In another research which was conducted by Terres et al. [40], numerical simulation results were shown to determine the heating in liquids when a solar cooker box type with internal reflector is used to this end. The data evaluated correspond to temperature values from bee honey, olive oil, milk and water when they are heated in the solar cooker. The maximum simulation temperatures reached are 91.8°C, 91.6°C, 86.2°C and 85.3°C that correspond to bee honey, olive oil, milk and water respectively. A comparative between simulation and experimental results also were shown. The values presented evidence the influence of the specific heat in each fluid considered. In the numerical simulation were used solar radiation and environment temperature values for February 26, 2006 in Mexico City.

Chen et al [41] investigated theoretically on the PCMs used as the heat storage media for box-type solar cookers. The selected PCMs are magnesium nitrate hexahydrate, stearic acid, acetamide, acetanilide and erythritol. For a two dimensional simulation model based on the enthalpy approach, calculations have been made for the melt fraction with conduction only. Different material such as glass, stainless steel, tin, aluminium mixed, aluminium and copper are used as the heat exchanger container materials in the numerical calculations. It is also found that the initial temperature of PCM does not have very important effects on the melting time, while the boundary wall temperature play an important role during the melting and has a strong effect on the melt fraction. The results also show that the effect of thickness of container material on the melt fraction is insignificant.

The results obtained in this paper show that acetamide and stearic acid, should be used as storage media in a box-type solar cooker to cook and/or to keep food warm in the late evening with different heat exchanger container materials. The large value of thermal conductivity of heat exchanger container material did not make a significant contribution on the melt fraction except for at very low thermal conductivities.

5. Energy and Exergy Efficiencies

Richard Petela [42] has been presented the theoretical exergy analysis of a solar cooker and the distribution of the exergy losses in the cooker. Equations for heat transfer between the three surfaces: cooking pot, reflector and imagined surface making up the system, were derived. The model allowed for theoretical estimation of the energy and exergy losses: unabsorbed insolation, convective and radiative heat transfer to the ambient, and additionally, for the exergy losses: the radiative irreversibilities on the surfaces, and the irreversibility of the useful heat transferred to the water.

The exergy efficiency of the SPC, was found to be relatively very low (~1%), and to be about 10 times smaller than the respective energy efficiency which is in agreement with experimental data from the literature. The influence of the input parameters (geometrical configuration, emissivities of the surfaces, heat transfer coefficients and temperatures of water and ambience) was determined on the output parameters, the distribution of the energy and exergy losses and the respective efficiencies.

The principles of radiative heat transfer applied in the present paper are presented e.g. by Holman[43]. Szargut and Petela [44] as well as Szargut et al. [45], present the concept of exergy and its application to the analysis of processes. Extensive review of the problems of radiation exergy is provided by bejan [46]. Some clarifications regarding exergy of thermal radiation are discussed by Petela [47].

The main reason of low efficiency of devices driven by solar radiation lies in the impossibility of full absorption of the insolation.

In relation to the exergy efficiency there is an additional reason which makes this efficiency significantly lower than the energy efficiency. A low exergy performance efficiency of SPC, and of other devices driven by solar radiation, is caused by the significant degradation of energy. The relatively high temperature (~6000 K) of solar radiation is degraded to the relatively low temperature e.g. to the temperature T_w of heated water, which is not much larger than the ambient temperature T_0.

The influence of the geometric configuration of the cooker on its performance was outlined. By applying the variation only of the "openness" (x2) and "depth" (y2) of the considered SPC it was shown that the energy efficiency of above 18%, and exergy efficiency of above 1.6%, could be reached. It can be confirmed by calculation that the determined optimal surface profile of the considered SPC can be scaled up, at the unchanged optimal efficiencies, to the SPC with the all dimensions changed proportionally. The scheme for calculations of the radiation shape factors illustrated in Figure 9.

Shukla and Gupta [48] presented an energy and exergy analysis of a concentrating solar cooker. The cooker was devised for community cooking and integrated with a linear parabolic concentrator which concentration ratio is 20. The experiments were carried out in both summer and winter conditions. Through the experimental results, the average efficiency of the solar cooker was determined to be 14%. Heat losses caused low efficiency were classified as optical losses (16%), geometrical losses (30%) and thermal losses (35%). The rest of the losses were due to edge losses, etc. The maximum temperature that the water in the cooker reached was 98°C during the tests.

In another research N. L. Panwar and et al. [49] presented an energy and exergy analysis of a domestic size parabolic solar cooker in actual use. The experimental time period was from 10:00 to 13:30 solar time. During the experiment, it was found that the maximum temperature of water was 368 K. The energy out of the cooker varied between 46.67 and 653.33 W, whereas its exergy output was in the range, 7.37-46.46 W. Over the time, both efficiencies were decreased because of the optical and thermal losses from the reflector and pot. By using properly insulated cooking pot, the considerable amount of conventional energy can be saved.

It is clear from Figure 10 that the ambient temperature was in the range of 301 K to 309 K. It was minimum at 10:00 h (301 K) and reached the maximum at 13:30 h (309 K).

It is clear from Figure 11 that energy and exergy efficiencies of cooker reduce with corresponding solar time. The maximum energy efficiency was evaluated 32.97% and it was observed at 10:30 h, whereas it was minimum at 13:30 h. As far as maximum exergy efficiency is concerned, it was evaluated 2.18% and it reduces as increasing solar time. Apart from increasing solar radiation, both energy and exergy efficiencies were decreased drastically and this may be due to high loss from pot as it is not insulated.

Ozturk [50,52,53] conducted several experimental researches on solar parabolic cookers and analyzed the performance parameters in terms of thermodynamic laws. Ozturk experimentally examined energy and exergy efficiencies of a simple design and the low cost parabolic cooker under the climatic conditions of Adana which is located in Southern Turkey (at 37_N, 35_E). The energy output of the parabolic cooker was determined to be 20.9–78.1 W, whereas its exergy output was in the range of 2.9–6.6 W. The results showed that the energy and exergy efficiencies of the parabolic cooker were calculated between 2.8–15.7% and 0.4–1.25%, respectively [52].

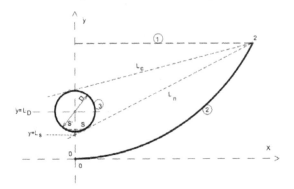

Figure 9. The scheme for calculations of the radiation shape factors

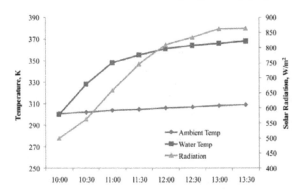

Figure 10. Temperature variation and solar radiation with time

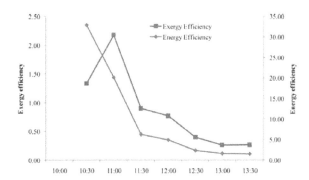

Figure 11. Energy and energy efficiencies with time

6. The Development Impact of Solar Cookers

Solar cooking has regularly been viewed as a solution looking for a problem, or a technological solution developed without sensitivity to user needs.[54]. Specific reference is made here to the activities of the DME/GTZ solar cooker field test executed in South Africa from 1996 which concluded, "Many advancements have been made in the technical advancement of solar cookers, but unfortunately, very little attention has been paid to the social context, as defined by the needs of the potential users." [54].

Different studies investigated solar cooker use rates since 1996 in South Africa. Based on[55-60] Solar cooker use rates can be accepted to be between 31% and 25%. Many factors influence solar cooker use rates and a change in use rates including: external conditions (weather conditions), change in interest and cooking patterns, solar stove characteristics (slow cooking), fuel saving, time saving and etc.

Since it is mainly women who do the cooking in the household, it is mainly their time that is being saved by using a solar cooker. Although most solar stoves cook slower than other stoves, they require very little attention once the food is in the stove [56].

Table 2 provides the results for the corresponding total average (over all users and all fuel types), during the first phase of the solar cooker field test, stating that the overall fuel savings were 38% [61].

Table 2. Average savings for all fuels by households

	Savings (%)	Weight
Parafin	33	0.28
Gas	57	0.16
Wood	36	0.56
Unweighted total average	42	-
Weighted total average	38.4	-

The ex-post purchase study [55] did not investigate savings specifically, but when asked why respondents had bought a solar cooker (independent of the model); the most cited reasons were monetary savings in fuel expenses and convenience (time savings, unattended cooking and having an additional "fuel" source) (Table 3) [55].

Table 3 Reseans for solar cooker acquisition

Resean	Entries
Savings	44
Convenience	29
Other	13

The investigation conducted by Market Research Africa also did not record specific savings (energy, monetary or time) although it was reported that users were motivated by cost savings/energy to cost purchase their solar cookers. Free energy, cost savings and no fuel costs were the most important perceived advantages of solar cookers [62].

7. Conclusion

Solar energy is free, environmentally clean, and therefore is recognized as one of the most promising alternative energy recourses options. In supplying the needed energy, solar cookers can fully or partially replace the use of firewood for cooking in many developing regions. In this paper, a review of the available literature on solar cookers is presented. The review covers a historic overview, classification, operation and thermodynamic analysis of different solar cookers as well as the reasons of why the solar cooking technology has never been able to gain any real extent of popularity.

References

[1] Abhishek Saxena and et al. A technical note on performance testing of a solar box cooker provided with sensible storage material on the surface of absorbing plate, Int. J. Renewable Energy Technology, Vol. 3, No. 2, 2012, 165-173.

[2] Riffat SB, Cuce E. A review on hybrid photovoltaic/thermal collectors and systems. Int J Low – Carbon Technol 2011;6 (3): 212-41.

[3] Cuce E, Bali T. Variation of cell parameters of a p-Si PV cell with different solar irradiances and cell temperatures in humid climates, Fourth international exergy, energy environment symposium, Sharjah, UAE; 19-23 April 2009.

[4] Cuce E, Bali T. A comparison of energy and power conversion efficiencies of m-Si PV celss in Trabzon, Fifth international advanced technologies symposium, Karabuk, Turkey; 13-15 May 2009.

[5] Cuce E, Bali T. Improving performanceparameters of silicon solar cells air cooling, Fifth international edge energy symposium and exhibition, Denizli, Turkey; 27-30 June 2010.

[6] Cuce E, Bali T. Swkucoglu SA. Effects of passive cooling on performance of silicon photovoltaic cells, Int J Low-Carbon Technol 2011; 6 (4): 299-308.

[7] Cuce PM, Cuce E. A novel model of photovoltaic modules for parameter estimation and thermodynamic assessment. Int J Low-Carbon Technol 2012;7 (2): 159-65

[8] Cuce PM, Cice E, Aygun C. Homotopy perturbation method for temperature distribution, efficiency and an effectiveness of conductive straight ns, Int J Low-Carbon Technol 2012.

[9] Johansson TB, Kelly H, Reddy AKN, et al. Renewable energy sources for fuels and electricity. Earthscan Publications Ltd. and Island Press; 1993.

[10] Lahkar PJ, Samdarshi SK. A review of the thermal performance parameters of box type cookers and identification of their correlations. Renew Sust Energy Rev 2010; 14: 1615-21.

[11] GHA, 2003. Global Health Alliance. Improving human and environmental health. http://www.glbhealth.org/ solarcooking".htm 13/ Feb/ 2003.

[12] Schwarzer, K., Krings, T., 1996. Demonstration und Feldtest von Solarkochern mit temporare Speicher Indien und Mali. Shaker, Aachen.

[13] Atul Sharma, C.R. Chen, V.V.S. Murty, Anant Shukla, Solar cooker with latent heat storage systems: A review, Renewable and Sustainable Energy Reviews 13 (2009) 1599-1605

[14] Pohekar SD, Dinesh Kumar M, Ramachandran. Dissemination of cooking energy alternatives in India-a review. Renewable and Sustainable Energy Reviews 2005;9 (4): 379-93.

[15] Nahar NM. Performance and testing of a hot box storage solar cooker. Energy Conversion and Management 2003; 44: 323-31.

[16] Halacy, B., Halacy, C. 19923 Cooking with the sun. Jack Howel, Lafayete, CA.

[17] GTZ and DME, 2002b. Solar cooker compendium volume 1. Scarcity of Household Energy and the rationale of solar cooking. GTZ, Pretoria.

[18] Laird, F. 2005. The society whose time had come. Solar Toda July/August, 36-39.

[19] Knudson, B. 2004. State of the art of solat cooking: A global survey of practices and promotion programs. SCI, Sacramento.

[20] S.K. Hannani, E. Hessari, M. Fardadi, M.K. JeddiMathematical modeling of cooking pots' thermal efficiency using a combined experimental and neural network method, Energy 31 (2006) 2969-2985

[21] Lof GOG. Recent investigation in the use of solar energy for cooking. Solar energy 1963; 7: 125-33.

[22] Funk PA, Larson DL. Parametric model of solar cooker performance. Solar Energy 1998; 62: 63-8.

[23] Saxena A, Varun, Pandey SP, Srivastav G. A thermodynamic review on solar box type cookers. Renew Sust Energy Rev 2011;15: 3301-18.

[24] R.M. Muthusivagami, R. Velraj, R. Sethumadhavan, Solar cookers with and without thermal storage—A review, Renewable and Sustainable Energy Reviews 14 (2010) 691-701

[25] Khan BH. Non-conventional energy resources. Tata McGraw Hill Publications; 2008.

[26] Kothari DP, Singal KC, Ranjan R. Renewable energy resources and emerging technologies. Prentice-Hill;2008.

[27] Klaus Kunhnke, Marianne Reuber, Detlef Schwefel, Solar Cookers in the Third World, Deutsche Gesellschaft für Technische Zusammenarbeit (GTZ) GmbH - 1990

[28] Harmim A, et al. Experimental investigation of a box-type solar cooker with a finned absorber plate. Energy 2010;35:3799-802.

[29] Ashok K. Areview of solar cooker designs. TIDE; 1998; 8: 1-37.

[30] Schwartzer K, Silva MEV. Solar cooking system with or without heat storage for families and institutions. Solar Energy 2003; 75: 35-41.

[31] Bazlar A, et al. A solar cooker using vacuum-tube collectors with integrated heat pipes. Solar Energy 1996; 58: 63-8.

[32] Felix Regin A, et al. Heat transfer characteristics of theral energy systems using PCM capsules: a review. Renewable and sustainble energy reviews 2008; 12: 2438-58.

[33] Sharma SD, Sagara K, Latent heat storae materials and systems: a review. International journal of green energy 2005; 2: 1-56.

[34] Zalba B, Marin JM, Cabeza LF, Mehling H. Review on thermal energy storage with phase change: materials, heat transfer analysis and applications. Applied Thermal Engineering 2003; 23: 251-83.

[35] Buddhi D, Sahoo LK. Solar cooker with latent heat storage: design and experimental testing. Energy Conversion and Management 1997; 38:4 93-8.

[36] Hussein HMS, El-Ghetany HH, Nada SA. Experimental investigation of novelindirect solar cooker with indoor PCM thermal storage and cooking unit. Energy Conversion and Management 2008; 49: 2237-46.

[37] Faghri A. Heat pipe science and technology. UK: Taylor and Frances; 1995.

[38] Hussein HMS, El-Ghetany HH, Nada SA. Performance of wickless heat pipe flat plate solar collectors having different pipes cross sections geometries and filling ratios. Energy Convers Manage 2006; 47: 1539.

[39] El-Sebaii AA. Thermal performance of a box-type solar cooker with outer inner reflectors. Energy 1997; 22 (10): 969-78.

[40] Terres H, Ortega JA, Gordon M, Morales JR, Lizard A. Heating of bee honey, olive oil, milk and water in a solar box type with internal reflectors. In: Energy sustainability conference, Long Beach, California, USA; 27-30 June 2007.

[41] Chen CR, Sharma A, Tyagi SK, Buddhi D. Numerical heat transfer studies of PCMs used in a box type solar cooker. Renew Energy 2008; 33 (5): 1121-29.

[42] Richard Petela, Exergy analysis of the solar cylindrical-parabolic cooker, Solar Energy 79 (2005) 221-233

[43] Holman, J.P., 1997. Heat Transfer, eighth ed. McGraw-Hill., Inc., New York.

[44] Szargut, J., Petela, R., 1965. Exergy. WNT, Warsaw (in Polish).

[45] Szargut, J., Morris, D.R., Steward, F.R., 1988. Exergy Analysis of Thermal, Chemical, and Metallurgical Processes. Hemisphere Publishing, New York.

[46] Bejan, A., 1997. Advanced Engineering Thermodynamics. Wiley, New York.

[47] Petela, R., 2003. Exergy of undiluted thermal radiation. Solar Energy 74, 469-488.

[48] Shukla SK, Gupta SK. Performance evaluation of concentrating solar cooker under Indian climatic conditions. In: Second international conference on energy sustainability, Jacksonville, Florida, USA; 10-14 August 2008.

[49] N. L. Panwar, S. C. Kaushik, and Surendra Kothari, Experimental investigation of energy and exergy efficiencies of domestic size parabolic dish solar cooker, J. Renewable Sustainable Energy 4, 023111 (2012).

[50] Ozturk HH. Second law analysis for solar cookers. Int J Green Energy 2004; 1 (2) 227-39.

[51] Ozturk HH, Oztekin S, Bascetincelik A. Evaluation of efficiency for solar cooker using energy and exergy analyses. Int J Energy 2003.

[52] Ozturk HH. Experimental determination of energy and exergy efficiency of solar parabolic-cooker. Solar Energy 2004; 77 (1): 67-71.

[53] Ozturk HH. Comparison of enerfy and exergy efficiency for solar box and parabolic cookers. J Energy Eng 2007; 133 (1): 53-62.

[54] Marlett Wentzel, Anastassios Pouris, The development impact of solar cookers: A review of solar cooking impact research in South Africa, Energy Policy 35 (2007) 1909-1919.

[55] Synopsis and Palmer Development Consulting, 2000. Long-term House- hold Acceptance of Solar Cookers. Ex-post Purchase Evaluation Study.

[56] Palmer Development Group, 1997a. Solar Cooker Field Test in South Africa. End-user acceptance Phase 1, Main Report, Volume 1. GTZ, Pretoria.

[57] Palmer Development Group, 1997b. Gender Review of the GTZ/DME Solar Cooker Field Test. GTZ, Pretoria.

[58] Kitzinger, X., 2004. Solar Cooker Usage and Lifetime of Solar Cookers in the Three Pilot Regions Huhudi, Pniel and Onseepkans Field report. Internal report. GTZ, Pretoria.

[59] Palmer Development Consulting, 2002a. End-user Monitoring Report. DME/GTZ Solar Cooker Field Test in South Africa. Department of Minerals and Energy Pretoria.

[60] Palmer Development Consulting, 2002 b. Internal Report Prepared for GTZ Evaluation Mission. Additional Inquiries into Use Rates Internal GTZ report.

[61] GTZ and DME, 2002 a. Solar Cooking Compendium. Challenges and Achievements of the Solar Cooker Field Test in South Africa. GTZ, Pretoria.

[62] Market Research Africa, 2003. Profile of Solar Cooker Purchasers Management report. GTZ, Pretoria.

6

Production of Bio-fuel (Bio-Ethanol) from Biomass (Pteris) by Fermentation Process with Yeast

Pradipsaha[1,*], Md. FakhrulAlam[1], Ajit Chandra Baishnab[1], MaksudurRahman Khan[1,2], M. A. Islam[1]

[1]Department of Chemical Engineering and Polymer Science, Shahjalal University of Science and Technology (SUST), Sylhet, Bangladesh
[2]Faculty of Chemical and Natural Resources Engineering, University Malaysia Pahang, Gambang, Kuantan, Pahang, Malaysia
*Corresponding author: pradip-cep@sust.edu

Abstract Bio-Ethanol is a renewable; eco-friendly energy source can be produced from bio-mass (hemicelluloses). Pteris (fern) is grown very fast and has not major economic importance but it is a reliable source of hemicelluloses can be converted to bio-Ethanol. In the hydrolysis of hemicelluloses the concentration of glucose and the reaction rate was observed with respect to the different parameters, like pH, temperature, substrate diameter, substrate loading. In this study we found optimal parameters, NH_4OH treatment as the best treatment among H_2SO_4, NH_4OH, and NaOH treatment. In addition, pH 7, temperature 35°C, substrate diameter 45μm-63μm, and substrate loading 0.25gm in 100ml working volume was found as optimum operation condition for hydrolysis reaction, which was carried out by *pseudomonas sp.*, isolated from cow dung. In this experiment sugar concentration was measured with UV spectrophotometer using DNS reagent finding the equilibrium time is 72hours for hydrolysis process and the maximum sugar concentration of 1.7625mg/l was achieved. Subsequently we study the fermentation process using yeast to produce Bio-Ethanol from reducing sugar solution obtained from the hydrolysis process on optimum reaction condition and yield the Ethanol concentration 0.333 mg/L, measured with UV spectrophotometer. Furthermore, it was also observed that the activity of yeast was sustain, in the reaction condition (pH = 7, temperature = 25°C), only for 50 hours. In sum, it could be conclude that 0.1332 mg of ethanol can be produce from 1gm of pteris where the conversion of reducing sugar to ethanol is 20% approximately in the optimum reaction condition.

Keywords: *Hemicellulose, pteris, hydrolysis, pseudomonas sp, reducing sugar, bio-ethanol, fermentation*

1. Introduction

With the developing of human life style and civilization, emphasis on energy consumption increases day by day. For this reason different nonrenewable energy sources such as coal, NG, petroleum based fuel etc. are the main targets of many countries. But elevation of CO_2 from this fuel is the most head cable reason considered as a main cause of global warming considered by Omer [1]. Beside this all are the nonrenewable energy source are declared not remain much in reservoir to fulfill the growing demand of the world. For this reason many countries of the world are trying to find out a suitable solution of the growing demand of the fuel with the best consideration of the environment. According to the Suleiman [2] renewable and eco-friendly energy sources with existing fuel energy, tidal energy, hydro energy, geothermal energy, solar energy, wind energy, and bio-ethanol energy from cellulosic materials etc., are the solution to overcome from this upcoming problem. However, bio-ethanol energy from cellulosic materials is ahead from other energy because of availability of the raw materials throughout the year with the as usual property of the other renewable

property such as less emission of greenhouse gas, biodegradable and less toxic etc., simultaneously improve the air quality informed by Charles [3]. DoKyoung et al. [4] found that fern (Pteris) is a celluloses plant which is economically less important and has the three common components (cellulose 30-50%, hemicellulose 20-40% and lignin 15-25%). The cellulose and hemicelluloses part can be hydrolyzed to produce reducing sugar since these are in amorphous state but the Lignin is a complex three-dimensional aromatic polymer hydrophobic in nature investigated by Jenni [5]. The hydrolysis process also depends on the particle size of substrate and loading of substrate. Mohammad. and Keikhosro [6] says that processing of lignocellulosic biomass to ethanol consists of four major unit operations (pretreatment, hydrolysis, fermentation and product separation or purification). Zheng et al [7] and Michelle [8] discussed about different types of pre-treatment techniques available such as chemical pretreatment (acid treatment, ammonia treatment, sulfuric acid treatment alkaline wet oxidation and ozone pretreatment), physical pretreatment (steaming, grinding and milling, blending, thermal, and irradiation), biological pre-treatment and the combine pretreatment. Chemical pretreatment overcome the recalcitrance of lignin present in the structure of hemicelluloses and physical pretreatment reduce the biomass physical size (increase the surface area). There are two methods generally used in hydrolysis of cellulose acid/base catalyzed hydrolysis and enzymatic hydrolysis. Alexander et al [9] investigated that Enzymatic hydrolysis is better than the acid/base catalyzed hydrolysis. Again there are two available ways of fermentation SSF (Simultaneous Saccarification and Fermentation) and SHF (Separate Hydrolysis and Fermentation). Again among different types of fermentatation it is observed by Kim et al [10] that SHF is better than the SSF in many causes. In the present study as chemical pretreatment acid treatment,ammonia treatment, sulfuric acid treatment and Physical pretreatment blending were performed from the available techniques. After pretreatment the hydrolysis performed to produce reducing sugar using cellulase produce from cellulytic bacteria *pseudomonas sp.* and then performing the SHF using *Saccharomy cescerevisiae* (Baker's yeast).

2. Materials and Methods

2.1. Isolation and Identification of Cellulytic bacteria

The cow dung a source of different bacteria was collected from the Toker bazaar, Sylhet and the study was performed in the research laboratory of Shahjalal University of Science and Technology, Sylhet, Bangladesh. The bacterial strain was *pseudomonas sp* identified by the Cowan and Steel's Manual [11] for the Identification of Medical Bacteria.

2.2. Extraction of Enzyme from Bacteria

A loop of bacteria cultured in nutrient Agar media was transfer into the production medium (KH_2PO_4:1.00gm/L; K_2HPO_4:1.145gm/L; MgSO4: 0.4gm/L; NH_4SO_4:5.0gm/L; $CaCl_2$:0.05gm/L; $FeSO_4$:0.00125gm/L; CMC:10gm/L) which was sterilized by autoclaving at 121°C for 15 min. 5 ml of the media was transferred into five screw cap test tubes and kept in a shaker incubator at 37°C for 24 hours at 100rpm, after that each of the 5ml seed culture test tube are poured into other 100ml conical flasks containing 35ml sterilized production media and kept in shaker incubator at 37°C for 24 hours at 100rpm. Subsequently the product was centrifuged for 15 minutes at 8000rpm maintaining 4°C temperatures in refrigerated centrifuge. The supernatant was collected in bayel as enzyme and kept in refrigerator for further use.

2.3. Pre-treatment of Pteris

Pre-treatment is required to achieve best enzyme performance, increasing reaction rate. Chemical pre-treatment were performed using H_2SO_4 acid, NaOH, NH_4OH, and Mechanical pre-treatment is performed by using a blender. The pteris was collected from the university area and washed vigorously with distilled water.

2.3.1. Chemical Pre-treatment

200ml of 1% (v/v) H_2SO_4, 1% (w/w) NaOH, and 4% (v/v) NH_4OH ware taken in separate beakers and 20gm of washed fern was mixed in each of the beaker. The mixture solutions was heated at 80-90°C in an oven for two hours, then neutralise with distilled water and dried at 105°C temperature for three hours.

2.3.2. Mechanical Pre-treatment

After different chemical pre-treated pteris leaves were dried and blended in a blender and desired particles sizes were separated using the sieve-shaker.

2.4. Experimental Procedure

The hydrolysis and fermentation reaction were performed in the rotary flask shaker (Model: LRD-750) with working volume 100ml in 250ml flasks. Based on the study purposes various amount of substrate with different particle size were taken in the flask. All the composition and parameter assumed same in the total reaction mixture due to continuous rotation. All experiments were run with different predetermined amount of enzyme in the hydrolysis environment with different amount of substrate, different size of particle, different pH, and different temperature to optimize the corresponding parameter. Samples were taken time to time and boiled with DNS solution after centrifuging the sample to destroy the enzymes activity which confirms the reaction ceasing. Then the sample was analysed for glucose in UV spectrophotometer according to the method by Miller [12]. When the hydrolysis reaction reached at its equilibrium then the reaction mixture were separated by filtration method and performing fermentation to produce bio ethanol using *Saccharomyces cerevisiae (*Baker's yeast) and detect the ethanol concentration using $K_2Cr_2O_7$ reagent and UV spectrophotometric method proposed by Adran and Prifysgol [13].

3. Result and Discussion

Though the reaction rate depends on different parameters, but in this study we assumed that the rate is affected only by Pre-treatment, Substrate loading, Enzyme loading, Particle size, Temperature, pH. To understand the effect of one parameter other parameters were kept constant and some run with differing the parameter which wanted to investigate.

3.1. Effect of Pre-treatment and Particle Size

Dried equal amount (pteris) substrate were taken by maintaining the other parameters same (particle size 45-63µm; pH 7;temperature 298K; [S] 2.5mg/L;[E] 10ml/L) for all three batches. The study shows that NH_4OH treatment has highest amount of reducing sugar concentration among all other treatment presented in Table 1. From the experiment done by Mark and Arthur [14] this consequence may be occurred due to higher removal of lignin, no formation of barrier for the hydrolysis and easily product recovery as ammonia gas. Again in the Figure 1 it is seen that the reducing sugar concentration is increased up to 72 hours from the starting for different particle size (45-63µm, 63-125µm, and 125-250µm). But for the particle size 45-63µmhigh amount of reducing sugar is obtained because for that size of particle the surface area of substrate was higher than the others. From this study it can be considered that the optimum size is 45-63µm and pre-treatment is NH_4OH treatment.

Table 1. Investigation of the effect of pre-treatment

Treatment name	Reducing sugar con.(mg/ml)
NH₄OH treatment	0.642
H₂SO₄ treatment	0.604
NaOH treatment	0.612

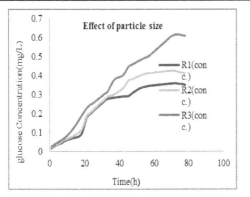

Figure 1. Investigation of the effect of particle size[R1 = 45-63μm; R2 = 63-125μm; and R3 = 125-250μm]

3.2. Effect of Substrate and Enzyme Loading

Different amount of NH₄OH pre-treated fern (pteris) substrate (1.25gm/L, 2.5gm/L, 3.75gm/L) having 45-63μm particle size and different amount of enzyme (10ml/L; 20ml/L; 30ml/L; 40ml/L) was inserted to investigate the effects The both study was performed maintaining pH 7; temperature 298K. The study shows that 2.5gm/L is optimum though for 3.75gm/L substrate the glucose concentration is slightly higher than for 2.5gm/L but the reaction rate is comparatively slow due to substrate to enzyme concentration ratio increased investigated by Kristensen [15] shown in Figure 2.

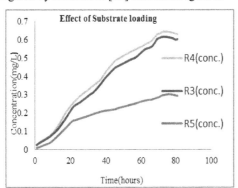

Figure 2. Effect of substrate loading.[R3 = 2.5gm/L, R4 = 3.75gm/L, R5 = 1.25gm/L]

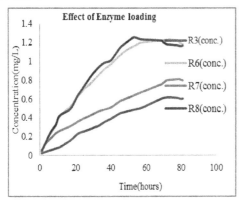

Figure 3. Effect of Enzyme loading.[R3 = 10ml/L R6 = 30ml/L, R7 = 20ml/L, R8 = 40ml/L]

Again Figure 3 shows that the concentration of the reducing sugar increased up to 71 hours for all enzymes loading. It is observed that 30ml/L enzyme loading is optimum because the reaction rate is too slow for the lower enzyme concentration again the 40ml/L enzyme comparatively less-efficient than 30ml/L enzyme loading.

3.3. Effect of pH and Temperature

Hydrolysis reactions are carried out simultaneously at different pH (4. 5, 6, 7, and 8) and at different temperature(25°C, 30°C, 35°C, 40°C, 45°C) with NH₄OH pre-treated fern (pteris) but the other parameters were kept constant, 10ml/L enzyme and 2.5gm/L substrate having particle size 45-63μm.The result were given in the Table 2 which indicates the optimum pH is 7 and the Optimum temperature is 35°C.

Table 2. Detection of optimum pH and temperature

pH	Glucose concentration (mg/L)	Temperature (°C)	Glucose concentration (mg/L)
4	0.450mg/L	25	0.615mg/L
5	0.489mg/L	30	1.023mg/L
6	0.589mg/L	35	1.612mg/L
7	0.615mg/L	40	1.351mg/L
8	0.577mg/L	45	1.033mg/L

3.4. Study at Optimum Condition

After optimizing the conditions (NH₄OH treatment as pre-treatment; particle size 45-63 μm; substrate loading 2.5gm/L; enzyme loading 30ml/L; pH 7; and temperature 35°C) for Enzymatic hydrolysis of preris a further reaction was studied and found the optimum sugar concentration 1.7625mg/L.

3.5. Fermentation of the Reducing Sugar Solution

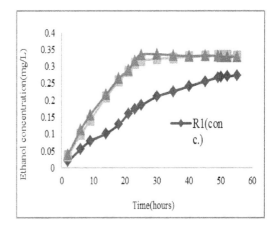

Figure 4. Observation the fermentation reaction rate of the reducing sugar.[R1 = 0.025gm ; R2 = 0.05gm ; R3 = 0.075gm]

After separating the sugar solution from the remaining unreacted particle using filter and decolourising with activated carbon the fermentation reaction was carried out with the *Saccharomyces cerevisiae* (Baker's yeast). For this purposes 0.025gm, 0.05gm, and 0.075gm of the *Saccharomyces cerevisiae was* added in three reactor flask having 75ml of the 1.7625mg/L concentrated reducing sugar solution. The other parameters were maintained constant pH = 7, and Temperature = 25°C. The data collected time to time and found the total concentration of the ethanol at 50 hours later is 0.333mg/L, which was the maximum concentration of ethanol for the study found for 0.075gm of yeast, and the others are 0.27mg/L for 0.025gm yeast and 0.329mg/L for 0.05gm yeast which were seen in the Figure 4.

4. Conclusion

This work is very significant because in the future the world's need much renewable energy to maintain the world's energy crisis and to protect the Environment. In this study a new hemicellulosic biomass source (pteris) is introduced into the renewable energy section which is economically less important and mostly available. This work suggests that the NH_4OH pre-treatment is the best among the NH_4OH, NaOH, and H_2SO_4 treatment for pteris. Again at the optimum condition of hydrolysis we get the optimum sugar concentration 1.7625 mg/L. and then performing fermentation with *Saccharomyces cerevisiae* (Baker's yeast) at 25°C temperature, PH = 7, we get 0.333mg/L ethanol which indicates that 0.1332mg of ethanol can be produce from 1gm of pteris where the conversion of reducing sugar to ethanol is 20% approximately in the optimum reaction condition.

Acknowledgement

This work was done under the full financial support of SUST research centre under the research grant 2013.

References

[1] Omer AM, the Energy Crisis, the Role of Renewable and Global Warming Greener, "*Journal of Environment Management and Public Safety,*" Vol. 1 (1), 038-070, 2012.

[2] Suleiman LN, Renewable Energy as a Solution to Nigerian Energy Crisis, "*VUAS,*" 13-19, 2010.

[3] Charles. EW, Biomass ethanol: technical progress, opportunities and commercial challenges, "*Annul. Rev. Energy Environ,*" 24:189-226, 1999.

[4] DoKyoung L, Vance N, Owens AB, Peter J , Composition of Herbaceous Biomass Feedstock, SDSU, "*SGINCCR*",07, 2007.

[5] Jenni R, "Isolation and Characterization of Lignin from steam pre-treated spruce and its utilization to study cellulase adsorption on lignin" University of Helsinki, 2009, 14.

[6] Mohammad. J.T, and Keikhosro. K, Enzyme biased hydrolysis process for ethanol from lignocellulosic materials: A Review, "*ncsu.edu/bioresources*" 2(4), 707-738.

[7] Zheng Y, Zhongli P, Ruihong Z, Overview of biomass pretreatment for cellulosic ethanol Production, "*Int J Agric & Biol Eng,*" 2(3), 51-68, 2009.

[8] Michelle LM, Factors Effecting Ethanol Fermentation via Simultaneous Saccharification and Fermentation, Shanghai Jiao Tong University, (2011) 10.

[9] Alexander D, Richard D, Scott G, Ryan P, Enzymatic Hydrolysis of Cellulosic Biomass for the Production of Second Generation Biofuels, "*WPI.*"May , 2009.

[10] Kim O, Magnus B and Gunnar L, A short review on SSF – an interesting process option for ethanol production from lignocellulosic feedstock, "*Biotechnology for Biofuels*" 1-7,2008.

[11] Cowan, S. T. & Steel, K. J. "*Cowan and Steel's Manual for the Identification of Medical Bacteria,*" 2nd edn. Revised by S. T. Cowan. Cambridge, UK: Cambridge University Press. 1974.

[12] MILLER, G.L. "*Analytical Chemistry,*" 1959, vol. 31, 426-428.

[13] Adran C, Prifysgol C, "concentration of ethanol in beer and wine, Experimental Handbook,"University of Wales BANGOR, 2006.

[14] Mark T. H, Arthur E. H., The Effect of Organosolv Pretreatment on the Enzymatic Hydrolysis of Poplar, "*Biotechnology and Bioengineering.*" Vol.26, 1984.

[15] Kristensen JB, Enzymatic hydrolysis of lignocellulose, Substrate interactions and high solids loadings, "*Forest & Landscape,*" Research No. 42-2008, 2.

7

Fermentable Sugar Production and Separation from Water Hyacinth Using Enzymatic Hydrolysis

Pradip saha[1,*], Md. Fakhrul Alam[1], Ajit Chandra Baishnab[1], Maksudur Rahman Khan[1,2], M. A. Islam[1]

[1]Department of Chemical Engineering and Polymer Science, Shahjalal University of Science and Technology, Sylhet, Bangladesh
[2]Faculty of Chemical and Natural Resources Engineering, University Malaysia Pahang, Pahang, Malaysia
*Corresponding author: pradip-cep@sust.edu

Abstract Water hyacinth containing a remarkable amount of cellulose which is found throughout the world as unusable material that can be used as one of the promising source for the production of glucose, initial step to produce bio ethanol. In this current study different types of treatments highest glucose concentration was obtained by hot water treated method. In addition, glucose was produced from water hyacinth using cellulytic enzyme pseudomonas sp., isolated from cow dung. Glucose concentration and production rate increases with the increasing of substrate concentration and enzyme loading, particle size 45µm and a pH 6.00 and temperature 40°C are the optimum for glucose production. A kinetic model rate expression has been developed for enzymatic hydrolysis of water hyacinth based on the Michaelis – Mentens model and parameters are determined. 0.15gm Reducing sugar are separated from the mixture of glucose water solution using 2 gm of water hyacinth. 0.531mg/l Cellulytic composition in water hyacinth are determined.

Keywords: hydrolysis, water hyacinth, pseudomonas sp., Michaelis – Mentens, kinetic model

1. Introduction

Energy requirements increasing day by day all over the world and oil, gas, crude oil are the considerable weapon to fulfill the growing demand throughout the world. Infact, that not going to possible to fulfill required the demand by these sources [1,2]. For these growing demand of fuel, worlds every country are trying to find out the other way and among them renewable energy such as solar energy, biodiesel production from biomass, bio ethanol production from biomass are the most appropriate outcome. Ligno cellulosic biomass presents an attractive substrate for bio ethanol production. Ethanol production based on corn [3], wheat [4] and sugarcane, lingo cellulosic biomass can be found in residues from the agriculture and forest industry or lignocellulosic rich energy crops, such as switch grass and elephant grass, can be grown on marginal land unfit for cultivation of crops for human or animal consumption [5] and researchers converted many lignocellulosic materials in to fermentable sugar initial step to bioethanol .In the present work water hyacinth was used for the production of fermentable sugar initial step to produce ethanol. Water hyacinth has been marked as the world's worst water weed and has garnered increasing international attention as an invasive species [5]. Water hyacinth has been identified by the International Union for Conservation of Nature (IUCN) as one of the 100 most aggressive invasive species [6] and recognized as one of the top 10 worst weeds in the world [7]. Beside this water hyacinth has contain about 20% of cellulose, 10%lignin, and 33% hemicelluloses [6]. So this high content of cellulose got strong favorability to the worlds scientists for sugar production. The conversion of lignocellulosic materials to bioethanol needs some important issues may be difficult for many situations [8]. Processing of lignocellulosic biomass to ethanol consists of four major unit operation; pretreatment, Hydrolysis, fermentation and product separation or purification [9]. In this work we produced fermentable sugar which is the initial step to produce bioethanol by using first two step noted above followed by separation. Generally two basic conventional approaches are followed for converting biomass into fermentable sugars. Saccharification can be carried by acid or enzymatic hydrolysis. Enzymatic hydrolysis got its favorability than acid hydrolysis for many reasons [9]. Still now their some factors should be considered as a headache for the production of fermentable sugar such as making much accessible cellulose more to enzyme, Degree of polymerization, pretreatment severity[10].There are many goal of the work among them enzymatic hydrolysis gets highest priority and study the activity of the enzyme at different conditions. Separating fermentable sugar and developing Michaelis – Mentens model for the designing of an effective process for the production of final product.

2. Materials and Methods

2.1. Substrate Preparation

Water hyacinth were collected from the nearby source and washed with water to remove the dust or other dirty. After that WH dried at 105°C for 5-6 hours in the drier oven.

2.2. Pretreatment of Substrate

Pretreatment of lignocellulosics aims to decrease crystallinity of cellulose, increase biomass surface area, remove hemicellulose, and break the lignin barrier [9]. Substrate are pretreated physically and chemically to accessible the substrate into enzyme. There different type of treatment among them those are applied in this work are given below.

2.2.1. Hot water Treated

10gm of dried water hyacinth were measured and mixed with 300ml distilled water in a biker and heated to the boiling point for 15 minutes. Treated WH dried at 102-105°C for 3 hours and was blended to get the desired size of the particle.

2.2.2. H$_2$SO$_4$ Treated

20gm of dried water hyacinth were measured and mixed with 150 ml of 1 %(v/v) of H$_2$SO$_4$ solution and heated at 115°C in oven for 3 hour. After that, it was washed with continuous charge of distilled water until the neutralization of water hyacinth sample and dried in air dryer for 3 hours at 105°C and was blended to get the desired particle size.

2.2.3. NaOH Treated

10gm of dried water hyacinth were measured and mixed with100 ml of 1 %(w/v) of sodium hydroxide solution and heated at 112°C in oven for 3 hours. Then washed again with distilled water to neutralize the sample and dried in air dryer for 3 hours and blended and particles with desired size were used for further experiment.

2.2.4. NH4OH Treated

10gm of dried water hyacinth were measured and mixed with vigorously in 100 ml of 1 %(v/v) of ammonium hydroxide solution followed by heating in oven at 110°C for 2 hour. Then neutralization was done by distilled water and dried in air dryer for 3 hours and blended into particles of desired size.

3. Hydrolysis

The hydrolysis of pre-treated **water hyacinth** was performed in rotary flasks shaker (Model: LRD: 750) with working volume of 150mL in 250mL flasks. Various amount of treated **water hyacinth** was taken a in each flask prior to the experiment and treated with Sodium acetate buffer (0.05 M) to maintain the p^H of the hydrolysis environment. All the components and p^H were assumed to have a uniform distribution in the flask due to continual rotation. All experimental run were conducted with addition of various predetermined amount of enzyme in the degradation environment with different amount of **water hyacinth**, different size of **water hyacinth**,

different enzyme loading, different pH and different temperature to delineate the corresponding respect . Samples were taken at every 1 hour interval; boiled for 5 min to destroy the enzyme, thus confirming the ceasing of the reaction. Then the samples were centrifuged, and analyzed for glucose concentration in **UV-spectrophotometer** according to the method by Miller [10].

4. Result and Discussion

4.1. Effect of Substrate Pretreatment

Pretreatment is considered one of the most crucial steps in bio ethanol production since it has a large impact on all other steps in the conversion process. Different types of treatments were used to increase the substrate activity to produce reducing sugar. mineral acid such as sulfuric acid ,base treatment such as sodium hydroxide, ammonium hydroxide and hydraulic treatment such as cold water or hot water treatment. Among them highest concentration was observed for hot water treatment.

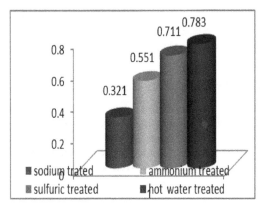

Figure 1. Effect of various types of pretreatment used for substrate (pH: 5.5 (acetate buffer), water hyacinth: 5 gm/L, enzyme: 15ml/ 1L mixture) on glucose yield

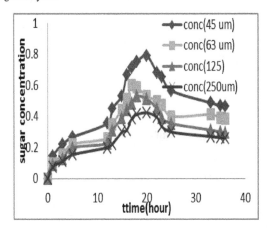

Figure 2. Effect of particle size on sugar production (WH: 3.3gm/l, enzyme: 30ml/l, pH: 5.5)

4.2. Effect of Substrate Particle Size

An efficient conversion of lignocelluloses into fermentable sugars is a key step in producing bioethanol in a cost effective and environmentally friendly way. Different sizes of particle ranging from 45 micrometer to 500 micrometer were taken for particle effect. Among them 45 micrometer shows greater production of reducing sugar.

4.3. Effect of Enzyme Loading

1ml to 18 ml enzyme was used to observe the effect of enzyme loading. Among them lower concentration was observed for 1ml enzyme and higher for 18 ml enzyme. It shows higher concentration of glucose for higher enzyme loading. This may be due to higher enzyme-substrate-active complex formation due to increased active sites by both charges

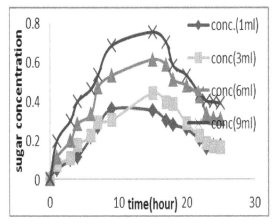

Figure 3. Effect of Enzyme loading (pH: 5.4 &WH loading: 3.33gm/L, temp: 24)

4.4. Effect of Substrate Loading

0.25mg to 1.5gm substrate was used to observe the effect of substrate. Among them lower concentration was observed for 25mg substrate and higher for 1.5gm substrate The effect of substrate concentration using water hyacinth is shown in Figure 4 given below.

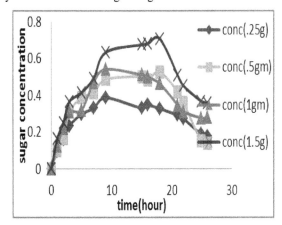

Figure 4. Effect of initial substrate concentration on sugar production (pH: 5.5, temperature: 26, enzyme: 30 ml/L)

In this observation it shows that the concentration of the glucose production increases with increasing of substrate concentration.

4.5. Effect of Temperature

Figure 5 represents the outcome of the consequence of temperature on the extent of bioconversion of glucose production where, the alteration of enzymatic efficiency of pseudomonas sp. for glucose production with different temperature is showed and the optimum temperature of 40 ± 2°C was required to attain the most excellent water hyacinth conversion to glucose.

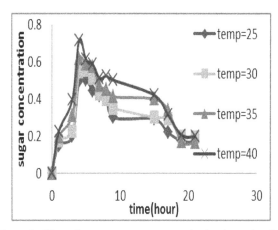

Figure 5. Effect of temperature on sugar production (water hyacinth loading: 3.33gm/L, pH: 5.5, enzyme loading: 30 ml/L).

4.6. Effect of pH

The ph effect for the activity of cellulytic bacteria from ranging 5 to 8 were observed the outcome of p^H alteration on glucose liberation is shown I Figure 6. The pH range between 6.0±0.2 gave the optimum yield of glucose.

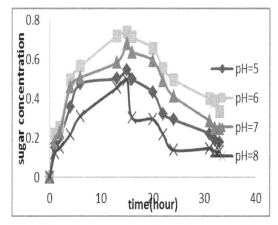

Figure 6. Effect of pH on sugar production (enzyme loading 36ml/L, WH loading: 3.33gm/L, temperature: 25 ± 2°C).

5. Development of Kinetic Modeling

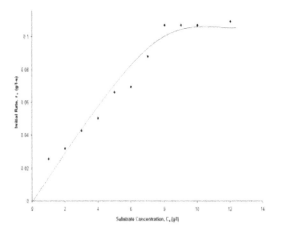

Figure 7. The initial rate as a function of substrate concentration (a typical curve)

A typical kinetic results by graphical Differentiation of curves using the initial rates method, afforded the plot of initial rates at different levels of substrate initial concentrations as shown in Figure 7.

From this curve, it can be observed that the reaction rate is proportional to the substrate concentration (that is, first order reaction) when the substrate concentration is in low range is also evident that the rate of reaction move toward a steady value as the substrate concentration becomes elevated. So we can say, the reaction rate alters progressively from first order to zero order as the substrate concentration was amplified. This form of conduct is commonly described by the Michaelis-Menten kinetic expression such as:

$$V_p = \frac{V_{max}[S]_o}{K_m + [S]_o} \quad (1)$$

Where, V_{max} (the maximum reaction rate) and K_m (rate constant) are the kinetic parameters, which are needed to be experimentally determined and $[S]_O$ is substrate concentration Applying the Line weaver – Burk method to linearize the rate expression by inverting equation (1) yields:

$$\frac{1}{V_p} = \frac{1}{K_m} + \frac{K_m}{V_{max}} \cdot \frac{1}{[S]_0} \quad (2)$$

So, from this straight line plot, the corresponding parameters (K_m & V_{max}) can be determined from intercept $1/V_{max}$ and the slope K_m/V_{max}.

The equation (2) is plotted by the data obtained with differentiation of initial glucose production by time for various initial substrate concentrations.

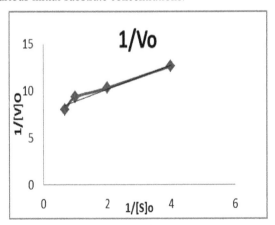

Figure 8. Linearized kinetic modeling plot

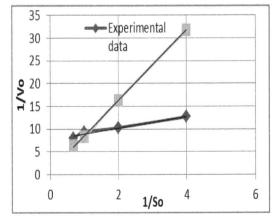

Figure 9. comparison between theoretical and experimental data

The data best fits a straight line with slope is 1.284 and intercept is 7.617 from this plot the kinetic parameters (V_{max} and K_m) were estimated as 0.102 gm/ (L. hour) and 0.778 gm/L respectively.

Based on the evaluated kinetic parameters, the model equation is given as:

$$V_o = \frac{0.778[S]_o}{5.932 + [S]_o}$$

6. Conclusion

Pretreatment is an important step to make cellulose more accessible to the enzyme that converts the carbohydrate polymers into fermentable sugar. In maximum time among different pretreatment acid pretreatment are used such as The acid pretreatment and enzymatic hydrolysis were used to evaluate to produce more sugar, to be fermented to ethanol .we have observed different acid pretreatment such as sulfuric acid,, base pretreatment such as ammonium oxide, sodium hydroxide and hydrothermal pretreatment such as hot water(at 105 for 15 minutes).We have highest result in hot water treatment among all other treatment and these treatment are better than other in many cases such as it is regarded as safer-equipment corrosion is reduced- and more environmentally friendly, as often no chemicals are required.

Another mentionable work we have done; the hydrolysis process were done by isolating the cellulytic bacteria from cow dung and separated the highest activity shower cellulytic enzyme *pseudomonas sp*. In this study fermentable sugar is separated and testified for sugar which is the unique work done. This is very important to know that how much fermentable sugar can be possible to produce using how amount water hyacinth whether for producing bioethanol or others work.

Another unique work done by measuring the cellulose concentration by passing of time; It should be noted that cellulytic concentration has strongly proof on behalf of the production of the fermentable sugar. Kinematic study has done and unknown parameters are calculated for designing the more economical scarification process and respectively for bioethanol production. It can be concluded that pre-treatment of water hyacinth enhances the rate of glucose production, while particle size of 45μm was found to be more favorable among the analyzed sizes. Operating temperature of 40°C and pH of 6.0 gave the best activity within the range of time investigated

The kinetics parameters of the reaction were obtained as $V_{max} = 0.778$ gm/ (L. hour) and $K_m = 5.932$ gm/L respectively.

Acknowledgement

This work is done under financial support of Ministry of Science and Technology, Government of the People's Republic of Bangladesh.

References

[1] Bentley, R W, Global oil and gas depletion: an overview. Energy Policy.30: 189-205, 2002.

[2] Cavallo, A J, "Predicting the peak in world oil production," *Nat Resour.Res.* 11: 187-195, 31: 426-428,2002.

[3] Kadam, K L, Rydholm, EC, and McMillan JD, "Development and Validation of a Kinetic Model for Enzymatic Saccharification of Lignocellulosic Biomass," *Biotechnol.Prog*, 20, 698-705, 2004.

[4] Y. Zhang, Jing, LX, Hui, JX, Zhen, H.Y, and Ying G. "Cellulase deactivation based kinetic modeling of enzymatic hydrolysis of steam-exploded wheat straw," *Bioresource Technology*, 101 8261-8266, 2010.

[5] Zhang, Y., Zhang, D., and Barrett S. "Genetic uniformity characterises the invasive spread of water hyacinth (Eichhorniacrassipes), a clonal aquatic plant," *Molecular Ecology,* 19: 1774-1786, 2010.

[6] Téllez, T., López E., Granado G., Pérez E., López R., and Guzmán J, "The water hyacinth, Eichhorniacrassipes: an invasive plant in the Guadiana River Basin (Spain)," *Aquatic Invasions* 3, 42-53. 2008.

[7] Shanab, S,, Shalaby, E., Lightfoot, D. and El-Shemy, H., "Allelopathic effects of water hyacinth (Eichhorniacrassipes)*," PLoS One* 5(10):e13200, 2010.

[8] Rubin, EM. "Genomics of cellulosic biofuels," *Nature* 454, 841-845,2008.

[9] Balat M, "Production of bioethanol from lignocellulosic materials via the biochemical pathway: A review," Energy conversion and Management 52 858-875, 2011.

[10] Hagerdal, HB, Galbe M, Gorwa-Grauslund MF, Liden G, and Zacchi G, "Bioethanol – the fuel of tomorrow from the residues of today," *Trends Biotechnol*; 24:549-56, 2006.

8

Biogas Production Using Geomembrane Plastic Digesters as Alternative Rural Energy Source and Soil Fertility Management

Seid Yimer[1], Bezabih Yimer[2], Omprakash Sahu[1,*]

[1]Department of Chemical Engineering, KIOT, Wollo University, Kombolcha, Ethiopia
[2]Mersa Agricultural TVET College, Woldya, Ethiopia
*Corresponding author: ops0121@gmail.com

Abstract The aim of work is to evaluate amount of gas production, economical feasibility and quality of slurry for geomembrane plastic biogas plants constructed below and above the ground surface and fixed-dome biogas plant of 3 m^3 capacity. Amount of gas and slurry were measured using calibrated biogas burner and weight balance respectively. The qualities of the slurry were analyzed in the laboratory using Kjeldahl and ash method respectively. Economic evaluation and comparison of the biodigester was carried out using cost-benefit analysis. Gas production and total-N was higher for a single layered and above ground plastic biodigester than others. Fermented slurry contained larger nitrogen content than fresh cow dung in both models of biodigester. The geomembrane plastic biogas plant gave higher net benefit than fixed-dome biogas plant. The biogas technology was found to increase income generation through increased crop production with the use of nutritive slurry as organic fertilizer and solve the problem of fuel shortage in the rural areas. Environmental impact assessment of the technology was studied and found that from the use of a geomembrane plastic biodigester, 360 m^3 of CO_2 and 600 m^3 CH_4 was prevented from emitting in to the atmosphere and save 0.562 hectare of forest per year from being deforested and hence attributed towards the decrease in global warming. Generally, the geomembrane cylindrical film biodigester technology was found cheap and simple way to produce gas in the study area.

Keywords: economical, feasibility, geomembrane, slurry

1. Introduction

The continuous depletion of fossil fuel is sticking the concern into the search for new energy sources to be used. The potential energy sources have been emerged as renewable energy resources. For a long time multifarious sources of renewable energy are being investigated to meet the increasing energy consumption rate. To counteract with the growing demand researchers are exploring the new sources. The developing countries are going ahead to face the shortage of available energy. Dependency from biomass such as fuel wood, charcoal, dried cow dung cake and crop residue in Ethiopia amounts to 95% (Benjamin, 2004). According to MOA (2000), on average each rural household spends ten hours per week searching for fuel wood. Females & children are engaged to search fire wood for about 5-6 hours journey [1]. When all forest uses are included, the deforestation rate in Ethiopia is around 1.1% per year [2]. According to Bech et al. [3], the forest cover of North Wollo and Habru district is 37,183.58 hectare and 1614 hectare respectively. According to FAO (2000) [4], the combustion of fossil fuels has caused serious air pollution problems, likewise the excessive consumption of fire wood results in deforestation on a large scale. IUCN (1990) [5] estimated that high forests covered 16% of the land area of Ethiopia in the early 1950 s, 3.6% in the early 1980 s and 2.7% in 1989. Biogas digestion was introduced into developing countries as a low-cost alternative source of energy to partially alleviate the problem of acute energy shortage for households, reduces deforestation and soil erosion, avoids scarcity of firewood, benefits environment globally and provides excellent fertilizer [6].

Biogas plants are a closed container in which anaerobic fermentation of cellulose containing organic material takes place so as to produce biogas and slurry. There are three basic designs of biogas plant popular in the world. These are the floating-drum type, fixed-dome type and bag digester. The floating-drum and fixed-dome type biogas plants have been introduced to Ethiopia. However, they became an obstacle to the rapid diffusion of biogas technology, because it takes a relatively long time (3-5 week) to build a plant, high initial cost of investment, shortage of adequate skilled person who can undertake construction & installation of the plant and transportation problem of the prefabricated steel drum from the urban areas to the interior rural regions of Ethiopia [7]. The introduction of the geomembrane plastic bag digester have not yet been experimented in Ethiopia. Considering the problem of biogas technology dissemination with the existing biogas plants, the study was conducted on alternative biogas plants constructed using geomembrane plastic in cylindrical shape. In this regard's an effort has been made to introduce geomembrane plastic biogas plant, comparisons of gas and slurry yield and economic feasibility analysis with the fixed-dome biodigester should be done and accordingly, the outcome of the study may have some contribution to set remedy to the problem.

2. Materials and Methods

2.1. Description of the Study Area

The experiment was conducted in Mersa Agricultural T.V.E.T College, Ethiopia at 11°35'N latitude and 39°38'E longitude with an altitude of 1557 m above sea level. The area is classified under moist warm climatic zone with mean annual rainfall of 1090 mm and with an average daily temperature of 21.12°C.

2.2. Geomembrane Plastic Biodigester Design Parameters

A. Selection of materials.
Construction materials: geomembrane plastic 0.5 mm in thickness, PVC pipes. Input materials are cow dung and water.
B. Temperature: Mesophilic (21.4°C).
C. Mixing ratio of substrate 1:1. I.e. 75 kg of cattle dung was mixed with 75 liters of water. Total dung required per day was calculated as [8].

$$\text{Total dung required / day}$$
$$= \frac{Gas production / day}{Gas / Kg of fresh dung} \tag{1}$$

D. Hydraulic Retention Time (HRT).
For mesophilic temperature of Mersa locality, HRT is selected as 40 days.

Number of cattle required was calculated by using the formula,

$$\frac{Total dung}{Dung / animal} \tag{2}$$

The volume of the digester was calculated as

Quantity of daily dung required / digester
$$= \frac{Weight of (dung + water)}{Density of slurry} \tag{3}$$

Minimum digester volume (D_O) = Volume of daily charge * Retention time
Actual digester volume = $D_O + 0.1 D_O$.
The volume of gas holder was calculated as,

$$V = (\text{volume of gas production}) \tag{4}$$

Total volume for geomembrane biodigester = Volume of digester + Volume of gas holder.
Hydrostatic pressure due to the slurry acting on the inner surface which exerts tensile load on the digester wall was calculated as = depth of slurry * density of slurry. According to Santra [9], a ratio of 4 in length to one in diameter was used to produce much gas and quality fertilizer in horizontal biogas plants. Where D = diameter and L= length and accordingly the dimension of the cylindrical plastic biodigester was calculated as

$$V = D2L / 4 \tag{5}$$

2.3. Experimental Design and Layout

Four biodigester were made from single & double layered cylindrical geomembrane plastic film, constructed below the ground on a trench excavated at a dimension of 7 m * 1.5 m * 0.5 m and above the ground surface on a concrete block wall platform constructed at a dimension of 7 m*1.5 m*(0.75 m and 0.5 m) with a slope of 2°. One Chinese model fixed-dome biogas plant was also another treatment. The capacities of the digesters were 3 m³. The experiment duration was from Nov. 2011 to July 2012. The treatments of the experiment were:

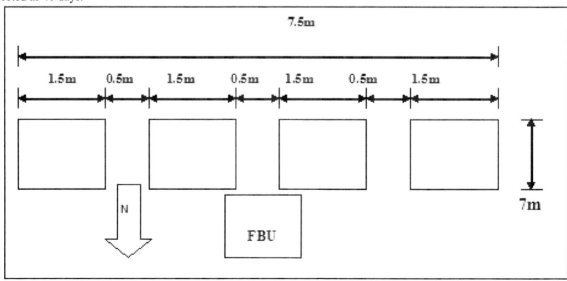

Figure 1. Layout of the Experimental Site

- PSA: plastic biodigester, single layered and constructed above ground.
- PDA: plastic biodigester double layered and constructed above ground.
- PSU plastic biodigester single layered and constructed under ground.
- PDA plastic biodigester double layered and constructed underground.
- FDU: Fixed dome biodigester constructed underground.

The manure was collected from Mersa Agricultural TVET College dairy farm and the nearby private cattle shed.

2.4. Geomembrane Plastic Construction Methodology

Materials used for the construction of geomembrane plastic biodigester were geomembrane plastic, PVC pipes, gate valves, GI caps, sockets, nipples, neoprene rubber hose and biogas stoves. Construction was done with the help of electrical geomembrane welding machine and CM-43 adhesive with other mixtures. The geomembrane plastic was cut at a dimension of 7 meter length and 4.50 meter width as per the design. Then, it was welded with the help of electrical plastic welding machine across the length and the circular part of the cylinder was fitted with the help of CM-43 adhesive and with other chemical mixtures. The biodigester has an inlet for entry of input materials, gas outlet for exit of produced gas and slurry outlet for disposal of fermented slurry. The digester and gas holder was made as one unit in a cylindrical shape.

Table 1. Technological Parameters of the Experimental geomembrane plastic biodigester

Constants	Value
Plastic width, m	2
Circumference, m	4
Internal diameter m	1.24
Plastic length, m	6.2
Loading rate, Kg ODM/m^3/day	10.36
Retention time, days	40
Quantity of daily dung required/digester, m^3	0.136
Minimum digester volume, m^3	5.44
Actual digester volume, m^3	5.984
Daily volume of gas production, m^3	3
Total volume of geomembrane plastic biodigester	7.484
Hydrostatic pressure due to the slurry, Kg/m^2	1023

2.5. Data Collection Procedures

Different data which were pertinent to the study objectives were collected following standard procedures. The following variables were measured and analyzed

during the study. Amount of gas production, quantity of input and output slurry, temperature of the air, pH of the fresh cow dung and digested slurry, total–N and organic matter content of the substrate and the slurry. Daily rainfall, temperature of the air and slurry was also measured.

2.5.1. Input to the Digesters

The type of input material which was found feasible and available in the study area was cow dung as there was dairy farm at a distance of 25 meter from the experiential site. Manure inputs were measured using bucket of 25 liter. One bucket of cattle dung was mixed with 1 bucket of water [8]. Thus, as per the design three bucket of dung (75 Kg) were mixed with three bucket of pure water (75 litres) so as to produce 3 m^3 gas per day on May 5, 2007. After the slurry mixture has been fed in to the digester, 15 liters starter material prepared from cattle dung and water in 1:1 ratio and allowed to ferment for one month in a closed barrel were added at equal amount to all five biodigester to initiate and facilitate the fermentation process.

2.5.2. Measurement of Gas Production

The quantity of gas produced from the two models of biodigester were measured with the help of standard sized biogas burner which is, certified by ISO and manufactured by gas crafters in Bombe, India. The burner has a capacity of 0.45 m^3 per hour and it was adjusted with the help of its air shutter until blue flame comes to burn with its maximum capacity. Time to burn was taken by stop watch for consecutive hours of one day. So the daily gas production from the digester (s) is the sum total of the hours run by each burner and its gas consumption rate.

2.5.3. Temperature of the Air and Slurry

The temperature of the air was measured via mini-max thermometer in a metrological station found around the study area and the daily temperature of the slurry in the biodigester was measured using ordinary electronic thermometer.

2.5.4. Total-Solids (DM) Content

The dry matter content of the substrate and spent slurry was measured by drying a sample at 70°C in an oven and weighing the residue on a precision electronic balance.

2.5.5. The Organic Dry Matter (ODM)

Only the organic or volatile constituents of the feed material are important for the digestion process. For this reason, only the organic part of the dry matter content was considered and this was analyzed in the laboratory.

2.5.6. PH of the Fresh Cow Dung and Digested Slurry

The pH of the fresh cow dung and fermented manure was measured by pH-meter in the laboratory.

2.5.7. Quality of Output Slurry

The compositions of digested slurry produced by anaerobic fermentation in the two models of biogas plants were determined in the laboratory. Five digested slurry samples from all five biogas plant and two fresh manure samples from the cow dung fed to the two types of biogas plant were taken by random sampling technique.

According to AOAC (1990) [10], total-N was determined by Foss-Tecator Kjeldahl procedures after taking 1 gram of manure sample and digested with sulpheric acid and salicylic acid and estimated by Kjeldahl method. The process of digestion took about three and half hours in the laboratory.

Organic matter content was determined by the use of Toffle furnace by ash method. 10 gram of well mixed manure in dry nickel crucible or silica basin was weighted and was put in a low flame or hot plate till the organic matter begins to burn. The crucible in a muffle furnace was placed at about 550°C for 8 hours. The crucible with a grayish white ash formed was removed from the furnace cool in a desicator and weigh. Therefore, the residue represents the ash and the loss in weight represents the moisture and organic matter [11]. Fresh cow dung samples were taken on April, 2011 and fermented slurry samples were taken after gas was fully generated and measured i.e. on June, 2011 from five biodigester and two cattle sheds.

2.5.8. The Efficiency of Biodigester

It was evaluated by comparing its gas and slurry production with the amount of substrate fed in relation to the volume of the digester and compared by calculating their specific gas production. Specific gas production was determined by dividing the daily volume of gas measured by the amount of cow dung loaded into the plant.

3. Results and Discussion

3.1. Biogas Production

Gas was burnt and measured after gas has completely produced within the designed HRT of 40 days with the help of calibrated biogas burner and stop watch (Figure 2). As can be seen in Table 2, gas production as the proportion of biodigester liquid volume was higher for a single layered and above ground plastic biodigester than others and very less amount of gas was measured from the fixed-dome biodigester. This was because more sun light temperature (27.65°C-32.7°C) was absorbed in a black geomembrane plastic sheeting digester than reinforced concrete fixed-dome biodigester (Figure 3). Temperature is very important factor which positively or negatively affects the activity of microorganisms in the production of biogas.

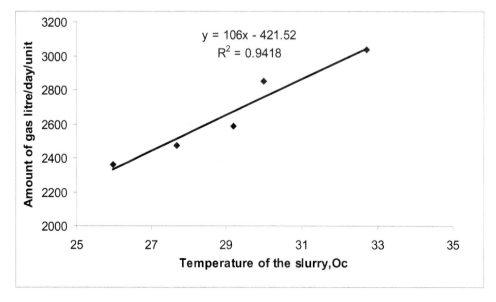

Figure 2. Burning and Measuring of Biogas with a Biogas Burner after Gas Generation

Figure 3. The Relationship between Temperature of the Slurry and amount of Gas Production for the Plastic and Fixed-Dome Biodigester

Table 2. Total Values for Biogas Production (9 Am to 4 Pm) in Biodigester with Different Types, Layers and Location of Installation

Biogas Production	BIODIGESTER TYPE					
	PDA	PSA	PDU	PSU	FBU	Average for plastic digesters
Hours recorded during burning of gas in a burner /plant	6.33	6.75	5.75	5.50	5.25	6.08
Liters/day/plant	2850	3037.5	2587.5	2475	2362.5	2737.5
Specific gas production per day (m³/kg)	0.0380	0.0405	0.0345	0.033	0.0315	0.0365
Average daily slurry temperature °C	29.99	32.7	29.18	27.65	25.96	29.88

PDA: plastic biodigester, double layered and constructed above ground
PSA: plastic Biodigester, single layered and constructed above ground
PDU: plastic biodigester, double layered and constructed underground
PSU: plastic biodigester, double layered and constructed underground
FBU: fixed-dome biodigester constructed underground.

3.2. Temperature of the Air and Slurry

The average atmospheric temperature during fermentation time between May 5, 2007 and June 14, 2007 was 24.55°C. According to Grewal et al. [8], temperature is one of the factor affecting the growth rate of micro-organisms involved in the production of biogas and effective anaerobic fermentation is carried out in mesophilic temperatures averaging between 24°C and 45°C. The local average atmospheric temperature of Mersa was 21.12°C according to 12 years of data which was very suitable for biogas production.

According to Hu Qichun (1991) [12], a biogas plant could perform satisfactorily only where mean annual temperatures are around 20°C or above or when the average daily temperature is at least 18°C. With the range of 20-28°C mean temperature, gas production increases over proportionally. If the temperature of the biomass is below 15°C, gas production will be so slow that the biogas plant is no longer economically feasible. So, the temperature recorded in the study area was very suitable for normal fermentation and higher amount of gas was produced by mesophilic bacteria. The average temperature

of the fermented material in the biodigester were in the range between 25.96°C and 32.7°C as described in (Table 3), which is greater than the critical 15°C. Thus, gas was completely produced within the designed HRT value of 40 days. As it is mentioned in Figure 3, R^2 is 0.945. Thus, 94.5% of the variability in the amount of gas produced by the biodigester was due to the variation in the slurry temperature.

The average temperature of the slurry measured in the geomembrane plastic biodigester exceed by 1.687°C to 6.737°C of the slurry in fixed-dome biodigester (Table 3). Thus, geomembrane bag was observed to have the best advantage of heating the digester contents easily and produce higher amount of gas (0.12-0.68 m³ of gas per m³ of digester per day) than fixed-dome biodigester made from reinforced concrete (Table 3). Since its walls are thin and black in color, it can be heated quickly with an external heat source, such as the sun radiation of the same degree in the study area. Similar results were reported by Bui Xuan and Preston [13] that found average temperatures in bag digesters, compared with dome types, are 2°C-7°C higher in the bag (0.235-0.61 m³ of gas per m³ of digester per day).

Table 3. Comparison of Average Slurry Temperature, °C and amount of Gas Produced, m³/Day of the Biodigester

Types of GPBD	Average Slurry temperature, °C of GPBD	Amount of gas produce, m³/day	GPBD, °C–FBU (25.963 °C)	GPBD, m³/day–FBU, 2.36 m³/day
PSA	32.700	3.04	6.737	0.68
PDA	29.988	2.85	4.025	0.49
PDU	29.175	2.59	3.212	0.23
PSU	27.650	2.48	1.687	0.12

GPBD: geomembrane plastic biodigester

3.3. Characteristics of Bio-Digested Slurry (Effluent) and the Influent

Organic substances passed through biogas plants not only produce a source of energy, but also a large quantity of digested slurry, which provides excellent organic fertilizer. The net weight of slurry discharged daily from the geomembrane plastic and fixed-dome biodigester were 123 Kg and 112.5 Kg respectively. Thus, the annual slurry output was 44895 Kg (45 tons i.e. 18% of the total substrate was lost) and 41062.5 Kg (41 tons i.e. 25% of the total substrate was lost), which was produced from 27375 kg of fresh dung fed to the digester annually, which in turn is equivalent to 13468.5 kg of dried dung cakes. Likewise, the loss in the amount of slurry have been

reported by UNESCO (1982) [14], and states that during digestion, about 20% of the total slurry is volatilized. Thus, geomembrane plastic biogas plant produced higher amount of slurry than fixed-dome biogas plant of the same capacity and using equal amount of input material. So, with the use of geomembrane plastic biodigester, the farmers could get 45 tones of fermented slurry which could be used to apply on the farmland as organic fertilizer. Thus, by conversion of cow-dung in to a more convenient and high-value fertilizer (biogas slurry), organic matter is readily available for agricultural purposes, thus protecting soils from depletion and erosion.

Therefore, the farmers should be advised to use geomembrane plastic biogas plant so as to utilize the dung for dual purposes such as the produced gas as fuel for cooking and the remaining large quantity of slurry as organic fertilizer which helps to increase crop production by preventing it from burning in the form of dry dung cake and ashing down the good manure creating unhygienic conditions in the kitchen.

3.4. Characteristics of Total-N in the Slurry and Influent

The average total nitrogen of the substrate and digested manure of the geomembrane plastic and fixed-dome biodigester were 0.37%, 1.13% and 1.15% respectively as described in Table 4. Similarly, Grewal et al. [8], reported that the total-N content of fresh dung as 0.242% which is 34.6% less than the result of this study. Thus, fermented slurry contained larger nitrogen content than fresh cow dung in both models of the biodigester because in an air tight biogas digester more organic acid such as acetic acid, prop ionic acid, butyric acid, ethanol and acetone was produced doting anaerobic fermentation of soluble simple organic substances which helps to absorb and fix ammonia and minimize the loss of nitrogen thus conserving the fertility of the manure. So, the higher quantity of nitrogen was converted in to the useful nitrate and ammonia which is the most important ingredient for plant growth.

There was also a greater conversion of organic substrate to nitrogen during 40 days of the experiment because microbial anaerobic degradation was facilitated with higher temperature.

According to Hu Qichun [12] (1991), during anaerobic fermentation, part of the total nitrogen is mineralized to ammonium and nitrate. Thus, it can be more rapidly taken up by many plants and in a number of applications, slurry from biogas plants is even superior to fresh dung especially when the slurry is spread directly on fields with a permanently high nitrogen demand (e.g. fodder grasses) or when using slurry compost to improve the structure of the soil.

Processing evidence suggests, however, that slurry is much more effective than dung when applied as fertilizer, French (1979) [15], discusses that slurry is 13 % more effective than dung, and Van Buren (1974) [16], reported that the ammonia content of organic fertilizer fermented for 30 days in a pit in china increased by 19.3%.

As can be seen in Table 4, the single layered geomembrane plastic biodigester constructed above ground produced higher amount of total nitrogen than others. This was due to higher amount of slurry temperature (Table 3) absorbed in the single layered above ground biodigester activated anaerobic micro-organisms to convert more simple organic substances of the substrate in to simple organic acid so as to fix more ammonia. Therefore, total-N was affected by material and position of biodigester construction. Mersa ATVET dairy farm fermented cow manure contains 1.0164%-1.2026% N and 50-75% Organic matter. Thus, the annual output of slurry equivalent to 13468.5 kg of dry dung cake converted to 152.2 kg of N and 8127 kg of organic matter on the average. In the slurry which has higher total nitrogen content there is higher proportion of ammonium which constitutes the more valuable form of nitrogen for plant nutrition [17].

Table 4. Effect of Material & Position of Biodigester Construction on the Composition of the Effluent

Components of the effluent and influent	Biodigester types and fresh cow dung fed to the biodigester							
	PSU	PDU	FBU	PSA	PDA	IFB	IPB	Average(For Plastic digesters)
Organic matter, %	50	50	66.7	75	60	80	85.7	58.75
Organic matter (Kg)	37.5	37.5	50.025	56.25	45	60	64.28	44.06
Organic carbon, %	29	29	38.686	43.5	34.8	46.4	49.706	34.08
Total N, %	1.148	1.1368	1.1536	1.2026	1.0164	0.3626	0.3822	1.126
Total N, Mg/kg	11480	11368	11536	12026	10164	3626	3822	11259.5
Dry matter, %	0.2	0.2	0.5	0.6	0.4	0.7	1.0	0.35
pH	6.8	6.9	7.2	7.1	7.6	6.8	7.0	7.1

IFB: influent of fixed-dome biodigester, IPB: influent of plastic biodigester
pH: hydrogen ion concentration.

3.5. Characteristics of Organic Matter in the Slurry and Substrate

The application of digested slurry to crop serves as a dual purpose; a soil conditioner as well as a source of plant nutrients. According to Table 4, the effluent in PSA showed larger organic matter content than the effluent in FBU. This could be because the amount of organic matter in the influent of plastic biodigester (IPB) was higher than that in the influent of fixed-dome biodigester (IFB). Higher value of organic matter was recorded from fresh cow dung than the slurry (Table 4). Even if higher value was recorded for fresh manure, the use of excreta of dairy cattle in biogas plants was found to be better than the farmyard manure in several ways. A part of nitrogen which is ammonia, found in the slurry becomes available to the plants. The ability of the wet digested slurry to aggregate soil particles immediately after application is also very unique. The digested manure was available in 40 days as compared to 4-6 months taken in the usual method of composting in a manure pit. According to Grewal et al. [8], almost all plant nutrients are retained in the digested slurry in such finely divided state that, it mixes up with the soil quickly and thoroughly, and the soil bacterial activity increases substantially. Thus the application of digested slurry gives better yields for all crops as compared to farmyard manure (FYM) made from the same quantity of cattle dung. The digested slurry in this study was thin and used directly to the crops through the irrigation channels and by direct splashing on the farmland.

Moreover, it was also put in a fish pond of Mersa agricultural T.V.E.T. College and fishes were nourished. Thus, it was proved that, the waste that comes out of the digestion process as slurry was very useful both as feed and organic fertilizer. Studies by Sokoine Agricultural University in Tanzania have shown that slurry from biogas improves productivity of land and maintains soil quality that can support crop production over a long period of time [18].

3.6. Characteristics of pH of Fermented Slurry

It is generally recommended that the pH inside the biodigester should be above seven for normal fermentation and maximum gas production [8]. In the present study, the pH of the fermented slurry and fresh cow-dung was 6.8-7.6 and (Table 4) with a hydraulic retention time of 40 days. Thus, the condition inside the digester was very comfortable for anaerobic micro-organisms to accomplish normal fermentation, higher gas production and fertile manure.

3.7. Efficiency of the Biodigester

The average specific gas production of the single layered above ground geomembrane plastic biodigester was greater than others (Table 2). That was due to higher amount of sun light was absorbed by the digester (32.7°C) that used to facilitate the activity of micro-organisms and increase the amount of gas production per weight of the substrate (Figure 2). Thus, the construction of single layered aboveground biodigester is better than other biodigester as efficiency of conversion of substrate organic matter to biogas was higher.

3.8. Economic Evaluations

In order to compute the value of biogas in terms of the value of traditional fuels saved by utilizing biogas, it was necessary to estimate the amount of dried dung cakes or fire wood needed to produce an equivalent amount of energy. Thus, according to UNISCO [14], Van Buren [16], used in this study assume that 1 m³ biogas substitutes for 3.47 kg of fire wood, 12.3 kg of dry dung fuel and 0.62 liter of kerosene oil.

Economic evaluation was also done for all biodigester using cost-benefit analysis which is the most commonly employed method used by many extension officers. It was able to evaluate the relative advantage of a geomembrane plastic biogas plant investment as compared to fixed dome biodigester on the basis of the anticipated minimum interest rate and economic life for the alternative designs. A discount rate of 18 % has been applied throughout the analysis according to Amhara regional state credit and saving institution (2007).

3.8.1. Market Price of Inputs

Dung: According to Senait Seyoum [19] (2007), the cost of dung was estimated in terms of.

Table 5. Summary of Market Value of Inputs and Outputs Used in the Analysis

Inputs	Market value (price in EB)
Dung (EB/kg)	0.50
Water (EB/5 m³)	2.00
Labor (EB/day)	15.00
Outputs	
Biogas (EB/litre)	3.47
Slurry (EB/tone)	90.62

EB: Ethiopian Birr (1$ = 18 Birr)

- Dung's value as fertilizer determined by the cost of an equivalent amount of commercial fertilizer, or.
- The market value of dung cakes, if dung is sold.
- 1 ton of DAP is 16 tone of dry manure.

Price of 1 tone of DAP according to 2006/2007 year was 1450 ET Birr. This was determined from a receipt voucher issued to Mersa agricultural T.V.E.T College purchasing office. The price of dried cow dung according

to Woldya market in 2006/2007 ranged from EB 5.00 to EB 8.00 per 50 kg sack, averaging EB 0.50 per kg of dried dung.

Water: This was valued according to the price charged by the water authority of Habru District, Mersa town. I.E. EB 2.00 per 5 m³.

Labor: The Labor used to collect water and spread slurry was valued at EB 15 day⁻¹.

3.8.2. Market Price of Outputs

Biogas: The biogas produced by the digestor was valued at the market value of fire wood or dried cow dung cakes, which it replaces. The observed price of fire wood in Mersa was between EB 15 and EB 20 per bundle weighing 15-20 kg. It was estimated that firewood averages EB 1.00 kg⁻¹.

Slurry: Output digested slurry was valued at the official price of DAP (diammonium phosphate). N and P contents of DAP roughly approximated the proportions of these nutrients in dried cow dung. According to Senait Seyoum [19], 1 ton of DAP is roughly equivalent to 16 ton of dry manure. The 2006/2007 price of DAP was EB 1450 tone⁻¹, thus each ton of dry cow dung which has less nutrient content than the fermented slurry is worth EB 90.62. Therefore, the cost of 1 ton of fermented slurry was assumed to be EB 90.62 to the minimum.

Three important technical assumptions were made with respect to gas production and use, and the efficient use of inputs and their conversion in the analysis.

Table 6. Summary of total costs and total benefits of geomembrane plastic and fixed-dome biogas plants

No	Biogas Models	Total Benefit	When used as fuel wood		When used as manure	
			Total Cost	Net Benefit	Total Cost	Net Benefit
1	PSA	7925.05	4427.90	3498.05	5035.45	2889.6
2	PDA	7687.57	4604.69	3082.88	5222.25	2465.32
3	PSU	7212.61	4127.78	3084.83	5562.47	1650.14
4	PDU	7355.10	4191.63	3164.10	5229.25	2125.85
5	FBU	6710.14	5662.96	1047.18	6389.32	320.82

1. The average daily gas production for a year from the geomembrane plastic and fixed-dome biodigester was assumed to be 2.74 m³ and 2.36 m³ as to the measurement taken once in drier months, but this could vary considerably with daily ambient temperature fluctuations in a year.
2. In the analysis it was assumed that all gas produced would be used, and it would have the same use as the dry dung or wood replaced by biogas.
3. The amount of slurry produced daily from the geomembrane plastic and fixed-dome biodigester was 123 tons and 112.5 tones.

As per the design, the total quantity of fresh dung required for 3 m³ size biogas plants in one year was 27375 kg or 27.375 tons but the quantity of fermented slurry collected after digestion and gas measurement was 45,000 kg and 41,000 kg in the geomembrane plastic and fixed-dome biodigester respectively. Thus, approximately 10 tons and 14 tons of digested slurry (18% and 25% of the input mixture) were lost during fermentation from the total of 55 tons of slurry mixture available in the geomembrane plastic and fixed-dome digesters respectively. The loss in the weight of the slurry was due to the loss of solids during fermentation. Similarly, UNESCO (1982), states that, during digestion, about 20% of the total slurry is volatilized. Thus, about 80% of the manure is collected from fresh dung. According to Grewal et al. [18], the loss of solids in the biogas plant rarely exceeds 27 percent even when maximum gas is generated.

3.8.3. Cost-Benefit Analysis of Biogas Plants

Therefore, the net benefit of geomembrane plastic biogas plants is greater than fixed-dome biogas plant and in particular the net benefit gained from PSA is greater than others. Thus, the use of geomembrane plastic biodigester is profitable than fixed-dome biodigester.

4. Conclusion

Generally, gas production and total nitrogen content as the proportion of biodigester liquid volume was higher for a single layered and above ground geomembrane plastic biodigester than the fixed-dome and other plastic biodigester. So the construction and use of single layered geomembrane plastic biodigester above the ground surface is preferable than other models and locations of installations of the biodigester. Fermented slurry contains larger nitrogen content than fresh cow dung in both models of biodigester. Thus fermented slurry has high fertilizer value in increasing the fertility of the soil than fresh cow dung. The geomembrane plastic biodigester gave higher net benefit than the fixed-dome biodigester. Thus, the geomembrane plastic biodigester is the cheapest model that an individual farmer could invest and acquire a better profit than the fixed-dome biodigester.

Considering the long-term benefit of plastic film biodigester technology both economically and environmentally, it is recommended to introduce the single layered above ground geomembrane plastic biogas plant to be used for the beneficiaries regardless of its higher net benefit, higher gas and fermented slurry production, simple construction and maintenance via extension education to promote its penetration and diffusion into rural areas. However, greater safety precaution during operation and usage of the plant and protection from damaging agents such as sharpened objects and rats is essential.

References

[1]   MOA (Ministry of Agriculture), 2000. Agro ecological Zones of Ethiopia, Natural Resource Management and Regulatory Department, Addis Ababa.

[2]   Wikipedia, 2006. Retrieved on Sep 3, 2007 available at http://en.wikipedia.org/wiki/Ethiopia.

[3]   Bech, N., Waveren and Evan, 2002. Environmental Support Project (ESP), Component 2: Environmental Assessment and Sustainable Land Use plan for North Wollo.

[4]   FAO (Food and Agricultural Organization of the United Nations), (2000). Statistical data base. FAO, Rome, Italy.

[5]   IUCN, 1990. Ethiopian National Conservation Strategy. Phase 1 Report. Based on the work of M.Stahl and A. Wood. IUCN, Gland, Switzerland Kristoferson LA and Bokhalders, 1991. Renewable Energy Technologies: Their application in developing countries. Intermediate Technology Publications, London, pp 112-117.

[6]   Vandana S, 2004. Alternative Energy. APH Publishing Corporation. New Delhi-11002.

[7]   Yacob Mulugeta, 2000. 'Renewable Energy Technologies and Implementation Mechanisms for Ethiopia', Energy Sources, 2 (1).

[8]   N.S.Grewal et al., 2000. Hand Book of Biogas Technology (A practical hand book). Punjab Agricultural University, Ludhiana. India.

[9]   S.C.Santra, 2001. Environnemental Science. New central book agency (p). l td. India.

[10]   AoAC, 1990. In: Helrick (Ed), official Methods of Analysis. Association of Official Analytical Chemists 15th Edition. Arilington. pp. 1230. Retrieved on December 12, 2007 available at www.mekarn.org/msc 2003-05/thesis 05/tram-p, pdf.

[11]   Tekalign Tadesse, L.Haque and E.A.Aduayi. 1991. International Livestock center for Africa. Addis Ababa, Ethiopia.

[12]   Hu Qichun, 1991. Systematical Study on Biogas Technology Application in Xindu Rural Area, China. Asian Institute of Technology, Bangkok/Thai.

[13]   Benjamin Jargstorf, 2004. Renewable energy and Development Brochure to accompany the mobile exhibition on Renewable energy in Ethiopia. Addis Ababa.

[14]   UNESCO (United Nations Educational, Scientific and Cultural Organization), 1982. Consolidation of information. Pilot edition. General information programme and universal system for information in science and Technology, UNESCO, Paris.

[15]   French D, 1979. The economics of renewable energy systems for developing countries.

[16]   Van Buren A (Ed), 1974. A Chinese Biogas Manual: Popularizing technology in the countryside. Intermediate Technology Publications, London.

[17]   Werner Kossmann, Uta Ponitz, 2007. Biogas Digest Volume 1. Information and Advisory Service on Appropriate Technology (ISAT). Retrieved on December 13, 2007 from www.gtz.de/de/documente/en-biogas_volume 1. pdf.

[18]   SURUDE (Foundation for sustainable Rural Development), 2002. Promotion of low cost biogas technology to resource poor farmers in Tanzania. Retrieved on July 8, 2007 from http://www.equator initiative.net.

[19]   Senait Seyoum, 2007. The economics of a biogas digester. Retrieved on Sep 15, 2007 from http://www.ilri.org/infoserv/webpub/Fulldocs/Bulletin30/economi. htm.

9

Thermal Performance Analysis of a Fully Mixed Solar Storage Tank in a ZEB Hot Water System

Mohammad Sameti[1], Alibakhsh Kasaeian[1,*], Seyedeh Sima Mohammadi[2], Nastaran Sharifi[3]

[1]Departmentof Renewable Energies, Faculty of New Sciences and Technologies, University of Tehran, Tehran, Iran
[2]Department of Mechanical Engineering, South Tehran Branch, Islamic Azad University, Tehran, Iran
[3]Department of Mechanical Engineering, University of Zanjan, Zanjan, Iran
*Corresponding author: akasa@ut.ac.ir

Abstract The intermittency is inherently affects the solar energy as the main hot water supplier in renewable systems. Therefore, the storage of the hot water is a vital part of a reliable energy supply system for buildings. Using a proper storage system makes it possible to fill the shortfall or emergency periods. This paper studies the daily thermal performance of a horizontal solar storage tank. It is assumed that the hot water load is similar to the Rand profile and the auxiliary heater is a biomass-fired boiler. The Rand profile assumes a daily hot water of consumption of 120 liters for a family of four. Furthermore, the tank is considered to be fully mixed because its height is relatively low. It means that the temperature in the solar zone of the tank is uniform. This assumption simplifies the analysis because the storage and the load temperature would be the same. The profiles for the storage water temperature, solar fraction and the useful energy gain was obtained and the effects of principal parameters are studied.

Keywords: solar energy, storage tank, Domestic Hot Water (DHW), Building Energy Simulation

1. Introduction

Zero Energy Building (ZEB) is a concept based on minimized energy demand and maximized harvest of local renewable energy resources. In these buildings, the electricity or heat from renewable energy resources can supply the same highly reduced energy demands [1,2].

Zero energy buildings are usually designed to utilize passive solar heat gain. For the purpose of space heating, the system is integrated with a thermal mass storage to fix diurnal indoor temperature variations. Along with that, the system should be able to provide hot water especially in domestic applications. The water heating systems are always inseparable parts of the residential buildings.

Approximately 18% of energy use in residential buildings and 4% in commercial buildings is for water heating. The hot water energy consumption is between about 220 *kWh* and1750 *kWh* per person and year for low energy demand and high energy demand patterns, respectively. The pattern consumption for the middle requirement range between 30 liters and 60 liters per person and day, with the warm water temperature of 45°C. The consumption is 440 *kWh* to 880 *kWh* per personequal to 1760 *kWh* to 3520 *kWh* for an average four-person household [3].

It is obvious that the *solar water heating systems* (SWH) can heat potable water only throughout the day and only when the sun is shining and the solar radiation is hitting the solar collector. Since the supply and the demand of thermal energy cannot usually be kept in balance, energy storage will play an important role for sustainable utilization. So it is essential for a solar water heating system to be accompanied with a thermal storage tank.

2. Solar Water Heating Components

As it can be seen from the schematics of a domestic solar water heating system depicted in Figure 1, a SWH is comprised of three main parts: *solar collector array*, *storage tank* and the *energy transfer system*.

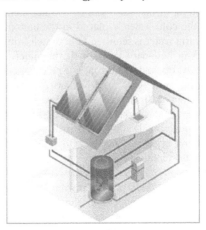

Figure 1. Schematics of solar water heating system components

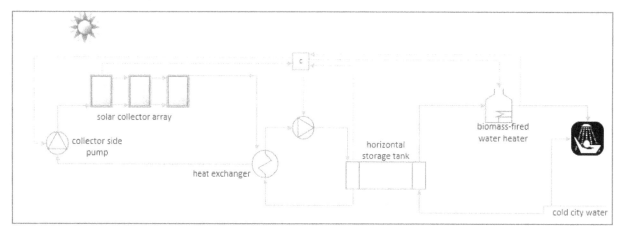

Figure 2. Schematics of a ZEB water heater system

2.1. Solar Collector Array

Solar collector arrays are normally mounted on the roof or placed on the ground. They transfer the absorbed thermal energy from the sun to a heat transfer fluid which can be water or a solution. Collectors can be *flat plate* or *evacuated tube*. Flat plate collectors are typically single glazed but may have an extra second glazing layer to improve their heat gain.

However, flat plate air collectors are mainly used for space heating purposes only. To produce both heat and electricity, flat plate air collectors are usually connected with photovoltaic panels. The more elaborate the glazing system, the higher the temperature difference that can be sustained between the absorber and the external wall. In general, most black paints which are widely used as the absorber plate color, reflect nearly 10% of the solar incident radiation.

2.2. Storage Tank

Storage tank is an insulated hot water tank that stores they gained thermal energy for later, shortfall or emergency uses. In *indirect systems*, the storage water in the tank receives the thermal energy from the heat transfer solution in the collector side loop through a heat exchanger which can be located inside or outside the storage tank or it can be served as its mantle. In *direct systems*, the potable existing is heated and used, directly.

Water storage tank systems designs are typically based on the exploiting of the warm and cold water tendency to stratify, i.e. the cold water is placed in the bottom of the tank while warm water is returned to or drawn from its top. A 9 in to 15 in boundary layer or thermoclineis formed between the warm and cold zones. Special mechanical designed diffusers or any array of nozzles assure the mixing phenomenon since the laminar flow is established within the tank. This laminar flow is necessary to increase stratification since the densities of return warm water and the supply water are 60°F and 40°F to 42°F respectively which are almost identical.

There are two types of solar storage tanks based on the thermal stratification: *fully stratified water tanks* and *fully mixed water tanks*. Comparison between these tanks employed in many solar water heating applications shows that the efficiency of the energy storage and the whole system may be increased up to 6% and 20%, respectively

[4]. The penalty associated with horizontal tanks is that the shallow tank depth degrades thermal stratification because of conduction through the walls of the tank and water [5].

2.3. Energy Transfer System

It uses liquid inside pipework which flows from solar energy collector array to the hot water storage tank. A *passive* transfer system does not involve mechanical devices such as pump or fan except for the energy needed to regulate dampers and/or controllers. *Active* solar uses electrical or mechanical equipment for the conversion of solar radiation into heat.

3. Model Equations

Model presented here is pertained to a system with schematics illustrated in Figure 2. At any hour j, the useful energy Q_u^j gained by the collector array with the area A_c can be expressed with equation (1).

$$Q_u^j = A_c F_R^{'}[G_t^j(\tau a) - U_{L,C}(T_i^j - T_a^j)] \qquad (1)$$

Where $F_R^{'}$ is the modified collector heat removal factor, $U_{L,C}$ is the overall heat loss coefficient from the collector (W/m²°C), T_i^j is the collector inlet temperature (°C) at hour j, °C), G_t is the irradiation (W/m²) which its values can be found in Figure 3, T_a^j is the ambient temperature (°C) at hour j and its values can be found in Figure 4 and τa is the product of the glass transmittance and the absorber absorptivity. It should be noted that the useful energy gain is always a positive. The factor $F_R^{'}$ is the consequence that occurs because the heat exchanger causes the collector side of the system to operate at a higher temperature than a similar system without a heat exchanger. The modified collector heat removal factor takes into account the presence of the heat exchanger and is given by

$$F_R^{'} = F_R \left\{1 + \frac{A_C F_R U_L}{(mC)_C}\left[\frac{(mC)_C}{\varepsilon(mC)_{min}} - 1\right]\right\}^{-1} \qquad (2)$$

where for the dimensionless capacitance rate $\zeta \neq 1$, we have

$$\varepsilon = \frac{1 - e^{NTU(1-\xi)}}{1 - \xi e^{NTU(1-\xi)}} \qquad (3)$$

$$NTU = \frac{UA}{(mC_p)_{min}} \qquad (4)$$

and

$$\xi = \frac{(mC_p)_{min}}{(mC_p)_{max}} \qquad (5)$$

In the above equations, C and m are the specific heat (*J/kg K*) and mass flow rate, respectively. Specifications for the heat exchanger can be found in Table 1.

For any instant of the time, energy balance for the storage tank can be written as the following.

$$m_t C \frac{dT_t}{dt} = Q_u - Q_i + Q_d \qquad (6)$$

Where m_t is the mass of storage tank water content (kg)with specific heat c which are both assumed to be fixed over the period of analysis (1 hour) and T_t is the storage tank water temperature (°C). The term Q_d is the demanded hot water energy per unit of the time (W) and Q_i is the system thermal loss (W). Using the finite difference approximation with a time step of 1 hour, Eq. (2) can be simplified into

$$m_t C(T_t^{j+1} - T_t^j) = Q_u^j - U_{L,t}(T_t^j - T_r)$$
$$- U_{L,p}(T_t^j - T_a^j) + m_d^j C(T_t^j - T_c) \qquad (7)$$

The extracted thermal energy corresponds to drawing of hot water with the mass m_d^j (kg) for the house at the top of the tank and replacing it with the cold make-up water from the mains inlet with temperature T_c (°C) and with the same mass at the below of the tank.

Furthermore, the system thermal loss can be expressed as the thermal loss from the tank $Q_{L,t}$ (Wh) and the thermal loss from the piping system $Q_{L,p}$ (Wh). $U_{L,t}$ and $U_{L,p}$ are the overall heat loss coefficients ($\frac{Wh}{°C}$) from the tank and the piping system, respectively, and T_r (°C) is the room temperature in which the tank is located. Values for different constants in equations (3) can be found in Table 2. Pipe surface temperature T_p^j is not uniform along the pipe length, therefore it is assumed that the heat loss from the piping is at its maximum rate by considering $T_p^j = T_t^j$.

Since the tank is fully mixed, the collector inlet temperature is assumed to be equal to the tank temperature, $T_p^j = T_t^j$. If the system is not able to provide the desired hot water outlet temperature which is $T_{set} = 50°C$, an auxiliary heater will be used to compensate the shortage in thermal energy. The *solar fraction* is defined as

$$f = \frac{Q - Q_d^j}{Q} \qquad (8)$$

Where Q is the required hot water energy for the building:

$$Q = Q_d^j + m_d^j C(T_{set} - T_t^j) \qquad (9)$$

or

$$f = \frac{T_{set} - T_t^j}{T_{set} - T_c} \qquad (10)$$

Using MATLAB to utilize a method developed by Kalogirou [5] or Duffie & Beckman [6] the hourly tank temperature and the solar fraction can be found. To start the algorithm, an initial condition is needed for the temperature in the tank as

$$initial\ condition : T_t^6 = 40°C \qquad (11)$$

The hot water daily consumption profile for m_d can be obtained by the Rand profile [7] which is shown in figure (4).

Table 1. Design parameters for heat exchanger

Parameter	Value
fluid	water glycol
c	3840 j/kg °C
m	1.35 kg/s
UA	5650 W/°C

Table 2. Values for design parameters used in the model

Parameter	Value
A_c	10m2
F_R	0.918
τa	0.9
$U_{L,C}$	7.5 W/m²°C
$U_{L,t}$	3.5 W/°C
$U_{L,p}$	20 W/°C
m_t	500 kg
T_r	22°C
T_c	18°C
c (water)	4180 j/kg °C
m (water)	0.95 kg/s

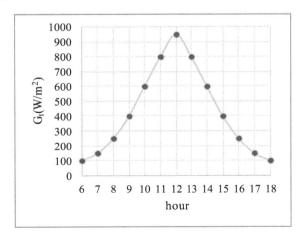

Figure 3. Distribution of Solar radiation in the analysis period

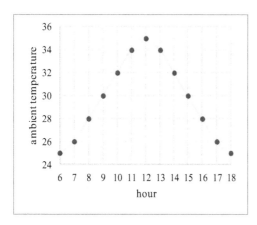

Figure 4. Ambient temperature in the analysis period

4. Results and Discussions

The useful energy gained throughout the day, temperature variation in the tank and the solar fraction are illustrated in Figures 6 to Figures 8. As it can be seen from Figures 7, the maximum energy gain is happened at noon because at this time the radiation is at its maximum level. The ambient temperature is relatively high and as a result, the losses are relatively low. The demand for hot water is descending between hours 10 and 15 so the temperature rises in the tank and reaches its maximum level at 15:00. The solar fraction was below 1 only for 6 hours a day (33% of the total time) and the modeled system is efficient enough to supply hot water for a small building however, its value never falls below 0.6. Note that the analysis is based on the Rand profile. The Rand profile assumes a daily hot water of consumption of 120 liters for a family of four. Another important point is the source which is used for the auxiliary heater in Figure 2. The fuel is biomass and the system is totally renewable and environment friendly.

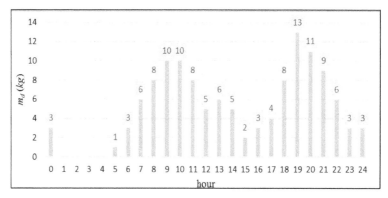

Figure 5. Rand profile for hot water energy consumption

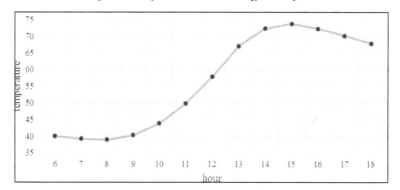

Figure 6. Hourly temperature in storage tank and supply for the load

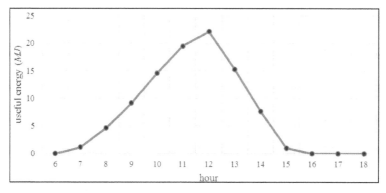

Figure 7. Useful energy gained by the solar heating system

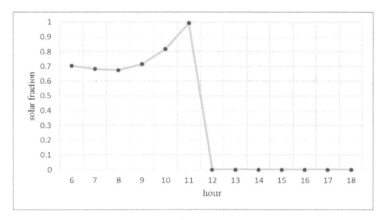

Figure 8. Solar fraction for the water heating systems

5. Conclusions

Zero energy buildings are usually designed to utilize passive solar heat gain. Temperature distribution in a fully mixed tank was modeled to obtain the temperature profile, solar fraction and useful energy gain. The solar fraction was below 1 only for 6 hours a day (33% of the total time) and the modeled system is efficient enough to supply hot water for a small building however, its value never falls below 0.6. The maximum in-tank temperature is happened in the low-demand hours when there is a chance for the tank to increase its temperature while no cold water can enter it. The minimum in-tank temperature happens in the morning while the demand is increasing and no enough radiation hits the collector. It is recommended to study the effects of the volume in the storage tank along with another thermal mass to achieve a complete hot water supply for the building. Also, it is recommended to propose a design for combined space heating and hot water.

References

[1] Sameti, M., Kasaeian, A., & Astaraie, F. R. (2014). Simulation of a ZEB Electrical Balance with a Hybrid Small Wind/PV. *Science and Education, 2* (1), 5-11.

[2] Kasaeian, A., Sameti, M., & Eshghi, A. T. (2014). Simplified Method for Night Sky Radiation Analysis in a Cool-Pool System. *Science and Education, 2* (1), 29-34.

[3] Eicker, U. (2009). *Low energy cooling for sustainable buildings.* Wiley. com.

[4] Han, Y. M., Wang, R. Z., & Dai, Y. J. (2009). Thermal stratification within the water tank. *Renewable and Sustainable Energy Reviews, 13* (5), 1014-1026.

[5] Kalogirou, S. A. (2013). *Solar energy engineering: processes and systems.* Academic Press.

[6] Duffie, J. A., & Beckman, W. A. (2013). *Solar engineering of thermal processes.* John Wiley & Sons.

[7] Bojić, M., Kalogirou, S., & Petronijević, K. (2002). Simulation of a solar domestic water heating system using a time marching model. *Renewable energy, 27* (3), 441-452.

10

Combination of Wind Catcher and Chimney for More Energy Efficient Architectural Buildings

Mohammadjavad Mahdavinejad[1], Sina Khazforoosh[2,*]

[1]Faculty of Art and Architecture, Tarbiat Modares University, Tehran, Iran
[2]Department of Architecture, University of Tehran-Kish International Campus, Iran
*Corresponding author: Sinakhazforoosh@ut.ac.ir

Abstract Energy, plays a crucial role in our everyday life. As energy supplies are limited, energy conservation is unquestionably one of the great importance to all of us. Although improving energy efficiency can be achieved in many ways, but energy saving in buildings plays one of the main roles in saving energy. Using natural ventilation systems can be effective to achieve this goal. This article tries to propose a pattern in which the combination of two common natural ventilation systems, Wind catcher and Chimney, will be use for optimum ventilation efficiency so that the energy efficiency will be increased.

Keywords: *energy efficiency, natural ventilation, wind catcher, chimney, evaporative cooling, Persiana*

1. Introduction

New technology is the fastest-growing and has a lot to do with energy efficiency. [1,2] Nowadays, this energy efficiency is employed in various fields such as engineering, applied sciences and construction industries. [3,4] Development of new materials helps other disciplines and has made a meaningful contribution to other fields especially energy efficiency in building construction. [5,6] Energy efficiency is going to find a more meaningful role in contemporary architecture. [7,8] Building construction industry is in need of energy efficiency to meet new customers' demands. The impacts of energy efficiency on architecture have been considered as the matter of significance in the current decades [9,10]. Improving energy efficiency can deliver a range of benefits to the economy and society. However energy efficiency programs are often evaluated only on the basis of the energy savings they deliver [11]. The International Energy Agency (IEA) estimates that residential, commercial and public buildings account for 30 percent to 40 percent of the world's energy consumption [12].

2. Research Method

The simulation research method is one of the well-known methods in quantitative researches. It also used in qualitative researches along with the modeling technique.

3. Literature Review

Regarding to the literature review one of the passive design methods is using and combining a wind catcher and a solar chimney as a mechanical eco concept which is a simple idea to increase natural ventilation in surrounding spaces. In past usually one of these items used in architectural designs to help natural ventilation but with this combination we can achieve to a more efficient ventilation system which mentioned in natural ventilation systems. Meir [13] explains that the air trap operates with the change of air temperature and the difference of weight of inside and outside the trap. The difference of weight of the air impels a suction process which causes the air to flow either to the bottom or to the top. It can be concluded that increasing the chimney width from 0.1 to 0.3 m with fixed air entrance size, increases the ventilation rate up to 25%. In addition, they found that the chimney width has more important effect on air flow pattern than the chimney inlet size. Mathur et al. [14] compared the natural ventilation rate from four different types of solar chimney experimentally and reported that when the absorber is inclined at 45°, the ventilation rate increases. In addition, by making the absorber more sloped, the ventilation rate will increase to 15.94% due to the increase of effective height of chimney.

4. Mechanism Analysis

A Wind catcher could come in various designs: uni-directional, bi-directional, and multi-directional. The construction of a wind catcher depends on the direction of airflow at that specific location: if the wind tends to blow from only one side, it is built with only one downwind opening and if there are two main directions for airflow it may have more than one downwind opening. The orientation of wind tower generally means the positions of the wind tower flank based on the four main geographical directions. [15] It is determined in view of function, use of wind power and the desired direction in which the wind blows.

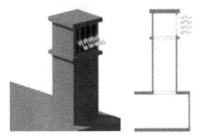

Figure 1. Uni-directional wind catcher

Figure 2. bi-directional wind catcher

The construction materials used for wind towers depend on climate. The choice of materials is made to ensure that the wind tower operates effectively as a passive cooling system. Wind towers in hot dry are built either of mud brick or more commonly of baked brick covered with mud plaster. Mud brick — adobe — passes heat at long time, because soil has got uncompressed volume and mud makes from water and soil [15].

Figure 3. wind catcher mechanism

Figure 4. wind catcher mechanism in hot climate. The shallow water pool helps to cool down the hot weather

A solar chimney is look like a tall and thin wind catcher in sky line of cities. They are vertical shafts which help the natural ventilation in buildings. In average a solar chimney is twice taller than a wind catcher. Its mechanism is based on second law of thermodynamics: heat transfer

always occurs from a hot body to a cold one. In other words, heat always flows to region of lower temperature, never to regions of higher temperature without external work beings performed on the system. A solar chimney usually place where it can absorb the most sun shine and appears in dark colors to raise the absorption range. The solar energy absorbed by chimney causes the air larger between two parallel planes of chimney to be heated so that the air of space in which the chimney entrance is located is sucked in. Therefore, the breeze inside the space lets the fresh air enter the space through openings [16].

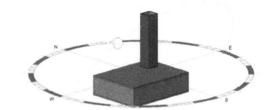

Figure 5. Orientation of a solar chimney based on the sun shine direction

Figure 6. The solar energy absorbed by chimney causes the air larger between two parallel planes of chimney to be heated so that the air of space in which the chimney entrance is located is sucked in

5. Proposing Pattern

Kashan is a city in and the capital of Kashan County, in the province of Isfahan, Iran. It lies in a desert at the eastern foot of the Central Iranian Range. Kashan is an ancient city and contains many houses which already use wind catcher as a natural ventilation system. This proposing pattern tries to presents an optimum combination between using a wind catcher and a chimney in one of these houses which is named as Alaghband house.

In order to find out the best place and orientation of the wind catcher and chimney, some base climate information of Kashan such as the direction of the main wind blow, the temperature and the solar path will be needed. Selected information illustrated below.

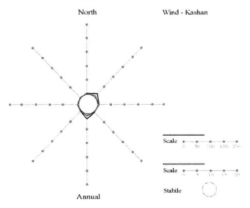

Figure 7. Wind speed by direction – Iran and stereographic Sun path Diagram – Iran – Kashan – 36N

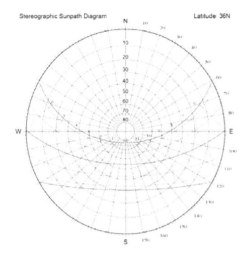

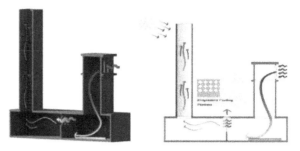

Figure 9. The proposing pattern mechanism

Figure 8. Wind speed by direction – Iran and stereographic Sun path Diagram – Iran – Kashan – 36N

This passing will cool down the weather again and will help to keep the temperature of interior spaces pleasurable. After this cool weather absorbed interior spaces heat and becomes warmer, it goes up and being led out of the building through the chimney which is hot because of the sun.

For optimum result, after hot weather entered through the wind catcher, it becomes cooler because of passing over a shallow water pool and moisturized. Then it enters interior spaces of the building and makes there cooler. After it becomes warmer it goes up through grating, placed at the top of the wall between wind catcher and chimney, which works as evaporative cooling Persiana [Figure 9].

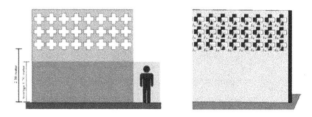

Figure 10. In order to provide interior spaces privacy and functions, the height of gratings are chose to be 2.10 meter above ground level

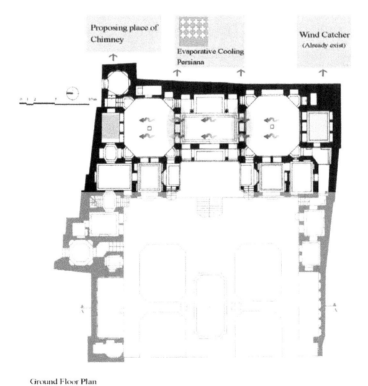

Ground Floor Plan

Figure 11. The Alaghband house Plan with the proposing pattern of combination of the wind catcher and the chimney as a passive natural ventilation system

This proposing pattern will work as a passive natural ventilation and will help to make the interior temperature much more comfortable. The proposing place of the chimney for this house, based on two main parameters: 1- The local climate information 2-The architectural plan of the house. This pattern is illustrated below in Figure 11.

6. Summary

In fact this proposing pattern is a combination of two common natural ventilation systems: 1-The wind catcher

2-The chimney. It will increase the efficiency of the ventilation system and in additional the energy efficiency. In past usually one of these items used in designs as a natural ventilation item but with this combination we can achieve to a more efficient ventilation system. Actually the two main parameters — Wind catcher and Chimney— are supplement each other in this method to achieve to the optimum result in ventilation.

7. Future Perspective

This pattern will be inefficient in some seasons or months. It needs to be model and analysis with Energy Plus software in order to have numerical information and charts about this pattern efficiency.

References

[1]   Mahdavinejad, M., Ahmadzadeh Siyahrood, S., Ghasempourabadi, M., Poulad, M., Development of Intelligent Pattern for Modeling a Parametric Program for Public Space (Case study: Isfahan, Mosalla, Iran), *Applied Mechanics and Materials,* (220-223) 2930-2935, 2012.

[2]   M. Mahdavinejad, M. Bemanian, M. Hajian, N. Pilechiha, Usage of Indigenous Architectural Patterns for Manufacturing Industrial Housing, Case: Renovation Project of Odlajan of Tehran, Iran, *Advanced Materials Research,* (548) 875-879, 2012.

[3]   M. Mahdavinejad, S. Mansoori, Architectural Design Criteria of Socio-Behavioral Approach toward Healthy Model, *Procedia-Social and Behavioral Sciences* (35) 475-482, 2012.

[4]   M. Mahdavinejad, M. Bemanian, G. Abolvardi, S. M. Elhamian, Analyzing the State of Seismic Consideration of Architectural Non-Structural Components (ANSCs) in Design Process (based on IBC). *International Journal of Disaster Resilience in the Built Environment,* 3 (2)133-147, 2012.

[5]   M. Mahdavinejad, A. Moradchelleh, Problems and Tendencies of the Development of the Architectural Sciences: Culture Research Aspect, *Middle-East J. Sci. Res.,* 10 (6) 677-682, 2011.

[6]   M. Mahdavinejad, A. Doroodgar, A. Moradchelleh, The Impacts of Revivalist Trends on the Contemporary Architecture of Iran (1977-2011), *Middle-East J. Sci. Res.,* 11 (2) 176-183, 2012.

[7]   M. Mahdavinejad, S. Matoor, N. Feyzmand, A. Doroodgar, Horizontal Distribution of Illuminance with Reference to Window Wall Ratio (WWR) in Office Buildings in Hot and Dry Climate, Case of Iran, Tehran, *Applied Mechanics and Materials,* 110-116 72-76, 2012.

[8]   M. Mahdavinejad, A. Doroodgar, H. Pourmand, New Engineering Materials and Developing Countries Architecture, *4th International Conference on Computer Modeling and Simulation,* Singapore. 2012.

[9]   Nazari, M., Mahdavinejad, M., & Bemanian, M., Protection of High-Tech Buildings Facades and Envelopes with One Sided Nano-Coatings. *Advanced Materials Research* (829) 857-861, 2014.

[10] Mahdavinejad, M., Nazari, M., & Khazforoosh, S. Commercialization Strategies for Industrial Applications of Nanomaterials in Building Construction. *Advanced Materials Research,* (829) 879-883, 2014.

[11] Lisa Ryan and Nina Campbell, SPREADING THE NET: THE Multiple benefits of energy efficiency improvements, *International Energy Agency,* OECD/IEA, 2012, 3.

[12] Schwarz, V., *Promoting Energy efficiency in buildings: Lessons Learned from International Experience,* United Nations Development Programme, 2009, 12.

[13] Meir, I and S. Roaf, Between Scylla and Charibdis: In search of the sustainable design paradigm between Vernacular and Hi-Tech, Procs, PLEA Conference, Santiago, Chile, 2003.

[14] J. Mathur, Anupma, and S. Mathur, Experimental investigation on four different types of solar chimneys, *Advances in Energy Research,* 151-156, 2006.

[15] Maleki, B. A. Wind Catcher: Passive and Low Energy Cooling System in Iranian Vernacular Architecture, *International Journal on Technical and Physical Problems of Engineering,* (3) 130-137, 2011.

[16] M. Mahdavinejad and M. Amini, Public participation for sustainable urban planning in Case of Iran, *Procedia Engineering,* (21) 405-413, 2011.

11

Synthesis of TiN@C Nanocomposites for Enhanced Electrochemical Properties

Danni Lei[1,2], Ting Yang[1,2], Baihua Qu[1,2], Jianmin Ma[1,2,*], Qiuhong Li[1,2], Libao Chen[1,2], Taihong Wang[1,2,*]

[1]Key Laboratory for Micro-Nano Optoelectronic Devices of Ministry of Education, Hunan University, Changsha, China
[2]State Key Laboratory for Chemo/Biosensing and Chemometrics, Hunan University, Changsha, China
*Corresponding author: nanoelechem@hnu.edu.cn (J.M. Ma); thwang@hnu.edu.cn (T.H. Wang)

Abstract TiN@C nanocomposites have been successfully synthesized by an annealing method using oleic acid as the carbon source. The as-prepared TiN@C nanocomposites were characterized by XRD, EDX, TEM techniques. When studied as anode materials for lithium-ion batteries, such unique structures endow composite electrodes with the long cycling ability and a high discharge capacity due to the existence of the carbon layer.

Keywords: TiN@C, nanocomposites, electrochemical properties, synthesis, lithium-ion batteries

1. Introduction

In recent years, titanium-based nanomaterials have been obtained unique attention as anode materials in lithium-ion batteries (LIBs) due to their electrochemical characteristics [1,2,3,4]. Among them, titanium nitride (TiN) is a semiconductor material with unusually high electrical conductivity, and relatively rarely studied as electrode materials for energy devices, such as supercapacitors and LIBs [1,5,6,7,8]. To date, it has been reported that TiN could act as a conductive additive for LIBs with minor capacity in literatures [1,7,8] Nevertheless, there are still no reports on improving its electrochemical performance in LIBs, especially under high potential range (1.0–3.0 V vs Li/Li+). Therefore, it is very meaningful to design novel TiN structures for studying their electrochemical properties in LIBs.

Over the past decades, much effort has been focused on the carbon coating electrode materials for improving their electrochemical properties in energy storage devices, which could possess high conductivity and a short distance of ion diffusion or mass transport [9-16]. Carbon coating techniques have been well developed. To date, many organic compounds including sucrose, glucose and acetylene gas have been used as the carbon sources [11,17,18]. However, there exist some disadvantages, such as inconsistent morphology, high temperature, complicated process and unsafety. Recently, organic acids or amines are found to be effectively used as carbon sources for coating other materials, such as oleic amine, oleic acid and other organic compounds [19-24]. Among them, oleic acid is a low-cost liquid, so it is a very suitable candidate for carbon coating reagent. Following this idea, we designed the TiN@C nonocomposites using oleic acid

as the carbon source for enhancing the electrochemical properties of TiN nanoparticles.

Herein, we have successfully prepared TiN@C nanocomposites using oleic acid as the effective carbon source, which absorbed on the surface of TiN could be easily carbonized, with carbon shell homogeneously coating on TiN nanocrystals. The synthetic route to prepare TiN@C nanocomposites is clearly given in Scheme 1. Moreover, the ultrathin carbon shell not only can also enhance the electronic conductivity of electrode material, resulting in the improved rate performance, but also achieve long cycling ability through effectively preventing TiN nanocrystals from pulverization during the charge-discharge process. Even at a current density of 2500 mA g^{-1} within the potential range of 1.0–3.0 V vs Li/Li+, the composite electrode still exhibits a specific capacity of 56 mA h g^{-1} (64 % of 87 mA h g^{-1} at 50 mA g^{-1}). Such TiN@C nanocomposites are demonstrated to be promising anode materials for high-performance LIBs.

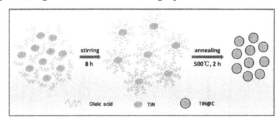

Scheme 1. An illustrative synthetic route to prepare TiN@C nanocomposites

2. Materials and Methods

2.1. Preparation of TiN@C Nanocomposites

All the chemicals were analytical grade and used without further purification. In a typical procedure, 0.5 g of commercial TiN was dispersed in 100 mL of oleic acid solution at room temperature. After being stirred for 8 hours, the precipitate was collected by centrifugation, and then washed with ethanol for two times. Finally, the product was dried completely in vacuum at 80°C, then annealed in a muffle furnace at 500°C for 2 h in Ar atmosphere with heating rate of 2°C min⁻¹. A brief summary of the synthesis procedure is shown in Scheme 1.

2.2. Characterization

The crystal structure of the products were determined by X-ray powder diffraction (XRD, Cu Kα radiation; λ = 1.5408 Å) with a SIEMENS D5000 X-ray diffractometer. The morphology and microstructure were characterized by transmission electron microscopy (TEM; JEOL, 2010) and scanning electron microscopy (SEM; Hitachi S4800) equipped with an energy-dispersive X-ray spectrometer (EDX).

2.3. Electrochemical Measurement

The assembly of the test cells (CR 2025-type) was performed in an argon-filled glove box with water and oxygen contents less than 1ppm before measurement. The anode electrode consisted of 80 wt% active material, 10 wt% conductivity agents (super carbon black), and 10 wt% binder (carboxyl methyl cellulose). One molar $LiPF_6$ in a mixture of ethylene carbonate (EC), dimethyl carbonate (DMC) and ethyl methyl carbonate (EMC) (EC : DMC : EMC = 1 : 1 : 1) was used as the electrolyte. The electrochemical measurements were carried out with a multichannel current static system (Arbin Instruments BT 2000, USA) after the cells were aged for 8 h. The electrochemical performances of the cells were evaluated within the potential range of 1.0 – 3.0 V versus Li/Li⁺.

3. Results and Discussion

3.1. Structural and Morphological Characterization

Figure 1a and b display low- and high-magnification TEM images of the commercial TiN nanoparticles. In Figure 1a and b, it can be found that the particles have a average size of about 20 nm. Figure 1c shows the low-magnification TEM image of TiN@C nanocomposites, which displays particle-like morphology with coating shells as indicated with black arrows. The high-magnification TEM image of TiN@C nanoparticle Figure 1d clearly exhibits that TiN nanoparticles are coated, and the coating carbon shell has a thickness of about 3 nm, as shown in the inset of Figure 1d. In Figure 1d, the intimate contact between the amorphous carbon layer and TiN nanocrystals could be observed, ruling out the possible increasing of contact resistance.

To confirm the crystal structure and phase of our TiN samples, the XRD characterization was performed. The XRD patterns of TiN and TiN@C samples are shown in Figure 2, respectively. In Figure 2a, the XRD pattern of commercial TiN nanoparticles can be indexed to the cubic crystal structure, which is well accorded with the standard card (JCPDS card No: 38-1420) without any obvious impurities. The XRD pattern of TiN@C nanocomposites in Figure 2b is similar with that of the commercial TiN sample. The carbon diffraction peaks could not be found in the pattern of TiN@C, which might be of the low

content of C in TiN@C and its inferior crystallinity. EDX characterization of TiN@C Figure S1 shows the peaks of C, Ti, N, O and Si (arising from the silicon substrate), which further confirm the existence of carbon element. The content of carbon element is about 20% in TiN@C nanocomposites.

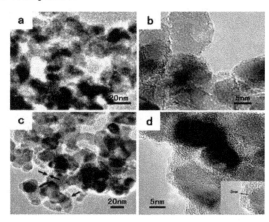

Figure 1. TEM images of the samples: (a and b) commercial TiN nanoparticles; (c and d) TiN@C nanoparticles

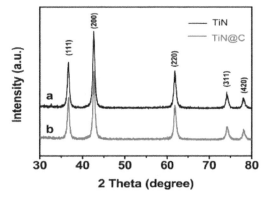

Figure 2. XRD patterns of the samples: (a) commercial TiN nanoparticles; (b) TiN@C nanoparticles

3.2. Electrochemical Properties

Inspired by the excellent characteristics of both titanium-based materials and carbon coating structures, we tested the electrochemical properties of TiN@C nanocomposites as anodes for LIBs. In this study, the specific capacity and cycling stability of electrodes were measured by constant current charge/discharge testing. The cycling stability for TiN and TiN@C electrodes are shown in Figure 3c. Compared with TiN electrodes, it is clear that the TiN@C electrodes deliver better cycling performance. The capacity of TiN@C has nearly no lost after 20 cycles while the capacity of TiN decrease gradually in all 200 cycles. The corresponding voltage profiles are plotted in Figure 3a and b. The initial capacity of TiN@C is 157 mA h g⁻¹, and decreases gradually to 76 mA h g⁻¹ after 200 cycles, while the capacity of TiN decreases from 127 to 54 mA h g⁻¹. The above results demonstrate that the cycling performance of TiN@C nanocomposite is much better than TiN.

Figure 3d shows rate performances of TiN and TiN@C electrodes at different current densities between 1.0 V and 3.0 V. It can be seen that TiN@C electrodes exhibit a stable capacity and good rate behaviors, while the capacity of TiN is lower than TiN@C electrodes. For the TiN@C electrodes, a specific discharge capacity of 87 mA h g^{-1} is obtained at a rate of 50 mA g^{-1} after 6 cycles; the discharge capacities of 80, 75, 70, 64, 56 mA h g^{-1} is observed at 100 mA g^{-1}, 200 mA g^{-1}, 500 mA g^{-1}, 1000 mA g^{-1}, 2500 mA g^{-1}, respectively. For comparison, the discharge capacity of TiN electrodes at the rate of 100 mA g^{-1}, 200 mA g^{-1}, 500 mA g^{-1}, 1000 mA g^{-1}, 2500 mA g^{-1}

are 85, 71, 60, 53, 45, 34 mA h g^{-1}, separately. When the rate was brought back to 50 mA g^{-1} after 6 cycles at 2500 mA g^{-1}, the discharge capacity of TiN@C electrode could recover to 80 mA h g^{-1} and remain. On the contrary, the commercial TiN electrodes show lower capacity (69 mA h g^{-1}), and their capacity displays the trend of decay. The above results demonstrate that TiN@C composites are excellent anode materials for LIBs.

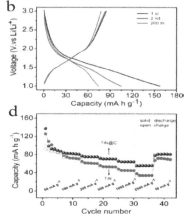

Figure 3. Electrochemical performance of TiN and TiN@C materials: (a) voltage profiles of TiN and TiN@C; (c and d) the corresponding cycling performance test at the 50 mA g^{-1} rate and rate capability

In order to understand the effects of carbon shells on the electrochemical properties of TiN nanoparticles, we examined their morphology change after electrochemical cycles. They were investigated on the copper foils by SEM. Figure 4a and 4c show the SEM photos of TiN before and after 200 cycles testing, respectively. The SEM images of TiN@C before and after 200 cycles are shown in Figure 4b and Figure 4d, respectively. Nearly all the TiN@C nanoparticles are unchanged on the battery electrodes and their SEI films are dense and slick, indicating that the composites are stable and strong enough to withstand the casting. On the other hand, an obvious change in morphology is deserved for TiN nanoparticles, which are coated by rough films Figure 4c. Hence, the SEI films on TiN@C are more stable than those on TiN nanoparticles. These results indicate that the carbon shells of TiN@C nanocomposites could effectively protect the active materials.

Here, the excellent lithium storage properties of TiN@C could be explained as follows: i) carbon shells enable the composite electrode to show a long cycle life

by preventing the TiN from volume change and pulverization during the charge-discharge process. ii) the carbon coating layer can effectively enhance the electronic conductivity of electrode material, which results in improved rate performance. The enhanced cycling and rate performance implies that this type of electrode can be a promising candidate for LIBs.

4. Conclusion

In summary, TiN@C nanocomposites have been synthesized by an annealing approach using the oleic acid as carbon resource. When used as the anode materials of rechargeable LIBs, TiN@C nanocomposites could exhibit a high specific capacity, the improved cycling performance and enhanced rate capabilities. The excellent properties could be due to the effective carbon coating. Our results suggest coating TiN nanoparticles with carbon is a reliable strategy to improve the electrochemical properties of TiN anode material.

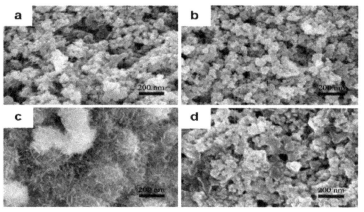

Figure 4. (a and b) SEM images of the commercial TiN and TiN@C materials as anode materials on copper foils; (c and d) SEM images of the commercial TiN and TiN@C in a fully lithiated state after 200 cycles

Acknowledgement

This work was supported by the National Natural Science Foundation of China (Grant No. 21103046, 21373081, 51302079 and 61376073) and the Young Teachers' Growth Plan of Hunan University (Grant No. 2012-118).

References

[1] Qiu Y., Gao L., "Novel polyaniline/titanium nitride nanocomposite: controllable structures and electrical/electrochemical properties", *J. Phys. Chem. B*, 109, 19732-19740, 2005.

[2] Armstrong G., Armstrong A. R., Bruce P. G., Reale P., Scrosati B., "TiO$_2$ (B) Nanowires as an Improved Anode Material for Lithium-Ion Batteries Containing LiFePO$_4$ or LiNi$_{0.5}$Mn$_{1.5}$O$_4$ Cathodes and a Polymer Electrolyte", *Adv. Mater.*, 18, 2597-2600, 2006.

[3] Shen L.F., Uchaker E., Zhang X. G. and Cao G. Z., Hydrogenated Li$_4$Ti$_5$O$_{12}$ nanowire arrays for high rate lithium ion batteries, *Adv. Mater.*, 24, 6502-6506, 2012.

[4] Seh Z. W., Li W. Y., Cha J. J., Zheng G. Y., Yang Y., McDowell M. T., Hsu P.C., Cui Y., "Sulphur-TiO$_2$ yolk-shell nanoarchitecture with internal void space for long-cycle lithium-sulphur batteries", *Nature Commun.*, 4, 1331, 2012.

[5] Choi D., Kumta P.N., "Nanocrystalline TiN derived by a two-step halide approach for electrochemical capacitors", *J. Electrochem. Soc.*, 153, A2298-A2303, 2006.

[6] Dong S. M., Chen X., Gu L., Zhou X. H., Xu H. X., Wang H. B., Liu Z. H., Han P. X., Yao J. H., Wang L., Cui G. L., Chen L. Q., "Facile Preparation of Mesoporous titanium nitride microspheres for electrochemical energy storage", *ACS Appl. Mater. Interfaces*, 3, 93-98, 2011.

[7] Snyder M. Q., Trebukhova S.A., Ravdel B., Wheeler M. C., DiCarlo J., Tripp C. P., DeSisto W. J., "Synthesis and characterization of atomic layer deposited titanium nitride thin films on lithium titanate spinel powder as a lithium-ion battery anode", *J. Power Sources*, 165, 379-385, 2007.

[8] Kim I., Kumta P.N., Blomgren G.E., "Si/TiN nanocomposites novel anode materials for Li-ion batteries", *Electrochem. Solid State Lett.* 3, 493-946, 2000.

[9] Ng S. H., Wang J. Z., Wexler D., Konstantinov K., Guo Z. P., Liu H. K., "Highly reversible lithium storage in spheroidal carbon-coated silicon nanocomposites as anodes for lithium-ion batteries", *Angew. Chem. Int. Ed.*, 45, 6896-6899, 2006.

[10] Konarova M., Taniguchi I., "Synthesis of carbon-coated LiFePO$_4$ nanoparticles with high rate performance in lithium secondary batteries", *J. Power Sources*, 195, 3661-3667, 2010.

[11] Ji X. X., Huang X. T., Liu J.P., Jiang J., Li X., Ding R. M., Hu Y. Y., Wu F. and Li Q., "Carbon-coated SnO$_2$ nanorod array for lithium-ion battery anode material", *Nanoscale Res. Lett.*, 5, 5649-5653, 2010.

[12] Kim J. H., Sohn H.J., Kim H. S., Jeong G. J., Choi W., "Enhanced cycle performance of. SiO-C composite anode for lithium-ion batteries", *J. Power Sources*, 170, 456-459, 2007.

[13] Yao W. L., Wang J. L., Yang J., Du G. D., "Enhancement of photoelectrochemical response by aligned nanorods in ZnO thin films", J. Power Sources, 176, 369-372, 2008.

[14] Lou X. W., Li C. M., Archer L.A., "Designed synthesis of coaxial SnO$_2$@carbon hollow nanospheres for highly reversible lithium storage", *Adv. Mater.*, 21, 2536-2539, 2009.

[15] Cui L. F., Yang Y., Hsu C. M., Cui Y., "Carbon-silicon core-shell nanowires as high capacity electrode for lithium ion batteries", *Nano Lett.*, 9, 3370-3374, 2009.

[16] Chen L. B., Yin X. M., Mei L., Li C.C., Lei D. N., Zhang M., Li Q. H., Xu Z., Xu C. M., Wang T. H., "Mesoporous SnO$_2$@carbon core-shell nanostructures with superior electrochemical performance for lithium ion batteries", *Nanotechnology*, 23, 035402, 2012.

[17] Li G., Wang X. L., Ma X. M., "Nb$_2$O$_5$-carbon core-shell nanocomposite as anode material for lithium ion battery", *J. Energy Chem.*, 22, 357-362, 2013.

[18] Seng K. H., Park M. H., Guo Z. P., Liu K. H., Cho J., "Self-assembled germanium/carbon nanostructures as high-power anode material for the lithium-ion battery", *Angew. Chem. Int. Ed.*, 51, 5657-5661, 2012.

[19] Jin Y.H., Seo S.D., Shim H.W., Park K.S., Kim D.W., "Synthesis of core/shell spinel ferrite/carbon nanoparticles with enhanced cycling stability for lithium ion battery anodes", *Nanotechnology*, 23, 125402, 2012.

[20] Gu F., Chen G., "Carbon Coating with Oleic Acid on Li$_4$Ti$_5$O$_{12}$", *Int. J. Electrochem. Sci.*, 7, 6168-6179, 2012.

[21] Xue D. J., Xin S., Yan Y., Jiang K. C., Yin Y. X., Guo Y. G., Wan L. J., "Improving the electrode performance of Ge through Ge@C core-shell nanoparticles and graphene networks", *J. Am. Chem. Soc.*, 134, 2512-2515, 2012.

[22] Kim K., Jeong J. H., Kim I. J., Kim H. S., "Carbon coatings with olive oil, soybean oil and butter on nano-LiFePO$_4$", *J. Power Sources*, 167, 524-528, 2007.

[23] Seo W. S., Lee J. H., Sun X. M., Suzuki Y., Mann D., Liu Z., Terashima M., Yang P.C., McConnell M. V., Nishimura D. G., Dai H.J., "FeCo/graphitic-shell nanocrystals as advanced magnetic-resonance-imaging and near-infrared agents", *Nature Mater.*, 5, 971-976, 2006.

[24] Xu C. J, Xu K. M., Gu H. W., Zheng R. K., Liu H., Zhang X. X., Guo Z. H., Xu B., "Dopamine as a robust anchor to immobilize functional molecules on the iron oxide shell of magnetic nanoparticles", *J. Am. Chem. Soc.*, 126, 9938-9939, 2004.

12

Introduction to Organic Solar Cells

Askari. Mohammad Bagher[*]

Department of Physics Azad University, North branch, Tehran, Tehran, Iran
*Corresponding author: MB_Askari@yahoo.com

Abstract Polymer solar cells have many intrinsic advantages, such as their light weight, flexibility, and low material and manufacturing costs. Recently, polymer tandem solar cells have attracted significant attention due to their potential to achieve higher performance than single cells. Photovoltaic's deal with the conversion of sunlight into electrical energy. Classic photovoltaic solar cells based on inorganic semiconductors have developed considerably [1] since the first realization of a silicon solar cell in 1954 by Chapin, Fuller and Pearson in the Bell labs. [2] Today silicon is still the leading technology on the world market of photovoltaic solar cells, with power conversion efficiencies approaching 15 – 20% for mono-crystalline devices. Though the solar energy industry is heavily subsidized throughout many years, the prices of silicon solar cell based power plants or panels are still not competitive with other conventional combustion techniques – except for several niche products. An approach for lowering the manufacturing costs of solar cells is to use organic materials that can be processed under less demanding conditions. Organic photovoltaic's has been developed for more than 30 years, however, within the last decade the research field gained considerable in momentum [3,4]. The amount of solar energy lighting up Earth's land mass every year is nearly 3,000 times the total amount of annual human energy use. But to compete with energy from fossil fuels, photovoltaic devices must convert sunlight to electricity with a certain measure of efficiency. For polymer-based organic photovoltaic cells, which are far less expensive to manufacture than silicon-based solar cells, scientists have long believed that the key to high efficiencies rests in the purity of the polymer/organic cell's two domains -- acceptor and donor.

Keywords: *organic solar cells, solar energy, photovoltaic, polymer*

1. Solar Energy

The amount of energy that the Earth receives from the sun is enormous: 1.75×1017 W. As the world energy consumption in 2003 amounted to 4.4×1020 J, Earth receives enough energy to fulfill the yearly world demand of energy in less than an hour. Not all of that energy reaches the Earth's surface due to absorption and scattering, however, and the photovoltaic conversion of solar energy remains an important challenge. State-of-the-art inorganic solar cells have a record power conversion efficiency of close to 39%, [6] while commercially available solar panels, have a significantly lower efficiency of around 15–20%. Another approach to making solar cells is to use organic materials, such as conjugated polymers. Solar cells based on thin polymer films are particularly attractive because of their ease of processing, mechanical flexibility, and potential for low cost fabrication of large areas. Additionally, their material properties can be tailored by modifying their chemical makeup, resulting in greater customization than traditional solar cells allow. Although significant progress has been made, the efficiency of converting solar energy into electrical power obtained with plastic solar cells still does not warrant commercialization: the most efficient devices

have an efficiency of 4-5%. [7] To improve the efficiency of plastic solar cells it is, therefore, crucial to understand what limits their performance.

2. Introduction

Organic solar cells can be distinguished by the production technique, the character of the materials and by the device design. The two main production techniques can be distinguished as either wet processing or thermal evaporation. Device architectures are single layer, bi layer hetero junction and bulk hetero junction, with the diffuse bi layer hetero junction as intermediate between the bi layer and the bulk hetero junction , Whereas the single layer comprises of only one active material, the other architectures are based on respectively two kinds of materials: electron donors (D) and electron acceptors (A). The difference of these architectures lays in the charge generation mechanism: single layer devices require generally a Scotty barrier at one contact, which allows separating photo excitations in the barrier field. The DA solar cells apply the photo induced electron transfer [5] to separate the electron from the hole. The photo induced electron transfer occurs from the excited state of the donor (lowest unoccupied molecular orbital, LUMO) to the LUMO of the acceptor, which therefore has to be a good electron acceptor with a stronger electron affinity. Subsequent to charge separation both the electron and the hole have to reach the opposite electrodes, the cathode and the anode, respectively. Thus a direct current can be delivered to an outer circuit. As the evidence of global warming continues to build-up, it is becoming clear that we will have to find ways to produce electricity without the release of carbon dioxide and other greenhouse gases. Fortunately, we have renewable energy sources which neither run out nor have any significant harmful effects on our environment. Harvesting energy directly from the sunlight using photovoltaic (PV) technology is being widely recognized as an essential component of future global energy production.

3. Organic Solar Cells

Organic materials bear the potential to develop a long-term technology that is economically viable for large-scale power generation based on environmentally safe materials with unlimited availability. Organic semiconductors are a less expensive alternative to inorganic semiconductors like Si; they can have extremely high optical absorption coefficients which offer the possibility for the production of very thin solar cells. Additional attractive features of organic PVs are the possibilities for thin flexible devices which can be fabricated using high throughput, low temperature approaches that employ well established printing techniques in a roll-to-roll process [8,9]. This possibility of using flexible plastic substrates in an easily scalable high-speed printing process can reduce the balance of system cost for organic PVs, resulting in a shorter energetic pay-back time. The electronic structure of all organic semiconductors is based on conjugated π-electrons. A conjugated organic system is made of an alternation between single and double carbon-carbon bonds. Single bonds are known as σ-bonds and are associated with localized electrons, and double bonds contain a σ-bond and a π-bond. The π-electrons are much more mobile than the σ-electrons; they can jump from site to site between carbon atoms thanks to the mutual overlap of π orbital's along the conjugation path, which causes the wave functions to delocalize over the conjugated backbone. The π-bands are either empty (called the Lowest Unoccupied Molecular Orbital - LUMO) or filled with electrons (called the Highest Occupied Molecular Orbital - HOMO). The band gap of these materials ranges from 1 to 4 eV. This π-electron system has all the essential electronic features of organic materials: light absorption and emission, charge generation and transport.

3.1. Structure of Organic Solar Cell

For organic solar cells based on polymer: fullerene bulk heterojunctions, the magnitude of JSC, VOC, and FF depends on parameters such as: light intensity [10], temperature [11,12], composition of the components [13], thickness of the active layer [14], the choice of electrodes used [15,16], as well as the solid state morphology of the film [17]. Their optimization and maximization require a clear understanding of the device operation and photocurrent, Jph, generation and its limitations in these devices. The relation between the experimental Jph and material parameters (i.e., charge-carrier mobility, band gap, molecular energy levels, or relative dielectric constant) needs to be understood and controlled in order to allow for further design of new materials that can improve the efficiency of this type of solar cells.

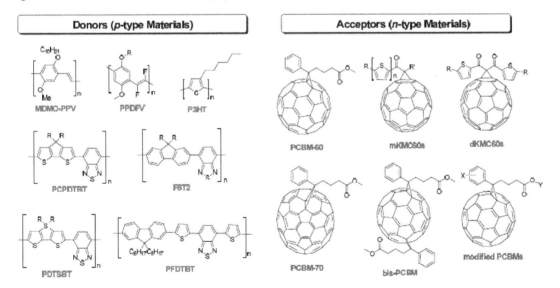

Figure 1. Chemical Structure of Organic solar cell Donor and Acceptor Materials

A first attempt to understand the physics behind the organic bulk hetero junction solar cells was done by using numerical models and concepts that are well established for inorganic solar cells, such as the p-n junction model. To improve the agreement of the classical p-n model with the experimental Jph of an organic bulk hetero junction cell, an expanded replacement circuit has been introduced [18,19,20]. This model replaces the photoactive layer by an ideal diode and a serial and a parallel resistance, which have an ambiguous physical meaning for an organic cell. However, different to classical p-n junction cells with spatially separated p- and n-type regions of doped semiconductors, bulk hetero junction cells consist of an intimate mixture of two un-doped (intrinsic) semiconductors that are nanoscopically mixed and that generate a randomly oriented interface. Moreover, due to the different charge generation, transport and recombination processes in bulk hetero junctions, the classical p-n junction model is not applicable to describe the Jph of these solar cells [21]. An alternative approach is to use the metal-insulator-metal (MIM) concept [22], where a homogenous blend of two unipolar semiconductors (donor/acceptor) is described as one semiconductor with properties derived from the two materials. This means that the photoactive layer is described as one 'virtual' semiconductor assuming that its conduction band is given by the LUMO of the acceptor

and its valence band is determined by the HOMO of the donor-type material. Under PV operation mode, the potential difference available in the MIM device, that drives the photo generated charge carriers towards the collection electrodes, is caused by the difference between the work functions of the metal electrodes.

As shown in Figure 1, several donor and acceptor materials are being reported, but none of them allegedly obtains over 3% efficiency except for P3HT/PCBM or PCPDTBT/PCBM.

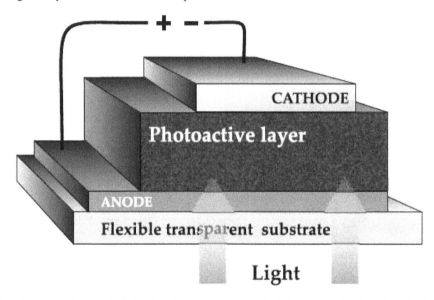

Figure 2. Schematic layout of an organic solar cell(Architecture of an organic photovoltaic device. The negative electrode is aluminum, indium tin oxide (ITO) is a common transparent electrode, and the substrate is glass. The schematic depicts a bulk heterojunction (BHJ) active layer where the donor and acceptor blend forms phase segregated domains within the active layer. The structure of the BHJ is critical to the performance of the solar device. - See more at: http://www.ssrl.slac.stanford.edu/content/science/highlight/2011-01-31/effects-thermal-annealing-morphology-polymer%E2%80%93fullerene-blends-organic#sthash.iE7FUkF8.dpuf)

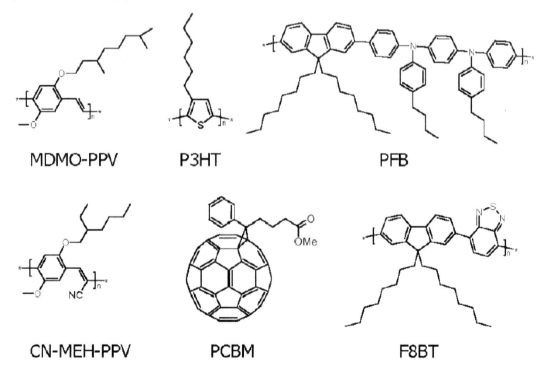

Figure 3. Several solution processible conjugated polymers and a fullerene derivative used in organic solar cells. Chemical structures and abbreviations of some conjugated organic molecules. From left: poly (acetylene) PA, poly(*para*-phenylene-vinylene) PPV, a substituted PPV (MDMO-PPV), poly(3-hexyl thiophene) P3HT, and a C60 derivative In each compound one can identify a sequence of alternating single and double bonds

3.2. Organic Solar Cell Application Field

We will summarize its application field by utilizing reports from Konarka and Plextronics. First, Konarka limits the application in 4 fields; 1) personal mobile phone charger, 2) small home electronics and mobile electronics attachment, 3) BIPV such as building's exterior wall, window, or blinder, and 4) power generation. Konarka predicts the market may be pioneered in each of these fields according to the module efficiency. In particular, the company predicts that the organic solar cell will be initially applied for special uses such as military market first due to low efficiency and high power generation unit cost.

Figure 4. Organic Solar Cell

4. Conclusions

Latest advances have shown a great potential for organic solar cells compared to conventional silicon cells. Their versatility in production methods, properties and applications looks very promising for the future of solar energy.

4.1. Organic Solar or Photovoltaic Cells (OPVs)

Organic or plastic solar cells use organic materials (carbon-compound based) mostly in the form of small molecules, dendrimers and polymers, to convert solar energy into electric energy. These semi conductive organic molecules have the ability to absorb light and induce the transport of electrical charges between the conduction band of the absorber to the conduction band of the acceptor molecule. There are various types of organic photovoltaic cells (OPVs), including single layered and multilayered structured cells. Both types are currently used in research and small area applications and both have their respective advantages and disadvantages.

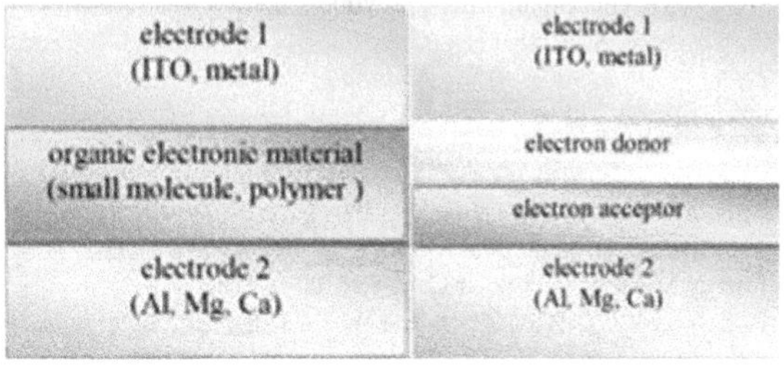

electrode 1 (ITO, metal)	electrode 1 (ITO, metal)
organic electronic material (small molecule, polymer)	electron donor
	electron acceptor
electrode 2 (Al, Mg, Ca)	electrode 2 (Al, Mg, Ca)

Figure 5. The Structure of a Single-Layer & a Multilayer Organic Solar Cell

4.2. Advantages of Flexible Organic Compared to Rigid Conventional Solar Cells

The latest advances in molecular engineering have uncovered a series of organic cell potential advantages that may eventually outbalance the benefits of silicon based solar cells. Although conventional solar cells currently dominate the existing market, the case may be quite different in the near future.

4.3. Manufacturing Process & Cost

Organic solar cells can be easily manufactured compared to silicon based cells, and this is due to the molecular nature of the materials used. Molecules are easier to work with and can be used with thin film substrates that are 1,000 times thinner than silicon cells (order of a few hundred nanometers). This fact by itself can reduce the cost production significantly.

Since organic materials are highly compatible with a wide range of substrates, they present versatility in their production methods. These methods include solution processes (inks or paints), high throughput printing techniques, roll-to-roll technology and many more, that enable organic solar cells to cover large thin film surfaces easily and cost-effectively. All above methods have low energy and temperature demands compared to conventional semi conductive cells and can reduce cost by a factor of 10 or 20.

4.4. Tailoring Molecular Properties

An important advantage of organic materials used in solar cell manufacturing is the ability to tailor the molecule properties in order to fit the application. Molecular engineering can change the molecular mass, bandgap, and ability to generate charges, by modifying e.g. the length and functional group of polymers. Moreover, new unique formulations can be developed with the combination of organic and inorganic molecules, making possible to print the organic solar cells in any desirable pattern or color.

4.5. Desirable Properties

The tailoring of molecular properties and the versatility of production methods described on the previous page enable organic polymer solar cells to present a series of desirable properties. These solar modules are amazingly lighter and more flexible compared to their heavy and rigid counterparts, and thus less prone to damage and failure. They can exist in various portable forms (e.g. rolled form) and their flexibility makes storage, installation, and transport much easier.

4.6. Environmental Impact

The energy consumed to manufacture a solar cell is less than the amount required for conventional inorganic cells. Consequently, the energy conversion efficiency doesn't have to be as high as the conventional cell's efficiency. An extensive use of organic solar cells could contribute to the increased use of solar power globally and make renewable energy sources friendlier to the average consumer.

4.7. Multiple Uses and Applications

The present situation indicates that organic solar cells cannot substitute for silicon cells in the energy conversion field. However their use seems to be more targeted towards specific applications such as recharging surfaces for laptops, phones, clothes, and packages, or to supply the power for small portable devices, such as cellphones and MP3 players.

Other than the domestic use, recent developments have shown a military application potential for organic solar modules. Research in the US (Konarka) has shown that organic cells can be used in soldier tents to generate electricity and supply power to other military equipment such as night vision scopes and GPS (global positioning system) receivers. This technology is thought to be extremely valuable for demanding missions.

4.8. The Current Situation

Organic Solar cells have certain disadvantages including their low efficiency (only 5% efficiency compared to the 15% of silicon cells) and short lifetime. Nonetheless, their numerous benefits can justify the current international investment and research in developing new polymeric materials, new combinations, and structures to enhance efficiency and achieve low-cost and large-scale production within the next years. A commercially viable organic solar cell production is the target of the next decade.

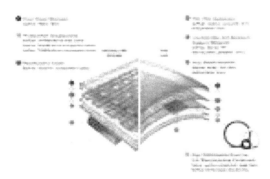

Figure 6. Schematic Comparison of a Rigid Crystalline Silicon to a Flexible Organic Solar Cell

Donor–acceptor based organic solar cells are currently showing power conversion efficiencies of more than 3.5%. Improving the nanoscale morphology together with the development of novel low band gap materials is expected to lead to power conversion efficiencies approaching 10%. The flexible, large-area applications of organic solar cells may open up new markets like "textile integration." Organic semiconductor devices in general and organic solar cells in particular can be integrated into production lines of packaging materials, labels, and so forth. Because there is a strong development effort for organic electronics integration into different products worldwide, the solar powering of some of these products will be desired. The next generation of microelectronics is aiming for applications of "electronics everywhere," and such organic semiconductors will play a major role in these future technologies. Combinations of organic solar cells with batteries, fuel cells, and so forth, will enhance their product integration. This inerrability of organic solar cells into many products will be their technological advantage.

The Si solar cell which has high manufacturing process expenses show delayed commercialization due to difficulties in overcoming its manufacturing cost limitation as Si wafer raw material supply shortage intensifies. On the other hand, the conjugated system organic/polymer material based organic solar cell is expected to reduce the manufacturing cost through new processes such as printing process. Therefore, the commercialization seems only possible by maximizing the energy conversion efficiency through a development of new conjugation system organic materials with reduced band gap.

References

[1]   A. Goetzberger, C. Hebling, and H.-W. Schock, Photovoltaic materials, history, status and outlook, Materials Science and Engineering R 40, 1 (2003).

[2]   D.M. Chapin, C.S. Fuller, and G.L. Pearson, A New Silicon p-n Junction Photocell for Converting Solar Radiation into Electrical Power, J. Appl. Phys. 25, 676 (1954).

[3]   H. Spanggaard and F.C. Krebs, A brief history of the development of organic and polymeric photovoltaic's, Sol. Energy Mater. Sol. Cells 83, 125 (2004).

[4]   H. Hoppe and N.S. Sariciftci, Organic solar cells: an overview, J. Mater. Res. 19, 1924 (2004).

[5]   N.S. Sariciftci, L. Smilowitz, A.J. Heeger, and F. Wudl, Photo induced electron transfer from a conducting polymer to buckminsterfullerene, Science 258, 1474 (1992).

[6]   M. A. Green, K. Emery, D. L. King, Y. Hishikawa, and W. Warta, Prog. Photovoltaic's 14, 455 (2006).

[7]   G. Li, V. Shrotriya, J. Huang, Y. Yao, T. Moriarty, K. Emery, and Y. Yang, Nature Mater. 4, 864 (2005).

[8]   S. E. Shaheen, R. Radspinner, N. Peyghambarian, G. E. Jabbour, *Fabrication of bulk hetero junction plastic solar cells by screen printing*, Applied Physics Letters 79 (2001), 2996.

[9]   G. Gustafsson, Y. Cao, G. M. Treacy, F. Klavetter, N. Colaneri, A. J. Heeger, *Flexible light-emitting-diodes made from soluble conducting polymers*, Nature 357 (1992), 477.

[10] P. Schilinsky, C. Waldauf, C. J. Brabec, *Recombination and loss analysis in polythiophene based bulk heterojunction photodetectors*, Applied Physics Letters 81 (2002), 3885.

[11] V. Dyakonov, *Electrical aspects of operation of polymer-fullerene solar cells*, Thin Solid Films 451-52 (2004), 493.

[12] I. Riedel, J. Parisi, V. Dyakonov, L. Lutsen, D. Vanderzande, J. C. Hummelen, *Effect of temperature and illumination on the electrical characteristics of polymer-fullerene bulk- heterojunction solar cells*, Advanced Functional Materials 14 (2004), 38.

[13] J. K. J. van Duren, X. N. Yang, J. Loos, C. W. T. Bulle-Lieuwma, A. B. Sieval, J. C. Hummelen, R. A. J. Janssen, *Relating the morphology of poly(p-phenylene viny- lene)/methanofullerene blends to solar-cell performance*, Advanced Functional Materials 14 (2004), 425.

[14] H. Hoppe, N. Arnold, N. S. Sariciftci, D. Meissner, *Modeling the optical absorption within conjugated polymer/fullerene-based bulk-heterojunction organic solar cells*, Solar Energy Materials and Solar Cells 80 (2003), 105.

[15] C. J. Brabec, A. Cravino, D. Meissner, N. S. Sariciftci, M. T. Rispens, L. Sanchez, J. C. Hummelen, T. Fromherz, *The influence of materials work function on the open circuit voltage of plastic solar cells*, Thin Solid Films 403.(2002), 368.

[16] C. J. Brabec, S. E. Shaheen, C. Winder, N. S. Sariciftci, P. Denk, *Effect of LiF/metal electrodes on the performance of plastic solar cells*, Applied Physics Letters 80 (2002), 1288.

[17] S. E. Shaheen, C. J. Brabec, N. S. Sariciftci, F. Padinger, T. Fromherz, J. C. Hummelen, *2.5% efficient organic plastic solar cells*, Applied Physics Letters 78 (2001), 841.

[18] C. J. Brabec, *Organic photovoltaics: technology and market*, Solar Energy Materials and Solar Cells 83 (2004), 273.

[19] P. Schilinsky, C.Waldauf, J. Hauch, C. J. Brabec, *Simulation of light intensity dependent current characteristics of polymer solar cells*, Journal of Applied Physics 95 (2004), 2816

[20] C. Waldauff, P. Schilinsky, J. Hauch, C. J. Brabec, *Material and device concepts for or- ganic photovoltaics: towards competitive efficiencies*, Thin Solid Films 451-52 (2004), 503.

[21] L. J. A. Koster, V. D. Mihailetchi, R. Ramaker, P. W. M. Blom, *Light intensity depen- dence of open-circuit voltage of polymer:fullerene solar cells*, Applied Physics Letters 86 (2005), 123509.

[22] C. J. Brabec, N. S. Sariciftci, J. C. Hummelen, *Plastic solar cells*, Advanced Functional Materials 11 (2001), 15.

13

Development of a Computational Interface for Hydropower Plants

Ajao K.R[*], Olabode O.F., Sule O.

Department of Mechanical Engineering, University of Ilorin, Ilorin, Nigeria
*Corresponding author: ajaomech@unilorin.edu.ng

Abstract The computational interface for hydropower plants was developed with Microsoft Visual Studio 2010 and tested using some available data from Jebba hydropower station to validate its functionality due to non availability of data for small hydropower plants which the interface is designed for. Hitherto several other methods have been used in hydropower computation, these methods range from manual computation with the use of pen and paper to the use of Microsoft Excel spreadsheet, some others utilize Visual Basic to commercial software. This interface was designed to compute flow duration values, plant capacity, power duration and the available energy. The program also have the project report feature, where user can view a concise report of the computation, save and print report sheets.

Keywords: computational interface, manual computation, flow duration values, plant capacity, report sheets

1. Introduction

Hydroelectric power is the power generated from the energy of falling water. When water fall due to gravity, the kinetic energy thereof can be converted to mechanical energy and then to electrical energy which is the most usable form of energy [1].

Small hydropower plant does not have a universal definition due to differences in what is accepted to be 'small' in different countries, but generally a small hydropower plant can be described to be a plant that can generate between 1MW to 50MW [2]. Thus, any plant capacity above the stated value may be considered as large hydropower plant.

It is essential to make necessary computations to determine the operating conditions of any hydroelectric station. It is from such computation that some details of the plant such as plant capacity, flow duration values and curves, renewable energy available can be obtained [3].

Before a hydropower plant design and construction is fully started, there is a great need to obtain the flow duration data which is used to generate the flow duration curve, this is referred to as hydrological modeling [4], and the power duration curve is also of equal importance. At this stage the outcome of the calculation helps to determine whether the whole project is worthwhile or not, especially when it is compared with cost and power demand.

It is necessary to make hydropower calculations because:
i. It can be used to predict the expected output of a hydropower plant

ii. It is efficient in determining the future flow characteristics of the site
iii. For existing sites that are not performing optimally, the parameters can be checked to study why there is drop in output and how best to boost the output.
Furthermore the computational interface is even more important because:
i. It presents a fast and more accurate means of design calculation when compared with manual estimation which on the other hand involves dealing with large figures that can easily induce human error.
ii. It is possible to easily compare values to study which one gives a closer output to the desired output without having to go through the rigorous activity of manual calculation.
iii. It presents an organized printable report which will help in construction and is also useful for future reference.

2.Materials and Methods

The Hydropower computation modules development process follows the general way of developing applications in Visual Basic express edition [5]. The process is as follows:
i. Planning the project: It is at this stage that the conceptual view of the program is gotten, its structure, target users and ultimately the purpose for which the program will be written.
ii. Creating the project: this is the creation of all files necessary for the application which is known as the project and when more than one is referred to, it is known as solution.

iii. Designing the User interface: it involves dragging various controls onto the design surface or form. Then the properties that define the appearance and behaviour of the form and the contents are set.

iv. Writing the code: this involves writing of the visual basic code that defines how the application behaves and interacts with the user.

v. Testing the code: it is possible to have bugs in the application, bugs are errors that disallow the application from producing desired results, it is from this stage that the application is tested and errors are removed or debugging.

vi. The program is then compiled in an executable format (.exe) extension which runs on windows operating system. The application is then ready for distribution.

Each interface has its own particular use, some of the interfaces are passive, and they neither accept data nor contribute to the actual computation, they are only necessary to aid the use of the application or give useful information about the application, while the other set of interfaces are active, they are indispensable during computation, without some of them, computations cannot be made while others aid the accuracy of the computations.

2.1. Description of Program Interface

There is a need for a deeper design understanding than the architectural design and configuration specification as it contains the procedural methods with which the inputs are transformed to output. The algorithm for the program is also developed, which contains step by step solution to the problem.

2.1.1. The Flow Interface

The flow form was designed to have a data grid view which displays data loaded into the application, it has two buttons, one for opening the open file dialog which helps in selecting the file to load while the other is the OK button for accepting the loaded data. The strength of this interface lies in its codes and after data has been loaded to the memory, they are sorted in descending order using the Bubble Sort Method [6], and a rank is given to each value,

the ranking was programmed to give the same rank to equal values and skip the next rank depending on how many values are equal. This and the equation for calculating the percentage equaled or exceeded are available on the web [7]. The equation for calculating percentage equaled or exceeded is given as:

$$P = 100 \; x \; \frac{M}{n+1} \qquad (1)$$

where,

P = the probability that a given flow will be equalled or exceeded (% of time)

M = the ranked position on the list

n = the number of events.

3.1.2. The Design Flow/Gross Head, Percentage Specified, Residual Flow And Losses Interface

All these forms are designed and function in the same way, they have textboxes into which the user enters the needed data, and a button which when clicked accepts the input data and stores at for use on other forms. Design flow is the maximum flow the turbine can use, the Gross head is the height of the falling water from the turbine, the percentage specified is a percentage value whose corresponding available flow value is the firm flow, and the residual flow is the flow left in the river throughout the year for environmental reasons. The losses include [8]: transformer losses which is the loss that occur due to the matching of the generator voltage to that of the transmission line, this value varies from 1-2%, parasitic electricity losses which is the loss due to the use of some of the energy generated to power auxiliary equipments it varies from 1-3%, maximum hydraulic losses which is the energy that is lost as water flows through the water passages and varies from 2-7% for most small hydropower plants. The maximum tail water which is the maximum reduction in available gross head that will occur during time of high flows in the river and annual downtime losses which is the loss due to the plant downtime and it is used in the computation of annual renewable energy available, it varies from 4-6%.

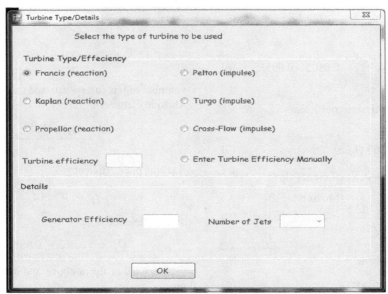

Figure 1. Turbine types interface

2.1.3. The Turbine Types Interface

This interface is responsible for accepting the user input of the type of turbine used, it has six radio buttons for the following types of turbine: Francis, Kaplan, Propeller, Pelton, Turgo and Cross flow Turbines. It also has a textbox for accepting manual efficiency which is enabled if the user decides not to use the model calculated efficiency and checks the enter efficiency manually radio button, the form also has a textbox that accepts the generator efficiency and a combo box for number of Jets which is used in the efficiency computation of Pelton and Turgo turbines.

2.2. Design Calculations

2.2.1. Turbine Efficiency

The Formulae for calculating the efficiency of the turbines are as given below [2]:
Reaction turbine runner size

$$d = kQ_d^{0.473} \qquad (2)$$

where: d = runner throat diameter in m
$k = 0.46$ *for* $d < 1.8$
$k = 0.41$ *for* $d \geq 1.8$
Q_d = Design flow (flow at rated head and full gate opening in m³/s)
Specific speed

$$n_q = kh^{-0.5} \qquad (3)$$

where:
n_q = specific speed on flow
k = 800 for Kaplan and propeller turbines
k = 600 for Francis turbine
H = rated head on turbine in m
for Francis turbine:
Specific speed adjustment to peak efficiency:

$$^\wedge e_{nq} = ((n_q - 56)/256)^2 \qquad (4)$$

Runner size adjustment to peak efficiency:

$$^\wedge e_d = (0.081 + {}^\wedge e_{nq})(1 - 0.789d^{-0.2}) \qquad (5)$$

Turbine peak efficiency:

$$e_p = (0.919 - {}^\wedge e_{np} + {}^\wedge e_{ed}) - 0.0305 + 0.005 R_m \qquad (6)$$

where:
R_m = turbine manufacture/design coefficient
Peak efficiency flow:

$$Q_p = 0.65 \, Q_d \, n_q^{0.05} \qquad (7)$$

Efficiencies at flows below peak efficiency flow:

$$e_q = \left\{ 1 - \left[1.25 \left(\frac{Q_p - Q}{Q_p} \right)^{(3.94 - 0.0195 n_q)} \right] \right\} e_p \qquad (8)$$

Drop in efficiency at full load:

$$^\wedge e_p = 0.0072 n_q^{0.4} \qquad (9)$$

Efficiency at full load:

$$e_r = (1 - {}^\wedge e_p) e_p \qquad (10)$$

Efficiencies at flows above peak efficiency flow:

$$e_q = e_p - \left[\left(\frac{Q - Q_p}{Q_d - Q_P} \right)^2 (e_p - e_r) \right] \qquad (11)$$

for Kaplan and Propeller turbines:
Specific speed adjustment to peak efficiency:

$$^\wedge e_{nq} = \left\{ (n_q - 170)/700 \right\} \qquad (12)$$

Runner size adjustment to peak efficiency:

$$^\wedge e_d = (0.0905 + {}^\wedge e_{nq})(1 - 0.789d^{-0.2}) \qquad (13)$$

Turbine peak efficiency:

$$e_p = (0.0905 - {}^\wedge e_{nq} + {}^\wedge e_d) - 0.0305 + 0.005 R_m \quad (14)$$

where:
R_m = turbine manufacture/design coefficient
for Kaplan turbines:
Peak efficiency flow:

$$Q_p = 0.75 Q_d \qquad (15)$$

Efficiencies at flows above and below peak efficiency flow:

$$e_q = \left[1 - 3.5 \left(\frac{Q_p - Q}{Q_p} \right)^6 \right] e_p \qquad (16)$$

for Propeller turbines:
Peak efficiency flow:

$$Q_p = Q_d \qquad (17)$$

Efficiencies at flows below peak efficiency flow:

$$e_q = \left[1 - 1.25 \left(\frac{Q_p - Q}{Q_p} \right)^{1.13} \right] e_p \qquad (18)$$

for Pelton turbines:
Rotational speed:

$$n = 31 \left(h \frac{Q_d}{j} \right)^{0.5} \qquad (19)$$

i = number of jets (user – selected value from 1-6)
Outside diameter of runner:

$$d = \frac{49.4 h^{0.5} j^{0.02}}{n} \qquad (20)$$

Turbine peak efficiency:

$$e_p = 0.864 d^{0.04} \qquad (21)$$

Peak efficiency flow:

$$Q_p = (0.062 + 0.001 j) Q_d \qquad (22)$$

Efficiencies at flows above and below peak efficiency flow:

$$e_q = \left[1 - \left\{ \left(1.31 + 0.025j \right) \left| \left(\frac{Q_p - Q}{Q} \right) \right|^{(5.6 + 0.4j)} \right\} \right] e_p \quad (23)$$

for Turgo turbines:

$$\text{Efficiency} = \text{pelton} - 0.03 \quad (24)$$

for Cross flow turbines:

Peak efficiency flow:

$$Q_p = Q_d$$

Efficiency:

$$e_q = 0.79 - 0.15 \left(\frac{Q_d - Q}{Q_p} \right) - 1.37 \left(\frac{Q_d - Q}{Q_p} \right)^{1.4} \quad (25)$$

All these formulae were written into codes and during runtime the values they generate are sent to the home form for them to be accessed by other forms, the generator efficiency is stored as well.

2.2.2. The Plant Capacity

The computed capacity of the plant is done using the following equations [2]:

$$P_{des} = pgQ_{des}H_g \left(1 - l_{hydr} \right) e_{t,des} \\ \times e_g \left(1 - l_{trans} \right) \left(1 - l_{para} \right) \quad (26)$$

where:

$P_{des} = Plant\ capacity$

$\rho = density\ of\ water$

$g = acceleration\ due\ to\ gravity$

$Q_{des} = Design\ head$

$H_g = gross\ head$

$l_{hydr} = maximum\ hydraulic\ losses$

$e_{t,des} = turbine\ efficiency\ at\ design\ flow$

$e_g = generator\ efficiency$

$l_{trans} = transformor\ losses$

$l_{para} = parasitic\ electricity\ losses$

The power duration values are computed using each available flow duration value in the equation as Q

$$P_{des} = pgQ[H_g - (h_{hydr} + h_{tail})] \\ \times e_t e_g \left(1 - l_{trans} \right) \left(1 - l_{para} \right) \quad (27)$$

where:

$$h_{hydr} = H_g l_{hydr,max} \frac{Q^2}{Q_{des}^2} \quad (28)$$

$l_{hydr,max}$ = maximum hydraulic losses specified by the user

$$h_{tail} = h_{tail,max} \frac{\left(Q - Q_{des} \right)^2}{\left(Q_{max} - Q_{des} \right)^2} \quad (29)$$

$h_{tail,max} = max\ imum\ tail\ water\ effect$

$Q_{max} = max\ imum\ water\ flow$

The renewable energy available is the area under the power duration curve and it is computed using the 21 values of power duration in the formula [2]:

$$E_{avail} = \sum_{k=1}^{20} \left(\frac{P_{s(k+1)} + P_{sk}}{2} \right) \frac{5}{100} 8760 \left(1 - l_{dt} \right) \quad (30)$$

where: l_{dt} is the is annual downtme losses.

3. Results and Discussion

The program interface requires real data from a functioning hydropower plant to ascertain its correctness. At this stage of the development of this interface program, data from Small Hydropower Plants (SHPs) would have been appropriate, however data from SHPs in Nigeria are either difficult to come by or are not available at all. The data of Jebba hydropower plant, Nigeria were obtained and used for the testing [9]. Although Jebba power plant is a large hydropower plant, using its data for testing is safe because the formulae for calculating the output of other sizes of hydropower plants are similar to that of small hydropower plant but additional losses need to be considered in the computation.

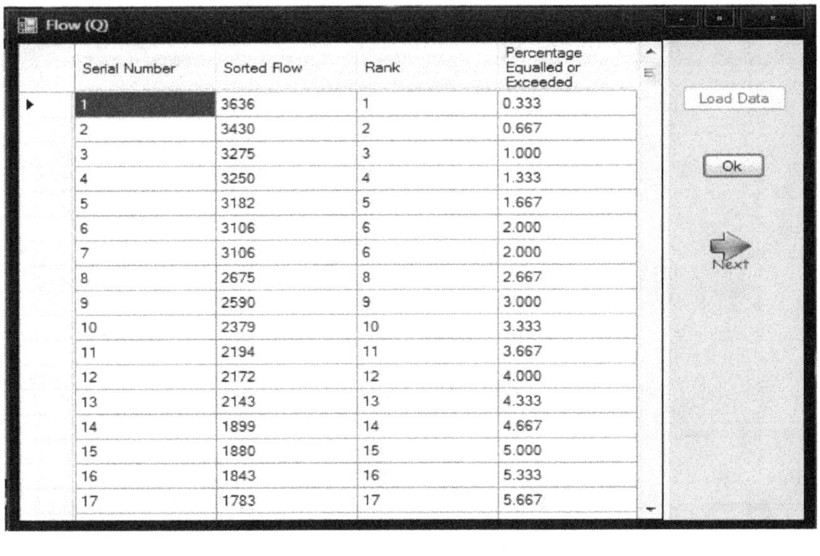

Flow (Q)				
	Serial Number	Sorted Flow	Rank	Percentage Equalled or Exceeded
▶	1	3636	1	0.333
	2	3430	2	0.667
	3	3275	3	1.000
	4	3250	4	1.333
	5	3182	5	1.667
	6	3106	6	2.000
	7	3106	6	2.000
	8	2675	8	2.667
	9	2590	9	3.000
	10	2379	10	3.333
	11	2194	11	3.667
	12	2172	12	4.000
	13	2143	13	4.333
	14	1899	14	4.667
	15	1880	15	5.000
	16	1843	16	5.333
	17	1783	17	5.667

Figure 2. The loaded flow form

The input data used for the testing are:

i. The monthly flow data of the river Niger from 1984-2008

ii. Design flow of $380m^3/s$ [10]

iii. Head of 27.6m [11]

iv. Generator efficiency of 91%

v. Fixed blade (propeller) type turbine

vi. Latitude = 9.138, Longitude = 4.7883 [12].

The losses, residual flow and percentage specified for firm flow were however not available. Nonetheless the program interface was however tested without these values because although they are necessary for accuracy, estimation can be made without them, moreover subsequently assumptions of the unavailable values within

acceptable range was used to show that if the values were known, the result would have been closer to the actual value.

With all the necessary input obtained, the program was run and tested. From the flow form shown in Figure 2 below, it can be seen that a flow of 3275 m^3/s is obtainable from the site only at 1% of the time in a year; also a flow of 1880 m^3/s is obtainable at 5% of the time and so on.

It can be seen from the Figure 3 that the flow values have been arranged in 5% increment of percentage equaled or exceeded, these are the values needed for further computation.

Serial Number	Sorted Flow	Percent Equalled or Exceeded
1	3842.00	0.0
2	1843.00	5.0
3	1533.00	10.0
4	1409.00	15.0
5	1354.00	20.0
6	1282.00	25.0
7	1209.00	30.0
8	1093.00	35.0
9	1049.00	40.0
10	1006.00	45.0
11	960.00	50.0
12	886.00	55.0
13	851.00	60.0
14	809.00	65.0
15	759.00	70.0
16	726.00	75.0

Figure 3. The flow duration values

Because residual flow was not entered it was passed as 0 to the application and that was what was subtracted from the flow duration values, hence the same values for the flow duration are for the available flow duration as well.

3.1. Flow Duration Curve

The flow duration and available flow duration have the same line because they share the same value. The flow duration curve obtained is depicted in Figure 4 below.

The following data were assumed with restriction to their range.

Percentage specified = 50%

Residual flow = $200m^3/s$

Transformer losses = 0.0125

Parasitic electricity losses = 0.02021

Maximum hydraulic losses = 0.07

Maximum tail water effect = 1

Annual downtime losses = 0.04

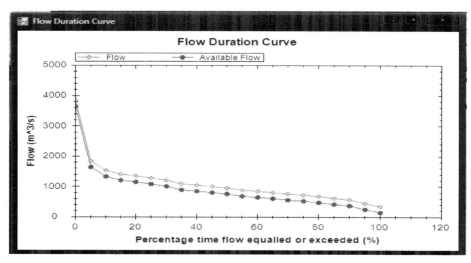

Figure 4. The flow duration curve

3.2. The Power Duration Curve

The power values were equal except the last value because any value larger than the design flow will not be used, rather the design flow will be used in its stead, this is because the turbine cannot utilize more than it is designed to use. This application plotted the curve using the power duration values as shown in Figure 5, and the area under this curve is the renewable energy available.

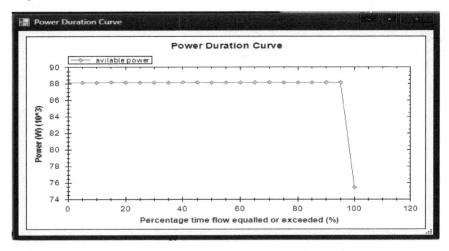

Figure 5. Power duration curve

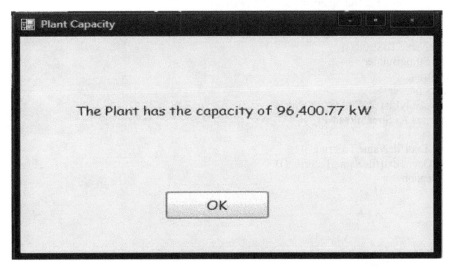

Figure 6. Plant capacity

As shown in Figure 6 below, a plant capacity of 96,400.77kW was obtained which is similar to the 96.4MW that a unit turbine generates in Jebba.

The performance of the program interface was satisfactory as it was able to compute the flow duration values and curve, available flow duration value and curve, plant capacity, power duration values and curve, renewable energy available etc. When compared with other commercially available computational interface for hydropower plants, such as the No Outage Com LLC interface output, it offers comparable details and it is user friendly like both RETScreen and No Outage Com LLC. However the program interface has some limitations which include its inability to be used to compute the energy demand for isolated grid.

4. Conclusion

Hydropower computation interface is essential for any hydropower installation and accurate estimation is needed for detailed information on the expected output of the plant. This program interface has the capability of computing the output of hydropower plants in a short time, develop a reliable expected flow and power characteristics of the site and generate a comprehensive report of the computation that is printable. The limitations of the program interface include its inability to compute the renewable energy delivered to an off grid load.

References

[1] Castaldi, D.,Chastain, E., Windram, M., Ziatyk, L. ,(2003). *A Study of Hydroelectric Power* : From a Global Perspective to a Local Application .Accessed on 10/04/2012,available online at http://www.ems.psu.edu/vikingpaper pdf , pp 4-5.

[2] Bennett, K., (1990), *Clean Energy Project Analysis*. Canada: Ret Screen

[3] Wang Z., Hongliang H. (2012), Hydropower Computation Using Visual Basic for Application Programming. China: ICAPIE. Accessed on 12/04/2012, available online at http://www.sciencedirect.com.

[4] Goran S. (2009), A Practical Guide to Assessment and Implementation of Small Hydropower.Australia: Tasmania. Accessed on 5/05/2012, available online at http://www.docstoc.com/docs/19906479/a-practical-guide-to-assesment-and-implementation-of-small , pp 1-5.

[5] The Visual Studio Combined help collection, Microsoft Visual Studio 2008 Documentation, (2007), Microsoft Corporation

[6] A visual representation of how bubble sort works. Accessed on 20/12/2011, available online at http://en.m.wikipedia.org/wiki/bubble_sort.

[7] Analysis Techniques: Flow Duration Analysis Tutorial (2002), Oregon State University. Accessed on 08/12/2011, available online at http://water.oregonstate.edu/streamflow/.

[8] RETScreen® International Clean Energy Decision Support Centre (1997), RETScreen® Software Online User Manuals. Canada: Ret Screen.

[9] Olukanni, D. O. & Salami, A. W. (2008). Fitting probability distribution functions to reservoir inflow at hydropower dams in Nigeria. Journal of Environmental Hydrology, USA, Vol. 16 Paper 35, pp. 1-7.

[10] Jebba Hydro Power Plant Brief History. Accessed on 12/11/2011, available online at http://jebbahydroelectricplc.org.

[11] Ajao K.R. & Sule B.F. (2011). Reduction of Carbon Dioxide (CO_2) in the Atmosphere, Hydropower as a Viable Renewable Energy Resource. Journal of Basic and Applied Scientific Research, TextRoad Publication, pp. 2127-2128.

[12] Jebba Hydroelectric power plant Nigeria-Geo. Accessed on 12/03/2012, available online at http://globalenergyobservatory.org/geoid/42544.

APPENDIX I

```
Flow form codes
Imports System.IO
Imports System.Text
Public Class FrmFlow
Public i As Integer
Private Sub Button3_Click(ByVal sender As System.Object, ByVal e As System.EventArgs) Handles Button3.Click
Dim openDialog As New OpenFileDialog openDialog.ShowDialog()
Dim fileName As String
fileName = openDialog.FileName.ToString()
Debug.WriteLine(fileName.ToString())
Dim builder As New StringBuilder
Dim singleChar As String
Dim intSingleChar As Integer
Dim itemlist As New ArrayList
Dim outWrite, outWrite2 As StreamReader
Try
outWrite = File.OpenText(fileName.ToString())
    outWrite2 = File.OpenText(fileName.ToString())
    Catch ex As Exception
    Beep()
    End Try
    'counting the entries
    Try
    While outWrite.Peek <> -1
    intSingleChar = outWrite.Read()
    singleChar = Chr(intSingleChar)
    " Debug.WriteLine(singleChar)
    If (singleChar.ToString() = "1") Then
        ElseIf (singleChar.ToString() = "2") Then
        ElseIf (singleChar.ToString() = "3") Then
        ElseIf (singleChar.ToString() = "4") Then
        ElseIf (singleChar.ToString() = "5") Then
        ElseIf (singleChar.ToString() = "6") Then
        ElseIf (singleChar.ToString() = "7") Then
        ElseIf (singleChar.ToString() = "8") Then
        ElseIf (singleChar.ToString() = "9") Then
        ElseIf (singleChar.ToString() = "0") Then
        ElseIf (singleChar.ToString() = vbNewLine Or singleChar.ToString() = vbCr) Then
    Else
    FrmHome.counter += 1
    End If
    End While
    Catch ex As Exception
    Debug.WriteLine(ex.ToString())
    End Try
    Dim counter2 As Integer
    counter2 = 0
```

```
Try
While outWrite2.Peek <> -1
    intSingleChar = outWrite2.Read()
    singleChar = Chr(intSingleChar)
    '' Debug.WriteLine(singleChar)
    If (singleChar.ToString() = "1") Then
        builder.Append("1")
    ElseIf (singleChar.ToString() = "2") Then
        builder.Append("2")
    ElseIf (singleChar.ToString() = "3") Then
        builder.Append("3")
    ElseIf (singleChar.ToString() = "4") Then
        builder.Append("4")
    ElseIf (singleChar.ToString() = "5") Then
        builder.Append("5")
    ElseIf (singleChar.ToString() = "6") Then
        builder.Append("6")
    ElseIf (singleChar.ToString() = "7") Then
        builder.Append("7")
    ElseIf (singleChar.ToString() = "8") Then
        builder.Append("8")
    ElseIf (singleChar.ToString() = "9") Then
        builder.Append("9")
    ElseIf (singleChar.ToString() = "0") Then
        builder.Append("0")
ElseIf       (singleChar.ToString().Equals(vbNewLine)      Or      singleChar.ToString().Equals(vbCr)      Or
singleChar.ToString().Equals(vbCrLf)) Then
        builder.Append("")
    Else
    FrmHome.arrayflow(counter2) = Double.Parse(builder.ToString())
        counter2 += 1
        builder.Clear()
End If
End While
'bubble sort
Dim counter3 As Integer
counter3 = 1
Dim k, l As Integer
For k = 0 To FrmHome.arrayflow.Count - 2
For l = k + 1 To FrmHome.arrayflow.Count - 1
If FrmHome.arrayflow(k) < FrmHome.arrayflow(l) Then
swap(FrmHome.arrayflow(k), FrmHome.arrayflow(l))
End If
Next
Next
Dim n, m As Integer
FrmHome.rank1(n) = 1
For n = 0 To FrmHome.arrayflow.Count - 2
For m = n + 1 To FrmHome.arrayflow.Count - 1
If FrmHome.arrayflow(m) = FrmHome.arrayflow(n) Then
FrmHome.rank1(m) = FrmHome.rank1(n)
Else
FrmHome.rank1(m) = m + 1
End If
Next
Next
counter2 += 1
i = 0
DataGridView1.Rows.Clear()
Do While (i < FrmHome.counter)
FrmHome.calc = FrmHome.rank1(i) / (FrmHome.counter + 1)
FrmHome.calc = FrmHome.calc * 100
FrmHome.calcArray.Add(FrmHome.calc)
```

```
DataGridView1.Rows.Add(New       String()       {i   +   1,   FrmHome.arrayflow(i),   FrmHome.rank1(i),
FrmHome.calc.ToString("0.000")})
        Debug.WriteLine(builder.ToString())
        Debug.WriteLine(i)
        i += 1
    Loop
    FrmHome.FlowDurationValuesToolStripMenuItem.Enabled = True
    Button3.Enabled = False
    FrmFlowDuration.flowduration()
  Catch ex As Exception
MsgBox("No/Wrong file selected", MsgBoxStyle.Exclamation, "Selection not Identified")
    End Try
  End Sub
  Private Sub swap(ByRef a As Integer, ByRef b As Integer)
    Dim temp As Integer
    temp = a
    a = b
    b = temp
  End Sub
Private Sub Button1_Click(ByVal sender As System.Object, ByVal e As System.EventArgs) Handles Button1.Click
    Me.Hide()

  End Sub
  Private Sub FrmFlow_Load(ByVal sender As System.Object, ByVal e As System.EventArgs) Handles MyBase.Load
  End Sub
  Private    Sub    LinkLabel1_LinkClicked(ByVal   sender   As   System.Object,   ByVal   e   As
System.Windows.Forms.LinkLabelLinkClickedEventArgs) Handles LinkLabel1.LinkClicked
    FrmResidualFLow.Show()
    Me.Hide()
    FrmResidualFLow.Visible = False
    FrmResidualFLow.ShowDialog(FrmHome)
  End Sub
  Private    Sub    DataGridView1_CellContentClick(ByVal   sender   As   System.Object,   ByVal   e   As
System.Windows.Forms.DataGridViewCellEventArgs) Handles DataGridView1.CellContentClick
End Sub
End Class


Flow duration codes
Public Class FrmFlowDuration
  Public Sub availflow()
    FrmAvailFlow.DataGridView1.Rows.Clear()
    Dim i As Integer
    i = 0
    Dim val As Double
    val = 0
    Do While (i <= 20)
      Try
  FrmHome.arrayavailflow(i) = (FrmHome.resultarray.Item(i) - FrmHome.ResFlow)
      Catch ex As Exception

  MsgBox("please open the flow duration window first", , "Unable to load Data")
      Exit Sub
      End Try
      FrmAvailFlow.DataGridView1.Rows.Add(New  String()  {(i + 1), FrmHome.arrayavailflow(i).ToString("0.00"),
val.ToString("0.0")})
      i += 1
      val += 5
    Loop
  End Sub
  Public Sub flowduration()
    DataGridView1.Rows.Clear()
    Dim val As Double
    Dim val4, val5, val6, val7 As Double
    Dim i As Integer
```

Flow duration codes

```
      i = 1
      val = 0
      Do While (i <= 21)
        If val = 0 Then
          val4 = FrmHome.calcArray.Item(0)
          val5 = FrmHome.calcArray.Item(1)
          val6 = FrmHome.arrayflow(0)
          val7 = FrmHome.arrayflow(1)
          FrmHome.result = ((val - val4) * (val7 - val6)) / (val5 - val4)
          FrmHome.result = FrmHome.result + val6
        ElseIf val = 100 Then
          val4 = FrmHome.calcArray.Item(FrmHome.calcArray.Count - 1)
          val5 = FrmHome.calcArray.Item(FrmHome.calcArray.Count - 2)
          val6 = FrmHome.arrayflow(FrmHome.calcArray.Count - 1)
          val7 = FrmHome.arrayflow(FrmHome.calcArray.Count - 2)
          FrmHome.result = ((val - val5) * (val6 - val7)) / (val4 - val5)
          FrmHome.result = FrmHome.result + val7
        Else
          Dim val3 As Integer
          val3 = FrmHome.calcArray.IndexOf(val)
          If val3 < 0 Then
            For Each calc1 As Double In FrmHome.calcArray
              If val > calc1 Then
                val4 = FrmHome.calcArray.Item(FrmHome.calcArray.LastIndexOf(calc1))
                val5 = FrmHome.calcArray.Item(FrmHome.calcArray.LastIndexOf(calc1) + 1)
                val6 = FrmHome.arrayflow(FrmHome.calcArray.LastIndexOf(calc1) + 1)
                val7 = FrmHome.arrayflow(FrmHome.calcArray.LastIndexOf(calc1) + 2)
              Else
              Exit For
              End If
            Next


            FrmHome.result = ((val - val4) * (val7 - val6)) / (val5 - val4)
            FrmHome.result = FrmHome.result + val6
          Else
            Dim val13 As Double
            val13 = FrmHome.arrayflow(val3 + 1)
            FrmHome.result = val13
          End If
        End If
  FrmHome.resultarray.Add(FrmHome.result)
  DataGridView1.Rows.Add(New String() {i, FrmHome.result.ToString("0.00"), val.ToString("0.0")})
        i = i + 1
        val = val + 5
      Loop
      FrmHome.AvailableFlowDurationValuesToolStripMenuItem.Enabled = True
      availflow()
      FrmHome.FlowDurationCurveToolStripMenuItem.Enabled = True
    End Sub
    Private    Sub    DataGridView1_CellContentClick(ByVal    sender    As    System.Object,    ByVal    e    As
System.Windows.Forms.DataGridViewCellEventArgs) Handles DataGridView1.CellContentClick
    End Sub
    Private Sub FrmFlowDuration_Load(ByVal sender As System.Object, ByVal e As System.EventArgs) Handles
MyBase.Load
    flowduration()
    End Sub
  Private Sub Button1_Click(ByVal sender As System.Object, ByVal e As System.EventArgs) Handles Button1.Click
    Me.Hide()
End Sub
End Class

Plant capacity codes
Public Class FrmPlantCapacity
```

```
 Private Sub FrmPlantCapacity_Load(ByVal sender As System.Object, ByVal e As System.EventArgs) Handles
MyBase.Load
 Dim hhydr As Single
hhydr = FrmHome.grosshead * FrmHome.Lhydrmax * (FrmHome.desflow ^ 2 / FrmHome.desflow ^ 2)
        'for francis turbine
        Dim df As Single
        Dim nqf As Single
        Dim enqf As Single
        Dim edf As Single
        Dim epf As Single
        Dim Qpf As Single
        Dim depf As Single
        Dim erf As Single
        df = ((0.46) * ((FrmHome.desflow) ^ (0.473)))
        nqf = 600 * ((FrmHome.grosshead - FrmHome.Lhydrmax) ^ (-0.5))
        enqf = ((nqf - 56) / 256) ^ 2
        edf = ((0.081 + enqf) * (1 - ((0.789) * (df ^ (-0.2)))))
        epf = (0.919 - enqf + edf) - 0.0305 + (0.05 * 4.5)
        Qpf = 0.65 * FrmHome.desflow * (nqf ^ (0.05))
        depf = 0.007 * (nqf ^ (0.4))
        erf = (1 - depf) * epf
        'for kaplan turbine
        Dim dk As Single
        Dim nqk As Single
        Dim enqk As Single
        Dim edk As Single
        Dim epk As Single
        Dim Qpk As Single
        dk = ((0.46) * ((FrmHome.desflow) ^ (0.473)))
        nqk = 800 * ((FrmHome.grosshead - FrmHome.Lhydrmax) ^ (-0.5))
        enqk = ((nqk - 170) / 700) ^ 2
        edk = ((0.095 + enqk) * (1 - ((0.789) * (dk ^ (-0.2)))))
        epk = (0.905 - enqk + edk) - 0.0305 + (0.05 * 4.5)
        Qpk = 0.75 * FrmHome.desflow
        'propellor turbine
        Dim dp As Single
        Dim nqp As Single
        Dim enqp As Single
        Dim edp As Single
        Dim epp As Single
        Dim Qpp As Single
        dp = ((0.46) * ((FrmHome.desflow) ^ (0.473)))
        nqp = 800 * ((FrmHome.grosshead - FrmHome.Lhydrmax) ^ (-0.5))
        enqp = ((nqk - 170) / 700) ^ 2
        edp = ((0.095 + enqp) * (1 - ((0.789) * (dp ^ (-0.2)))))
        epp = (0.905 - enqp + edp) - 0.0305 + (0.05 * 4.5)
        Qpp = FrmHome.desflow
        'for pelton turbine

        Dim npel As Single
        Dim dpel As Single
        Dim eppel As Single
        Dim Qppel As Single
        npel = 31 * (((FrmHome.grosshead - FrmHome.Lhydrmax) * (FrmHome.desflow / FrmHome.j)) ^ (0.5))
        dpel = (49.4 * ((FrmHome.grosshead - FrmHome.Lhydrmax) ^ (0.5)) * ((FrmHome.j) ^ (0.02))) / npel
        eppel = 0.864 * (dpel ^ (0.04))
        Qppel = (0.662 + (0.001 * FrmHome.j)) * FrmHome.desflow
        'For turgo
        Dim ept As Single
        ept = eppel - 0.03
        'For crossflow
        Dim Qpc As Single
        Qpc = FrmHome.desflow
```

```
   If Frm_Turbine_Type.RadioButton7.Checked = True Then
      'manual plant capacity

FrmHome.Pdes = (1000 * 9.81 * FrmHome.desflow) * (FrmHome.grosshead) * (1 - FrmHome.Lhydrmax) *
(FrmHome.eff) * FrmHome.Eg * (1 - FrmHome.Ltrans) * (1 - FrmHome.Lpara)
      End If
      'francis plant capacity
      If Frm_Turbine_Type.RadioButton1.Checked = True Then
         Dim eff As Single
         If FrmHome.desflow = Qpf Then
            eff = epf
         ElseIf FrmHome.desflow < Qpf Then
eff = (1 - (1.25 * (((Qpf - FrmHome.desflow) / Qpf) ^ (3.94 - 0.0195 * nqf)))) * epf
ElseIf FrmHome.desflow > Qpf Then
eff = epf - ((((FrmHome.desflow - Qpf) / (FrmHome.desflow - Qpf)) ^ 2) * (epf - erf))
         End If
         FrmHome.Pdes = (1000 * 9.81 * FrmHome.desflow) * (FrmHome.grosshead) * (1 - FrmHome.Lhydrmax) * (eff)
* FrmHome.Eg * (1 - FrmHome.Ltrans) * (1 - FrmHome.Lpara)


      End If
      'kaplan plant capacity
      If Frm_Turbine_Type.RadioButton2.Checked = True Then
         Dim eff As Single
         If FrmHome.desflow = Qpk Then
            eff = epk
         Else
            eff = (1 - 3.5 * (((Qpk - FrmHome.desflow) / Qpk) ^ 6)) * epk
         End If
         FrmHome.Pdes = (1000 * 9.81 * FrmHome.desflow) * (FrmHome.grosshead) * (1 - FrmHome.Lhydrmax) * (eff)
* FrmHome.Eg * (1 - FrmHome.Ltrans) * (1 - FrmHome.Lpara)
      End If
      'propellor plant capacity
      If Frm_Turbine_Type.RadioButton3.Checked = True Then
         Dim eff As Single
         eff = epp
         FrmHome.Pdes = (1000 * 9.81 * FrmHome.desflow) * (FrmHome.grosshead) * (1 - FrmHome.Lhydrmax) * (eff)
* FrmHome.Eg * (1 - FrmHome.Ltrans) * (1 - FrmHome.Lpara)
      End If
      If Frm_Turbine_Type.RadioButton4.Checked = True Then
         Dim eff As Single
         If FrmHome.desflow = Qppel Then
            eff = eppel
         Else
            eff = (1 - ((1.31 + (0.025 * FrmHome.j)) * (Math.Abs((Qppel - FrmHome.desflow) / Qppel)) ^ (5.6 + (0.4 *
FrmHome.j)))) * eppel
         End If
         FrmHome.Pdes = (1000 * 9.81 * FrmHome.desflow) * (FrmHome.grosshead) * (1 - FrmHome.Lhydrmax) * (eff)
* FrmHome.Eg * (1 - FrmHome.Ltrans) * (1 - FrmHome.Lpara)
      End If
      If Frm_Turbine_Type.RadioButton5.Checked = True Then
         Dim eff As Single
         If FrmHome.desflow = Qppel Then
            eff = ept
         Else
            eff = ((1 - ((1.31 + (0.025 * FrmHome.j)) * (Math.Abs((Qppel - FrmHome.desflow) / Qppel)) ^ (5.6 + (0.4 *
FrmHome.j)))) * eppel) - 0.03
         End If
         FrmHome.Pdes = (1000 * 9.81 * FrmHome.desflow) * (FrmHome.grosshead) * (1 - FrmHome.Lhydrmax) * (eff)
* FrmHome.Eg * (1 - FrmHome.Ltrans) * (1 - FrmHome.Lpara)
      End If
      If Frm_Turbine_Type.RadioButton6.Checked = True Then
         Dim eff As Single
```

```
        eff = 0.79 - 0.15 * ((FrmHome.desflow - FrmHome.desflow) / Qpc) - 1.37 * (((FrmHome.desflow -
FrmHome.desflow) / Qpc) ^ 14)
        FrmHome.Pdes = (1000 * 9.81 * FrmHome.desflow) * (FrmHome.grosshead) * (1 - FrmHome.Lhydrmax) * (eff)
* FrmHome.Eg * (1 - FrmHome.Ltrans) * (1 - FrmHome.Lpara)
      End If
      If FrmHome.Pdes = 0 Then
        Label1.Text = "One or some of the parameters needed for this computation has not been entered, go to help for
assistance"
      Else
        FrmHome.Pdes = FrmHome.Pdes / 1000
        Label1.Text = "The Plant has the capacity of " & FrmHome.Pdes.ToString("#,###.##") & " " & "kW"
        FrmHome.PowerDurationValuesToolStripMenuItem.Enabled = True
        Frmpowerduration.powerduration()
      End If
    End Sub


    Private Sub Button1_Click(ByVal sender As Object, ByVal e As System.EventArgs) Handles Button1.Click
      Me.Hide()
    End Sub

End Class


```

Power duration codes
```
Public Class Frmpowerduration
  Public Sub powerduration()
    Dim hhydr, htail As Single
    DataGridView1.Rows.Clear()
    Dim c As Integer
    'for francis turbine
    Dim df As Single
    Dim nqf As Single
    Dim enqf As Single
    Dim edf As Single
    Dim epf As Single
    Dim Qpf As Single
    Dim depf As Single
    Dim erf As Single
    df = ((0.46) * ((FrmHome.desflow) ^ (0.473)))
    nqf = 600 * ((FrmHome.grosshead - FrmHome.Lhydrmax) ^ (-0.5))
    enqf = ((nqf - 56) / 256) ^ 2
    edf = ((0.081 + enqf) * (1 - ((0.789) * (df ^ (-0.2)))))
    epf = (0.919 - enqf + edf) - 0.0305 + (0.05 * 4.5)
    Qpf = 0.65 * FrmHome.desflow * (nqf ^ (0.05))
    depf = 0.007 * (nqf ^ (0.4))
    erf = (1 - depf) * epf
    'for kaplan turbine
    Dim dk As Single
    Dim nqk As Single
    Dim enqk As Single
    Dim edk As Single
    Dim epk As Single
    Dim Qpk As Single
    dk = ((0.46) * ((FrmHome.desflow) ^ (0.473)))
    nqk = 800 * ((FrmHome.grosshead - FrmHome.Lhydrmax) ^ (-0.5))
    enqk = ((nqk - 170) / 700) ^ 2
    edk = ((0.095 + enqk) * (1 - ((0.789) * (dk ^ (-0.2)))))
    epk = (0.905 - enqk + edk) - 0.0305 + (0.05 * 4.5)
    Qpk = 0.75 * FrmHome.desflow
    'propellor turbine
    Dim dp As Single
    Dim nqp As Single
    Dim enqp As Single
    Dim edp As Single
```

```
      Dim epp As Single
      Dim Qpp As Single
      dp = ((0.46) * ((FrmHome.desflow) ^ (0.473)))
      nqp = 800 * ((FrmHome.grosshead - FrmHome.Lhydrmax) ^ (-0.5))
      enqp = ((nqk - 170) / 700) ^ 2
      edp = ((0.095 + enqp) * (1 - ((0.789) * (dp ^ (-0.2)))))
      epp = (0.905 - enqp + edp) - 0.0305 + (0.005 * 4.5)
      Qpp = FrmHome.desflow
      'for pelton turbine

      Dim npel As Single
      Dim dpel As Single
      Dim eppel As Single
      Dim Qppel As Single
      npel = 31 * (((FrmHome.grosshead - FrmHome.Lhydrmax) * (FrmHome.desflow / FrmHome.j)) ^ (0.5))
      dpel = (49.4 * ((FrmHome.grosshead - FrmHome.Lhydrmax) ^ (0.5)) * ((FrmHome.j) ^ (0.02))) / npel
      eppel = 0.864 * (dpel ^ (0.04))
      Qppel = (0.662 + (0.001 * FrmHome.j)) * FrmHome.desflow
      'For turgo
      Dim ept As Single
      ept = eppel - 0.03
      'For crossflow
      Dim Qpc As Single
      Qpc = FrmHome.desflow

      If Frm_Turbine_Type.RadioButton7.Checked = True Then
        'manual power duration
        For c = 0 To 20
          If FrmHome.arrayavailflow(c) > FrmHome.desflow Then
            htail   =   FrmHome.Htailmax   *   (((FrmHome.resultarray.Item(c)   -   FrmHome.desflow)   ^   2)   /
((FrmHome.arrayflow(0) - FrmHome.desflow) ^ 2))
            FrmHome.arrayavailflow(c) = FrmHome.desflow
            hhydr = FrmHome.grosshead * FrmHome.Lhydrmax * (FrmHome.desflow ^ 2 / FrmHome.desflow ^ 2)
          ElseIf FrmHome.arrayavailflow(c) < FrmHome.desflow Then
            FrmHome.arrayavailflow(c) = FrmHome.arrayavailflow(c)
            hhydr   =   FrmHome.grosshead   *   FrmHome.Lhydrmax   *   ((FrmHome.arrayavailflow(c))   ^   2   /
FrmHome.desflow ^ 2)
            htail = 0
          End If
          FrmHome.pduration(c) = (1000 * 9.81 * FrmHome.arrayavailflow(c)) * ((FrmHome.grosshead) - (hhydr +
htail)) * (FrmHome.eff) * FrmHome.Eg * (1 - FrmHome.Ltrans) * (1 - FrmHome.Lpara)

        Next
      End If
      'francis plant capacity
      If Frm_Turbine_Type.RadioButton1.Checked = True Then
        Dim eff As Single
        For c = 0 To 20
          If FrmHome.arrayavailflow(c) > FrmHome.desflow Then
            htail   =   FrmHome.Htailmax   *   (((FrmHome.resultarray.Item(c)   -   FrmHome.desflow)   ^   2)   /
((FrmHome.arrayflow(0) - FrmHome.desflow) ^ 2))
            FrmHome.arrayavailflow(c) = FrmHome.desflow
            hhydr = FrmHome.grosshead * FrmHome.Lhydrmax * (FrmHome.desflow ^ 2 / FrmHome.desflow ^ 2)
          ElseIf FrmHome.arrayavailflow(c) < FrmHome.desflow Then
            FrmHome.arrayavailflow(c) = FrmHome.arrayavailflow(c)
            hhydr   =   FrmHome.grosshead   *   FrmHome.Lhydrmax   *   ((FrmHome.arrayavailflow(c))   ^   2   /
FrmHome.desflow ^ 2)
            htail = 0
          End If
          If FrmHome.arrayavailflow(c) = Qpf Then
            eff = epf
          ElseIf FrmHome.arrayavailflow(c) < Qpf Then
            eff = (1 - (1.25 * (((Qpf - FrmHome.arrayavailflow(c)) / Qpf) ^ (3.94 - 0.0195 * nqf)))) * epf
          ElseIf FrmHome.arrayavailflow(c) > Qpf Then
```

```
            eff = epf - (((((FrmHome.arrayavailflow(c) - Qpf) / (FrmHome.desflow - Qpf)) ^ 2) * (epf - erf))
         End If
         FrmHome.pduration(c) = (1000 * 9.81 * FrmHome.arrayavailflow(c)) * ((FrmHome.grosshead) - (hhydr +
htail)) * (eff) * FrmHome.Eg * (1 - FrmHome.Ltrans) * (1 - FrmHome.Lpara)
        Next

    End If
    'kaplan plant capacity
    If Frm_Turbine_Type.RadioButton2.Checked = True Then
       Dim eff As Single
       For c = 0 To 20
          If FrmHome.arrayavailflow(c) > FrmHome.desflow Then
             htail   =   FrmHome.Htailmax  *  (((FrmHome.resultarray.Item(c)   -   FrmHome.desflow)   ^   2)   /
((FrmHome.arrayflow(0) - FrmHome.desflow) ^ 2))
             FrmHome.arrayavailflow(c) = FrmHome.desflow
             hhydr = FrmHome.grosshead * FrmHome.Lhydrmax * (FrmHome.desflow ^ 2 / FrmHome.desflow ^ 2)
          ElseIf FrmHome.arrayavailflow(c) < FrmHome.desflow Then
             FrmHome.arrayavailflow(c) = FrmHome.arrayavailflow(c)
             hhydr   =   FrmHome.grosshead   *   FrmHome.Lhydrmax   *   ((FrmHome.arrayavailflow(c))   ^   2   /
FrmHome.desflow ^ 2)
                htail = 0
          End If
          If FrmHome.arrayavailflow(c) = Qpk Then
             eff = epk
          Else
             eff = (1 - 3.5 * (((Qpk - FrmHome.desflow) / Qpk) ^ 6)) * epk
          End If
          FrmHome.pduration(c) = (1000 * 9.81 * FrmHome.arrayavailflow(c)) * ((FrmHome.grosshead) - (hhydr +
htail)) * (eff) * FrmHome.Eg * (1 - FrmHome.Ltrans) * (1 - FrmHome.Lpara)

       Next
    End If
    'propellor plant capacity
    If Frm_Turbine_Type.RadioButton3.Checked = True Then
       Dim eff As Single
       For c = 0 To 20
          If FrmHome.arrayavailflow(c) > FrmHome.desflow Then
             htail   =   FrmHome.Htailmax  *  (((FrmHome.resultarray.Item(c)   -   FrmHome.desflow)   ^   2)   /
((FrmHome.arrayflow(0) - FrmHome.desflow) ^ 2))
             FrmHome.arrayavailflow(c) = FrmHome.desflow
             hhydr = FrmHome.grosshead * FrmHome.Lhydrmax * (FrmHome.desflow ^ 2 / FrmHome.desflow ^ 2)
          ElseIf FrmHome.arrayavailflow(c) < FrmHome.desflow Then
             FrmHome.arrayavailflow(c) = FrmHome.arrayavailflow(c)
             hhydr   =   FrmHome.grosshead   *   FrmHome.Lhydrmax   *   ((FrmHome.arrayavailflow(c))   ^   2   /
FrmHome.desflow ^ 2)
                htail = 0
          End If
          If FrmHome.arrayavailflow(c) = Qpp Then
             eff = epp
          Else
             eff = (1 - 1.25 * ((Qpp - FrmHome.arrayavailflow(c)) / Qpp) ^ 1.13) * epp
          End If
          FrmHome.pduration(c) = (1000 * 9.81 * FrmHome.arrayavailflow(c)) * ((FrmHome.grosshead) - (hhydr +
htail)) * (eff) * FrmHome.Eg * (1 - FrmHome.Ltrans) * (1 - FrmHome.Lpara)
        Next

    End If
    If Frm_Turbine_Type.RadioButton4.Checked = True Then
       Dim eff As Single
       For c = 0 To 20
          If FrmHome.arrayavailflow(c) > FrmHome.desflow Then
             htail   =   FrmHome.Htailmax  *  (((FrmHome.resultarray.Item(c)   -   FrmHome.desflow)   ^   2)   /
((FrmHome.arrayflow(0) - FrmHome.desflow) ^ 2))
             FrmHome.arrayavailflow(c) = FrmHome.desflow
```

```
        hhydr = FrmHome.grosshead * FrmHome.Lhydrmax * (FrmHome.desflow ^ 2 / FrmHome.desflow ^ 2)
      ElseIf FrmHome.arrayavailflow(c) < FrmHome.desflow Then
        FrmHome.arrayavailflow(c) = FrmHome.arrayavailflow(c)
        hhydr = FrmHome.grosshead * FrmHome.Lhydrmax * ((FrmHome.arrayavailflow(c)) ^ 2 /
FrmHome.desflow ^ 2)
          htail = 0
      End If
      If FrmHome.desflow = Qppel Then
        eff = eppel
      Else
        eff = (1 - ((1.31 + (0.025 * FrmHome.j)) * (Math.Abs((Qppel - FrmHome.desflow) / Qppel)) ^ (5.6 + (0.4 *
FrmHome.j)))) * eppel
      End If
        FrmHome.pduration(c) = (1000 * 9.81 * FrmHome.arrayavailflow(c)) * ((FrmHome.grosshead) - (hhydr +
htail)) * (eff) * FrmHome.Eg * (1 - FrmHome.Ltrans) * (1 - FrmHome.Lpara)
      Next


    End If
    If Frm_Turbine_Type.RadioButton5.Checked = True Then
      Dim eff As Single
      For c = 0 To 20
        If FrmHome.arrayavailflow(c) > FrmHome.desflow Then
          htail = FrmHome.Htailmax * (((FrmHome.resultarray.Item(c) - FrmHome.desflow) ^ 2) /
((FrmHome.arrayflow(0) - FrmHome.desflow) ^ 2))
          FrmHome.arrayavailflow(c) = FrmHome.desflow
          hhydr = FrmHome.grosshead * FrmHome.Lhydrmax * (FrmHome.desflow ^ 2 / FrmHome.desflow ^ 2)
        ElseIf FrmHome.arrayavailflow(c) < FrmHome.desflow Then
          FrmHome.arrayavailflow(c) = FrmHome.arrayavailflow(c)
          hhydr = FrmHome.grosshead * FrmHome.Lhydrmax * ((FrmHome.arrayavailflow(c)) ^ 2 /
FrmHome.desflow ^ 2)
          htail = 0
        End If
        If FrmHome.desflow = Qppel Then
          eff = ept
        Else
          eff = ((1 - ((1.31 + (0.025 * FrmHome.j)) * (Math.Abs((Qppel - FrmHome.desflow) / Qppel)) ^ (5.6 + (0.4 *
FrmHome.j)))) * eppel) - 0.03
        End If
          FrmHome.pduration(c) = (1000 * 9.81 * FrmHome.arrayavailflow(c)) * ((FrmHome.grosshead) - (hhydr +
htail)) * (eff) * FrmHome.Eg * (1 - FrmHome.Ltrans) * (1 - FrmHome.Lpara)
        Next


    End If
    If Frm_Turbine_Type.RadioButton6.Checked = True Then
      Dim eff As Single
      For c = 0 To 20
        If FrmHome.arrayavailflow(c) > FrmHome.desflow Then
          htail = FrmHome.Htailmax * (((FrmHome.resultarray.Item(c) - FrmHome.desflow) ^ 2) /
((FrmHome.arrayflow(0) - FrmHome.desflow) ^ 2))
          FrmHome.arrayavailflow(c) = FrmHome.desflow
          hhydr = FrmHome.grosshead * FrmHome.Lhydrmax * (FrmHome.desflow ^ 2 / FrmHome.desflow ^ 2)
        ElseIf FrmHome.arrayavailflow(c) < FrmHome.desflow Then
          FrmHome.arrayavailflow(c) = FrmHome.arrayavailflow(c)
          hhydr = FrmHome.grosshead * FrmHome.Lhydrmax * ((FrmHome.arrayavailflow(c)) ^ 2 /
FrmHome.desflow ^ 2)
          htail = 0
        End If
        eff = 0.79 - 0.15 * ((FrmHome.desflow - FrmHome.desflow) / Qpc) - 1.37 * (((FrmHome.desflow -
FrmHome.desflow) / Qpc) ^ 14)
          FrmHome.pduration(c) = (1000 * 9.81 * FrmHome.arrayavailflow(c)) * ((FrmHome.grosshead) - (hhydr +
htail)) * (eff) * FrmHome.Eg * (1 - FrmHome.Ltrans) * (1 - FrmHome.Lpara)
        Next


    End If
```

```
        'populating the datagrid
        Dim d, value As Integer
        d = 0
        value = 0
        Do While (d <= 20)
           FrmHome.pduration(d) = FrmHome.pduration(d) / 1000
           DataGridView1.Rows.Add(New String() {(d + 1), FrmHome.pduration(d).ToString("#,###.##"), value})
           d += 1
           value += 5
        Loop
        FrmHome.PowerDurationCurveToolStripMenuItem.Enabled = True
        FrmHome.RenewableEnergyAvailableToolStripMenuItem.Enabled = True
    End Sub

    Private Sub Frmpowerduration_Load(ByVal sender As System.Object, ByVal e As System.EventArgs) Handles
MyBase.Load
        If FrmHome.arrayavailflow(0) = 0 Then
           Dim result As DialogResult = MessageBox.Show("Power duration can not be computed because flow values have
not been loaded. Do you want to load flow data?", _
                                           "Unable    to    compute    power    duration",    MessageBoxButtons.YesNo,
MessageBoxIcon.Question, MessageBoxDefaultButton.Button1)
           If result = DialogResult.Yes Then
              Me.Close()
              FrmFlow.ShowDialog()
           ElseIf result = DialogResult.No Then
              Me.Close()
           End If
        Else
           powerduration()
        End If
    End Sub

    Private Sub Button1_Click(ByVal sender As System.Object, ByVal e As System.EventArgs) Handles Button1.Click
        Me.Hide()
    End Sub

End Class
```

Renewable energy available codes
```
Public Class FrmRenEnergy

    Private Sub RenEnergy_Load(ByVal sender As System.Object, ByVal e As System.EventArgs) Handles MyBase.Load
        If FrmHome.arrayavailflow(0) = 0 Then
           Dim result As DialogResult = MessageBox.Show("Renewable energy available can not be computed because
flow values have not been loaded. Do you want to load flow data?", _
                                           "Unable  to  compute  Renewable  energy  available",  MessageBoxButtons.YesNo,
MessageBoxIcon.Question, MessageBoxDefaultButton.Button1)
           If result = DialogResult.Yes Then
              Me.Close()
              FrmFlow.ShowDialog()
           ElseIf result = DialogResult.No Then
              Me.Close()
           End If
        Else
           Dim k As Integer
           Dim Db, Dd, Dc As Double
           k = 0
           Db = 0
           Do While (k <= 19)
              Dd = ((FrmHome.pduration(k) + (FrmHome.pduration(k + 1))) / 2) * (5 / 100) * 8760 * (1 - FrmHome.Ldt)
              Dc = Dd + Db
              Db = Dc
              k += 1
           Loop
```

```
        FrmHome.RenEnergy = Db
        Db = Db / 1000
        Label1.Text = ("The Renewable Energy Available is " & Db.ToString("#,###.##") & " kWh/yr")
      End If
    End Sub
    Private Sub Button1_Click(ByVal sender As System.Object, ByVal e As System.EventArgs) Handles Button1.Click
      Me.Hide()
    End Sub
End Class
```

14

Optimal Controller for Wind Energy Conversion Systems

Hossein Nasir Aghdam, Farzad Allahbakhsh[*]

Department of Engineering, Ahar Branch, Islamic Azad University, Ahar, Iran
*Corresponding author: annakulmalaa@gmail.com

Abstract Increasing of word demand load is caused a new Distributed Generation (DG) enter to power system. One of the most renewable energy is the Wind Energy Conversion System (WECS), which is connected to power system using power electronics interface or directly condition. In this paper an optimal Lead-Lag controller for wind energy conversion systems (WECS) has been developed. The optimization technique is applied to Lead-Lag optimal controller in order to control of the most important types of wind system with doubly Fed Induction Generator is presents. Nonlinear characteristics of wind variations as plant input, wind turbine structure and generator operational behavior demand for high quality optimal controller to ensure both stability and safe performance. Thus, Honey Bee Mating Optimization (HBMO) is used for optimal tuning of Lead-Lag coefficients in order to enhance closed loop system performance. In order to use this algorithm, at first, problem is written as an optimization problem which includes the objective function and constraints, and then to achieve the most desirable controller, HBMO algorithm is applied to solve the problem. In this study, the proposed controller first is applied to two pure mathematical plants, and then the closed loop WECS behavior is discussed in the presence of a major disturbance. Simulation results are done for various loads in time domain, and the results show the efficiency of the proposed controller in contrast to the previous controllers.

Keywords: WECS- HBMO, optimal controller, lead lag controller

1. Introduction

According to increasing consumption of fossil fuels in recent years and their limited resources, it is very vital to use unlimited natural energy resources such as water, wind and sun. A very common application of such resources is electrical energy generation. Water and wind energy are mostly used to rotate shafts of prime movers to generate energy [1,2].

WECS are very challenging from the control system. In fact, wind turbine control enables a better use of the turbine capacity as well as the alleviation of aerodynamic and mechanical loads, which reduce the useful life of the installation [3].

Various approaches are introduced in literature in order to control WECS. In [4] a combined adaptive supervisory control approach is introduced which has a radial basis Zfunction adaptive controller and a simple supervisor. In [5] a PID controller is proposed in which PID parameters are tuned via a wavelet neural network that is a single layer network with wavelets of type RASP1 as hidden nodes. This network proposes a pre specified model structure to identify the unknown plant. In [6,7] a self tuning control method for wind energy conversion system is introduced using wavelet networks.

Many predictive control algorithms were proposed during the last decades, such as dynamic matrix control (DMC), model algorithmic control (MAC) and generalized predictive control (GPC). Although they can obtain high performance of the controlled system, few of them have been realized in electrical machine or drive application due to the heavy calculation power required by them [8]. Improving the reactive power compensation and active filtering capability of a Wind Energy Conversion System is proposed in [9]. The proposed algorithm is applied to a Doubly Fed Induction Generator (DFIG) with a stator directly connected to the grid and a rotor connected to the grid through a back-to-back AC-DC-AC PWM converter. It has been shown also that the proposed strategy allows an operating full capacity of the RSC in terms of reactive power compensation and active filtering. Ref [10] describes a novel teaching experience using wind energy as the starting point for understanding power electronics and electrical machines. The results point out the wide variety of concepts involved in the course, the numerous competences that it can enhance, and its positive reception by learners. An analysis tool is proposed to estimate the theoretical control limit of the RSC in suppressing the short circuit rotor currents during grid faults in [11]. To execute the analysis, a simplified DFIG model with decoupled stator and rotor fluxes is presented; the low-voltage ride through (LVRT) problem can be formulated as an optimization problem, which intends to suppress the rotor winding currents with voltage constraints. The results are expected to help the manufacturers to assess and improve their RSC controllers or LVRT measures.

In this paper an optimal-tuning Lead-Lag controller for a WECS is proposed based on Honey Bee Mating Optimization (HBMO) algorithm. So far none of the mentioned approaches in the literature analyzed the WECS in the presence of external disturbances. But, in this study, a major disturbance is applied to the close loop WECS and then system performance is analyzed. Furthermore, it will be shown that the system response in the proposed controller is considerably faster while no fluctuations are mounted on the control or output signals.

This paper is organized as follows: in section 2 the wind energy conversion system is explained. In section 3 the HBMO algorithm is described. Proposed controller is applied to mathematical plants in section 4. Section 5 shows the advantages of the proposed controller over classic controller. Finally, section 6 contain conclusion of proposed algorithm.

2. Wind Energy Conversion System

Wind energy conversion systems have very interesting specifications such as simplicity, high reliability and low maintenance expenses. These systems are more cost-effective than all the other renewable energy resources. A WECS could be used in three different topologies: standalone, hybrid and grid-connected. The first topology is used in battery chargers for applications such as illumination, remote radio repeaters and sailboats. The second topology is used in combining wind turbines with solar and diesel generators [2].

In this study, we discuss a horizontal-axis wind turbine has been discussed for which the output power is defined as below [12]:

$$P_w = 0.5\rho_a C_p \, V_\omega^3 A_r \qquad (1)$$

Where ρ is the air density, A is the area swept by blades, V_ω is the wind speed and C_p is the power factor which is a nonlinear function of $\lambda = \dfrac{\omega R}{V_\omega}$ where R is the radius of the blades and ω is the angular speed of the turbine blades. C_p is usually approximated as $C_p = \alpha\lambda + \beta\lambda^2 + \gamma\lambda^3$ where α, β, γ are constructive parameters for a given wind turbine.

According to physical and mathematical equations of the system, the generation torque of the wind turbine could be obtained as below [12]:

$$T_t = 0.5\rho_a \left(\frac{C_p}{\lambda}\right) V_\omega^2 \pi R^2 \qquad (2)$$

The dynamics of the whole system (turbine plus generator) are related to the total moment of inertia. Thus ignoring torsion in the shaft, electric dynamics of the generator and other higher order effects, the approximate system's dynamic model is:

$$J\omega' = T_t(\omega, V_\omega) - T_g(\omega, \alpha) \qquad (3)$$

The input to our wind energy conversion system is a time-varying wind speed. In this paper the wind speed changes assumed is shown in Figure 1:

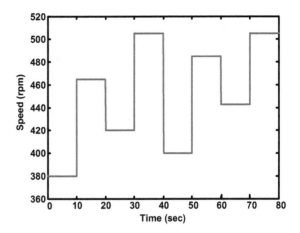

Figure 1. time-varying wind speed

Now the need for a controller which satisfies the following conditions arises. Reduce steady state error of the output signal Reject disturbance. The open loop system response depicted in Figure 2, shows the vital need of advanced controller for reducing error and convergence increment. One of the simplest controllers that seem to be capable of satisfying all the above conditions is a Lead-Lag controller.

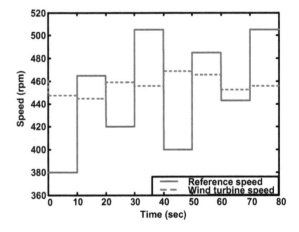

Figure 2. open loop response of WECS

3. HBMO Algorithm

The honey bee is a social insect that can survive only as a member of a community, or colony. The colony inhabits an enclosed cavity. A colony of honey bees consist of a queen, several hundred drones, 30,000 to 80,000 workers and broods during the active season. A colony of bees is a large family of bees living in one bee-hive. The queen is the most important member of the hive because she is the one that keeps the hive going by producing new queen and worker bees [13]. Drones' role is to mate with the queen. Tasks of worker bees are several such as: rearing brood, tending the queen and drones, cleaning, regulating temperature, gather nectar, pollen, water, etc. Broods arise either from fertilized (represents queen or worker) or unfertilized (represents drones) eggs. The HBMO Algorithm is the combination of several different methods corresponded to a different phase of the mating process of the queen. In the marriage process, the queen(s) mate during their mating flights far from the nest. A mating flight starts with a dance performed by the queen who then starts a mating flight during which the drones follow the queen and mate with her in the air. In each mating, sperm reaches the spermatheca and accumulates there to form the genetic pool of the colony. The queen's size of spermatheca number equals to the maximum number of mating of the queen in a single mating flight is determined. When the queen mates successfully, the genotype of the drone is stored. At the start of the flight, the queen is initialized with some energy content and returns to her nest when her energy is within some threshold from zero or when her spermatheca is full. In developing the algorithm, the functionality of workers is restricted to brood care, and therefore, each worker may be represented as a heuristic which acts to improve and/or take care of a set of broods. A drone mates with a queen probabilistically using an annealing function as [14]:

$$P_{rob}(Q,D) = e^{-\frac{\Delta(f)}{s(t)}} \qquad (4)$$

Where Prob (Q, D) is the probability of adding the sperm of drone D to the spermatheca of queen Q (that is, the probability of a successful mating); Δ (f) is the absolute difference between the fitness of D (i.e., f (D)) and the fitness of Q (i.e., f (Q)); and S (t) is the speed of the queen at time t. It is apparent that this function acts as an annealing function, where the probability of mating is high when both the queen is still in the start of her mating–flight and therefore her speed is high, or when the fitness of the drone is as good as the queen's. After each transition in space, the queen's speed, S(t), and energy, E(t), decay using the following equations:

$$S(t+1) = \alpha \times s(t) \qquad (5)$$

$$E(t+1) = E(t) - \gamma \qquad (6)$$

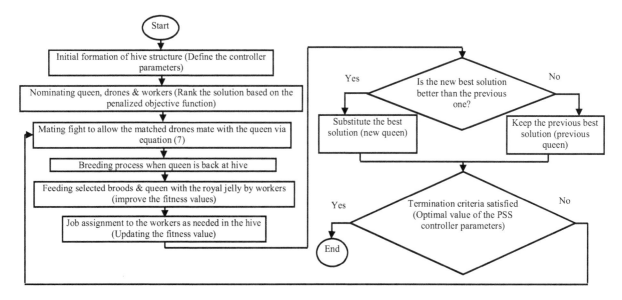

Figure 3. Algorithm and computational flowchart of HBMO

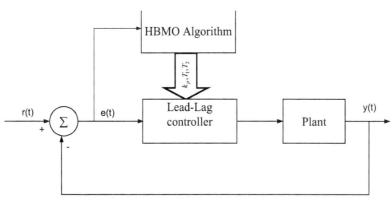

Figure 4. Schematic of proposed controller desining

where α is a factor and γ is the amount of energy reduction after each transition. Also, Algorithm and computational flowchart of HBMO method to optimize the WECS controller parameters is presented in Figure 4.

Thus, HBMO algorithm may be constructed with the following five main stages:

A. The algorithm starts with the mating–flight, where a queen (best solution) selects drones probabilistically to form the spermatheca (list of drones). A drone is then selected from the list at random for the creation of broods.

B. Creation of new broods by crossoverring the drones' genotypes with the queen's.

C. Use of workers (heuristics) to conduct local search on broods (trial solutions).

D. Adaptation of workers' fitness based on the amount of improvement achieved on broods.

E. Replacement of weaker queens by fitter broods.

4. Controller Designing Based on HBMO

Due to develop of system controllers, the conventioal contreollers are used widely in power system applications. making applicable of conventional controllers is simple against the new controllers of power systems [14]. The Lead-Lag controllers are widely used in most cases of power systems controllers which compensat very well. One of the most benefits of these controllers is the easily implemention in analog and digital systems. If these controllers are designed optimaly, indubitable they become one of the most implimanted controllers in modern systems. This paper introduce a new optimal Lead-Lag controller, which is used of HBMO algorithm to designing the controller of wind energy conversion system in order to control of output power. The overall controller schematic is shown in Figure 4.

Lead-Lag general controller is expressed in equation (7) which the controller k_p, T_1, T_2 parameters should be optimized using the proposed algorithm. In the load variations, it is obvious that the transient mode of the WECS system depends on the controller parameters. The conventional controller designing method are not viable to be implemented because this system is an absolute nonlinear system. So these methods would have not efficient performance in the system.

$$G_c(s) = k_p \frac{1+sT_1}{1+sT_2} \tag{7}$$

In order to design controller optimal controller using HBMO for the WECS from the speed vaiation curve, we consider the worst condition for load design controllers for these conditions. Figure 5, depicts the worst condition for wind variation in the system.

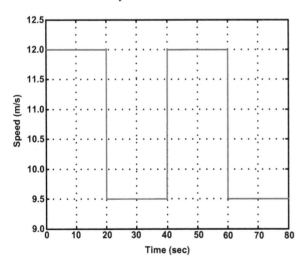

Figure 5. Worst case of wind speed variation

At first, problem should be written as an optimization problem and then by appliyning the proposed optimization method, the best Lead-Lag controller is achieved. Selecting objective function is the most important part of this optimization problem. Because, choosing different objective functions may completely change the particles

variation state. In optimization problem we considered the power error signal in order to achieved the best controller.

$$J = \int_{tstart}^{tsim} \left[\left| N_{ref} - N_{out} \right| \right] . dt \tag{8}$$

Where, *tsim* is the simulation time in which objective function is calculated, the N_{out} is the real speed of turbine and N_{ref} is the reference speed signal. We are reminded that whatever the objective function is a small amount in this case the answer will be more optimized. Each optimizing problem is optimized under a number of constraints. At this problem constraints should be expressed as .

Minimize J subject to
$$k_p^{min} \leq k_p \leq k_p^{max}$$
$$T_1^{min} \leq T_1 \leq T_1^{max} \tag{9}$$
$$T_2^{min} \leq T_2 \leq T_2^{max}$$

Where, T_1, T_2 are in the interval [0.01 10] and k_p in the interval [1 500].

In this problem, the number of particles, dimension of the particles, and the number of repetitions are selected 40, 3, 80, respectively. After optimization, results are determined as below:

$$k_p = 12.4 \ , T_1 = 0.063, T_2 = 6.7532 \tag{10}$$

5. Simulation Results

The conventional PID controller has initial and random values in proportional, integral and derivative coefficients as below:

$$k_p = 90 \ , k_i = 8, k_d = 2.5 \tag{11}$$

Closed loop system response to the input shown in Figure 1, in the absence of disturbance is plotted in Figure 6. According to Figure 6, Proposed PID controller gives a satisfactory closed loop response when there is no disturbance. In order to check the capability of the designed Proposed PID controller in the presence of disturbance, a major disturbance on the firing angle of the thyristor at 35 seconds is applied (See Figure 7).

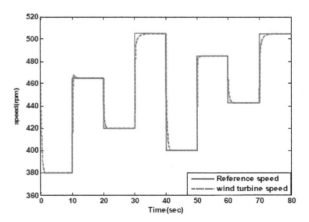

Figure 6. System response using conventional PID controller

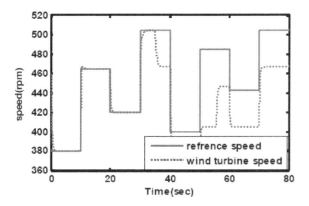

Figure 7. System response using conventional PID controller for thyristor disturbance

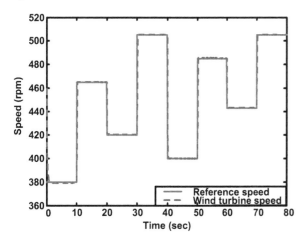

Figure 8. System response using proposed controller based on HBMO

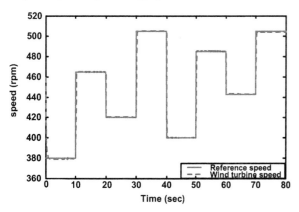

Figure 9. System response using proposed controller which disturbance has been applied to firing angle

In order to qualify the proposed controller, closed loop system responses to the input shown in Fig 3, in the absence of disturbance and uncertainty, with disturbance applied to firing angle of the thyristor, with disturbance applied to output speed of the given turbine and with parametric uncertainty are depicted in Figs 8 through 12 respectively. It is also desired to check whether the proposed controller is capable of rejecting disturbances and being robust to uncertainties simultaneously. Fig.13 depict corresponding control signal in the presence of both the disturbance and the uncertainty. According to Figs 8 through 13 one can say that the proposed controller is much more effective than conventional PID controller.

The proposed controller offers a fast response with almost zero overshoot and is very operational in rejecting disturbances. It could be easily seen that our controller behaves effectively and pretty fast when a disturbance is applied. It also operates pretty well when there are parametric uncertainties.

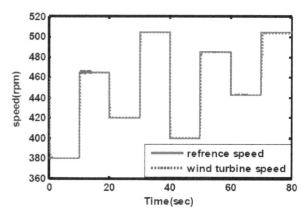

Figure 10. System response with proposed controller which disturbance applied to firing angle at 10

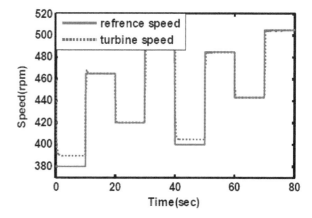

Figure 11. Uncertain system response with proposed controller

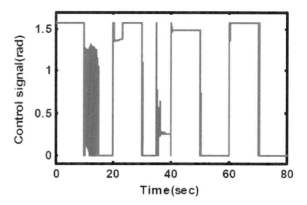

Figure 12. Control signal with both the disturbance and the uncertainty

6. Conclusion

In this paper, a novel method to control the WECS was proposed. It was shown that these proposed methods, can explain instinct nonlinear behavior turbine/ generator systems very well. At first, the conventional PID controller is used in order to control of WECS speed. It was seen that the proposed controller offered a desired closed loop response when the system was subjected to major external disturbance. Then, proposed method, which was used the HBMO optimization algorithm controller, was applied to control the WECS. Issues such as external disturbances which are associated with all practical systems ask for model-free. The proposed controller not only increases the closed system band width but also offers a desired response when the system is subjected to various disturbances. The high adaptation speed along with high stability and performance characteristics make this method a high quality control approach.

References

[1]   K.Ogawa, N.Ymammura, M.Ishda, "Study for Small Size Wind Power Generating System Using Switched Reluctance Generator", IEEE International Conference on Industrial Technology, pp. 1510-1515, 2006.

[2]   F. D. Bianchi, H. De Battista and R. J. Mantz, "Wind Turbine Control Systems Principles, Modeling and Gain Scheduling Design" Springer-Verlag London Limited 2007.

[3]   Miguel Angel Mayosky, Gustavo I.E.Cancelo "Direct Adaptive Control of Wind Energy Conversion Systems Using Gaussian Networks" IEEE Trans on Neural Networks, Vol. 10, pp.898-906, July 1999.

[4]   M.Sedighizade, A.Rezazadeh "Adaptive PID Control of Wind Energy Conversion Systems Using RASP1 Mother Wavelet Basis Function Networks" Proceedings of Academy of Science, Engineering and Technology Vol.27, pp. 269-273. February 2008.

[5]   M.Sedighizade "Nonlinear Model Identification and Control of Wind Turbine Using Wavelets" Proceedings of the 2005 IEEE Conference on Control Applications Toronto, pp. 1057-1062 Canada, 2005.

[6]   M. Kalantari, M. Sedighizadeh "Adaptive Self Tuning Control of Wind Energy Conversion Systems Using Morlet Mother Wavelet Basis Functions Networks"12th Mediterranean IEEE Conference on Control and Automation MED'04, Kusadasi, Turkey, 2004.

[7]   X. Zhang, D. XU and Y. LIU, "Predictive Functional Control of a Doubly Fed Induction Generator for Variable Speed Wind Turbines," 5th World Congress on Intelligent Control and Automation, June 15-19, Hangzhou. P.R. China, 2004.

[8]   G. A. Smith, Power electronics for recovery of wind and solar energy, Wind Eng., Vol. 19, no. 2, 1995.

[9]   M. Boutoubat, L. Mokrani, M. Machmoum "Control of a wind energy conversion system equipped by a DFIG for active power generation and power quality improvement" Renewable Energy 50 (2013) 378e386.

[10]   Noradin Ghadimi "Genetically Adjusting of PID Controllers Coefficients for Wind Energy Conversion System with doubly fed induction Generator" World Applied Sciences Journal 15 (5), 702-707, 2011.

[11]   Shuai Xiao, Hua Geng, Honglin Zhou, Geng Yang" Analysis of the control limit for rotor-side converter of doubly fed induction generator-based wind energy conversion system under various voltage dips" IET Renew. Power Gener., 2013, Vol. 7, Iss. 1, pp. 71-81.

[12]   P.Puleston, "Control strategies for wind energy conversion systems", Ph.D. dissertation, Univ. La Plata, Argentina, 1997.

[13]   Yannis M, Magdalene M, Georgios D (2011). "Honey bees mating optimization algorithm for the Euclidean traveling salesman problem," Infor. Sci. 181(20):4684-4698.

[14]   Niknam T (2011). An efficient multi-objective HBMO algorithm for distribution feeder reconfiguration. Expert Syst. Appl. 38(3): 2878-2887.

15

Optimization of Biodiesel Production from Waste Cooking Oil

Seid Yimer, Omprakash Sahu[*]

Department of Chemical Engineering, KIOT Wollo University, Kombolcha (SW), Ethiopia
*Corresponding author: ops0121@gmail.com

Abstract Energy is basic need for growth of any country. The world energy demand is increasing so rapidly because of increases in industrialization and population that limited reservoirs will soon be depleted at the current rate of consumption. Both the energy needs and increased environmental consciousness have stimulated the researching of an alternative solution. So an attempted has been made to investigation of biodiesel production using transesterification reaction with solid or heterogeneous catalyst at laboratory scale and to compare the physical properties with the standard biodiesel properties. The selected process parameters are temperature ranged from 318 K to 333 K, molar ratio of methanol to oil from 4:1 to 8:1, mass ratio of catalyst to oil from 3% to 5% and rotation speed at optimum biodiesel yield was produced at 600 rpm.

Keywords: alcohol, bioenergy, methanol, oil, waste

1. Introduction

"Sustainable" second generation biofuels don't yet exist on any useful scale but small quantities particularly from other existing waste streams - may be available by 2020. To meet the 10% target, and assuming the world carries out all the other measures above and if the quadruple waste counting comes into effect as proposed in the new commission proposal, an extra 22.7 Petajoules (PJ) of waste energy would be required to get us the rest of the way to the 10% target. In theory, there's quite a lot of waste energy out there. Food waste could be an aerobically digested to produce methane although this preferentially should be burnt to produce electricity. Much of this energy is in the form of wood, energy crops, forestry residues or agricultural straw, which isn't easy to transform into vehicle fuel and costly to transport. There's a risk that creating demand for waste as a fuel can reduce the incentive to reduce waste in the first place and reducing waste always saves more energy in the long term. As with all wastes, the primary focus should be on waste avoidance, then recovery, and finally disposal. In the waste management hierarchy, energy-from-waste comes fairly low down. Materials may have other more sustainable uses and have all sorts of unintended knock-on effects. So some solution should be required to minimize the cost like production of biodiesel [1]. Biodiesel is the name for a variety of ester based fuels (fatty ester) generally defined as monoalkyl ester made from renewable biological resources such as vegetable oils (both edible and non edible), recycled waste vegetable oil and animal fats [2]. This renewable source is as efficient as petroleum diesel in powering unmodified diesel engine. Today's diesel engines require a clean burning, stable fuel operating under a variety of conditions. Using biodiesel not only helps maintaining our environment, it also helps in keeping the people around us healthy [3,4,5]. Biodiesel is miscible with petrodiesel in all ratios. In many countries, this has led to the use of blends of biodiesel with petrodiesel instead of neat biodiesel [6]. There are different types of feed stocks that are used for the production of biodiesel. These includes linseed oil, palm seed oil, waste cooked vegetable oil, sunflower seed oil, cotton seed oil, cooking seed oil and animal fats [7,8,9]. Oilseed plants are used for the production of biodiesel through the process called transesterification reaction which is a process by which alcohol reacts with vegetable oil in the presence of catalyst. Triglycerides are major components of vegetable oils and animal fats. Chemically, triglycerides are esters of fatty acids with glycerol. Fatty acid ethyl esters are mostly involved because ethanol is the cheapest alcohol, but other alcohols, namely methanol, may be employed as well [10]. In this way, highly viscous triglycerides are converted in long chain monoesters presenting much lower viscosity and better combustion properties to enhance the burning. Homogeneous or heterogeneous catalysis are used to enhance the reaction rate [11,12,13].

However, the synthesis of biodiesel from these low quality oils is challenging due to undesirable side reactions as a result of the presence of FFAs and water. The pretreatment stages, involving an acid catalyzed pre-esterification integrated with water separation, are necessitated to reduce acid concentrations and water to below threshold limits prior to being processed by standard biodiesel manufacturing [14,15]. Besides catalyzing esterification, acid catalysts are able to catalyze TG transesterification, opening the door for the use of acid catalysts to perform simultaneous FFA esterification and TG transesterification [16,17]. The most common approach for processing waste oil in the biodiesel synthesis is a two-step acid-pretreatment before the successive base-catalyzed transesterification [18,19,20,21]. By using a two-step sulfuric acid-catalyzed pre-esterification, Canakci and Van Gerpen [20] were able to reduce the acid levels of the high FFA feedstocks (reaction mixtures containing 20-40 wt% FFA) below 1 wt% within 1 h, making the feedstocks suitable for the subsequent alkali-catalyzed transesterification.

Recently, the two-step catalyzed processes was shown to be an economic and practical method for biodiesel production from waste cooking oils where the acid values of 75.9 mg KOH/g were presented [18]. Employing a ferric sulfate catalyzed reaction followed by KOH catalyzed transesterification, a yield of 97.3% fatty acid methyl ester (FAME) was achieved within 4 h. By integrating the heterogeneous catalyst in the pretreatment process, several advantages have been introduced such as no acidic wastewater, high efficiency, low equipment cost, and easy recovery compared to corrosive liquid acids. Another reaction route for a two-step process was proposed and proven by Saka and his co-workers [22,23], where the first step involves hydrolysis with subcritical water at 270°C and subsequently followed by methyl esterification of the oil products at the same temperature. In this process, triglycerides were hydrolyzed with subcritical water to yield FFAs, which further reacted with supercritical methanol, resulting in a completed reaction within 20 min. Besides catalyzing the esterification, acid catalysts are able to perform TG transesterification; however, acid catalysts are 3 orders of magnitude slower than basic catalysts [24], thus allowing FFA esterification and TG transesterification to be catalyzed simultaneously [25]. The slow activity can be traded off with a decrease in process complexity, equipment pieces, and the amount of waste stream. For instance, Zhang et al. [3,8] have shown that, in biodiesel production using waste cooking oils, a one-step acid catalyzed process offered more advantages over the alkali-catalyzed process with regard to both technological and economical benefits. It was also proved to be a competitive alternative to a two-step acid pre-esterification process. Hence, it is imperative for this. The main aim of this work is to utilized the waste cooking oil for the production of biodiesel which is collected from the cafeteria, restaurant. The effect of temperature, effect of alcohol to oil molar ratio and effect of catalyst weight on the yield of biodiesel has also studied.

2. Material and Methods

2.1. Material

The waste cooking oil has been collected from the restaurant.

2.2. Method

2.2.1. Central Composite Design

Experimental design was analyzed and done by the Design Expert 7.0.0 software application. Experimental design selected for this study is CCD and the output measured is biodiesel yield gained. Process variables revised are reaction temperature, molar ratio of ethanol to oil and weight percentage of catalyst. To get maximum conversion; reaction period and rotation speed was set at 2 hours and 500 rpm respectively and at constant atmospheric pressure. The operating limits of the biodiesel production process conditions are reasons to choose levels of the variables Three level three factors CCD was made use of in the optimization study, needing 20 experiments to be done. Catalyst concentration, ethanol to oil molar ratio and reaction temperature were the independent variables selected to optimize the conditions for biodiesel production by using sodium hydroxide as main catalyst for performing transesterification reaction. Twenty experiments were done and the data was statistically analyzed by the Design Expert 7.0.0 software and to get suitable model for the percentage of fatty acid methyl ester as a function of the independent variables. The model was tested for adequacy by analysis of variance. The regression model was found to be highly significant with the correlation coefficients of determination of R-Squared ($R2$), adjusted R-Squared and predicted R-Squared having a value of 0.9966, 0.9936 and 0.9788, respectively. The yield of the transesterification processes were calculated as sum of weight of FAME produced to weight of cooking oil used, multiplied by 100. The formula is given as:

$$Yield\ of\ FAME = \left(\frac{Weight\ of\ fatty\ acid\ methyl\ ester}{Weight\ of\ fat\ used}\right) \times 100\% \quad (1)$$

2.2.2. Development of Model

The model equation that correlates the response to the transesterification process variables in terms of actual value after excluding the insignificant terms was given below. The predicted model for percentage of FAME content (R) in terms of the coded factors is given by Eq. 2.

$$Yield\ of\ FAME(\%) = 86.07 + 13.34\,xA + 0.75\,xB$$
$$+2.13\,xC - 0.66\,xAxB - 0.97\,xAxC - 0.68\,xBxC \quad (2)$$
$$-5.62\,xA^2 + 0.090\,xB^2 - 0.55\,xC^2$$

Table 1. Physicochemical property of biodiesel

S. No	Biodiesel properties	Measured values	Units
1	Density at 20°C	950	kg/m^3
2	Kinematic viscosity 40°C	38.7	mm^2/s
3	Flash point (°C)	150	°C
4	Acid value	12.6	mgKOH/g
5	Saponification value mgKOH/g	180	mgKOH/g
6	Moisture content	0.019	(%)w/w
7	Ash content	0.04	(%)w/w
8	Iodine value	90.6	I_2g/100g
9	Free fatty acidic	5.28	%
10	Melting point	30.2	°C

3. Result and Discussion

3.1. Physicochemical Properties

After transesterification biodiesel obtained, whose physical and chemical properties is mention in Table 1. The physical-chemical properties of generated biodiesel were found all satisfactoriness.

3.2. Effect of Methanol-to-Oil Molar Ratio

The methanol-to-oil ratio is one of the important factors that affect the conversion of triglyceride to FAME. Stoichiometrically, three moles of methanol are required for each mole of triglyceride, but in practice, a higher molar ratio is required in order to drive the reaction towards completion and produce more FAME as products. The results obtained in this study are in agreement with this. As shown in Figure 1, the methanol-to-oil ratio showed positive influence to the yield of methyl ester, but the yield started to decrease as the ratio much increased. The increase is due to the positive sign in the experimental model. The decrease in the yield contrary to increase in molar ratio may be due to the separation problem resulted from excessive methanol. Higher ratio of methanol used could also minimize the contact of access triglyceride molecules on the catalyst's active sites which could decrease the catalyst activity.

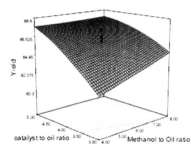 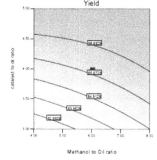

Figure 1. Interaction Effect of temperature and catalyst ratio versus yield (a) surface plot (b) contour plot

3.3. Effect of Catalyst Concentration

The RSM was used to optimize the conditions of biodiesel which is shown in Figure 2. It was observed that the catalyst concentration influenced the biodiesel yield in a positive manner up to a certain concentration. Beyond this concentration, the biodiesel yield decreased with increase in potassium hydroxide concentration. When the catalyst mount was improved, the interactive (active) site of the catalyst was increased; thus, the transesterification reaction was accelerated and biodiesel yield was increased.

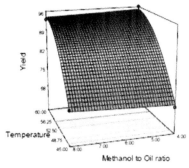 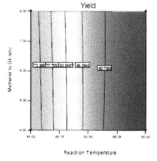

Figure 2. Interaction effect of catalyst ratio and methanol ratio versus yield (a) surface plot (b) contour plot

3.4. Effect of Temperature

The effect of temperature with biodiesel conversion is shown in Figure 3. Temperature increase clearly influences the reaction rate and biodiesel yield in a positive manner. The temperature increase affected the biodiesel yield in a positive manner till 60°C and after that it decreased. The increase in the yield of FAME at higher reaction temperature is due to higher rate of reaction. From the experimental model analysis the p-value of the temperature term was nearer to the p-value limit. Hence, its effect on the biodiesel is almost constant.

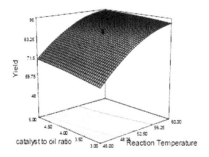 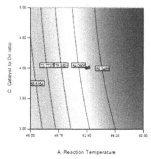

Figure 3. Interaction effect of temperature and methanol ratio versus yield (a) surface plot (b) contour plot

The results above have shown that the three transesterification process variables and the interaction among the variables affect the yield of FAME. Therefore, the next step is to optimize the process variables in order to obtain the highest yield using the model regression developed. Using the optimization function in Design Expert, it was predicted that at the following conditions; 60°C of reaction temperature, methanol to cooking oil ratio of 4 and 5 wt% of catalyst, an optimum FAME yield of 95.08% can be obtained. In order to verify this prediction, experiments were conducted and the results were comparable with the prediction. It was found that the experimental value of 94.98% of FAME content agreed well with the predicted value. Therefore, this study shows that tribasic sodium phosphate is a potential catalyst for the production of biodiesel from cooking oil cooking oil via heterogeneous transesterification. The optimization result also tells the same result as the ANOVA output. The ANOVA output shows that the transesterification process is highly and significantly affected by the temperature, catalyst weight and the interaction between the temperature and the catalyst.

4. Conclusion

Biodiesel is alternative option in place diesel, the factors like profitability, availability, low sulfur content, low aromatic content, biodegradability and renewability makes biodiesel more advantageous. The result shows that biodiesel production using heterogeneous catalyst, is a considerable potential in biodiesel production process, mainly because of catalyst regeneration (decrease of catalyst cost),Simplification of separation process (decrease of production cost) and decrease of wastewater (development of environmental friendly process). At 333 K of reaction temperature, methanol to waste cooking oil ratio of 4 and 5 wt% of catalyst, an optimum fatty acid methyl ester yield of 95.10% was obtained. The sodium phosphate has excellent activity during transesterification. As a solid catalyst, sodium phosphate can decrease the waste water treatment and the steps of purification. It has a potential for industrial application in the transesterification of waste cooking oil to biodiesel. Hence, sodium phosphate has good catalytic performance

Reference

[1] Chivers D., Rice T., Alternatives to Biofuels, Meeting the Carbon Budgets ‐ 2012 Progress Report to Parliament, Chapter 5, pages 176-187.

[2] Lotero, E., Liu, Y.J., Lopez, D.E., Suwannakarn, K., Bruce, D.A., and Goodwin, J.G., Jr., "Synthesis of biodiesel via acid catalysis" Ind. Eng. Chem. Res. 44 (2005) 5353.

[3] Zhang, Y., Dube, M.A., McLean, D.D., and Kates, M., "Biodiesel production from waste cooking oil: 2. Economic assessment and sensitivity analysis" Bioresour. Technol. 90 (2003) 229.

[4] Srivastava, A. and Prasad, R., "Triglycerides-based diesel fuels" Renewable & Sustainable Energy Reviews 4 (2000) 111.

[5] Ma, F.R. and Hanna, M.A., "Biodiesel production: a review" Bioresour. Technol. 70 (1999) 1.

[6] Schwab, A.W., Bagby, M.O., and Freedman, B., "Preparation and Properties of Diesel Fuels from Vegetable-Oils" Fuel 66 (1987) 1372.

[7] Ziejewski, M., Kaufman, K.R., Schwab, A.W., and Pryde, E.H., "Diesel engine evaluation of a nonionic sunflower oil-aqueous ethanol microemulsion" J. Am. Chem. Soc. 61 (1984) 1620.

[8] Zhang, Y., Dube, M.A., McLean, D.D., and Kates, M., "Biodiesel production from waste cooking oil: 1. Process design and technological assessment" Bioresour. Technol. 89 (2003) 1.

[9] National Renderers Association (2005), www.renderers.org

[10] Kusdiana, D. and Saka, S., "Effects of water on biodiesel fuel production by supercritical methanol treatment" Bioresour. Technol. 91 (2004) 289.

[11] Warabi, Y., Kusdiana, D., and Saka, S., "Reactivity of triglycerides and fatty acids of rapeseed oil in supercritical alcohols" Bioresour. Technol. 91 (2004) 283.

[12] Bunyakiat, K., Makmee, S., Sawangkaew, R., and Ngamprasertsith, S., "Continuous production of biodiesel via transesterification from vegetable oils in supercritical methanol" Energy & Fuels 20 (2006) 812.

[13] Saka, S. and Kusdiana, D., "Biodiesel fuel from rapeseed oil as prepared in supercritical methanol" Fuel 80 (2001) 225.

[14] Hsu, A.F., Jones, K.C., Foglia, T.A., and Marmer, W.N., "Continuous production of ethyl esters of grease using an immobilized lipase" J. Am. Chem. Soc. 81 (2004) 749.

[15] Fukuda, H., Kondo, A., and Noda, H., "Biodiesel fuel production by transesterification of oils" J. Biosci. Bioeng. 92 (2001) 405.

[16] Deng, L., Nie, K.L., Wang, F., and Tan, T.W., "Studies on production of biodiesel by esterification of fatty acids by a lipase preparation from Candida sp. 99-125" Chinese J. Chem. Eng. 13 (2005) 529.

[17] Chang, H.M., Liao, H.F., Lee, C.C., and Shieh, C.J., "Optimized synthesis of lipase-catalyzed biodiesel by Novozym 435" J. Chem. Technol. Biotechnol. 80 (2005) 307.

[18] Wang, Y., Ou, S.Y., Liu, P.Z., Xue, F., and Tang, S.Z., "Comparison of two different processes to synthesize biodiesel by waste cooking oil" J. Mol. Catal. A. 252 (2006) 107.

[19] Lepper, H. and Friesenhagen, L., "Process for the production of fatty acid esters of short-chain aliphatic alcohols from fats and/or oils containing free fatty acids" 1986 U.S.

[20] Canakci, M. and Van Gerpen, J., "Biodiesel production from oils and fats with high free fatty acids" Trans. ASAE 44 (2001) 1429.

[21] Zullaikah, S., Lai, C.C., Vali, S.R., and Ju, Y.H., "A two-step acid-catalyzed process for the production of biodiesel from rice bran oil" Bioresour. Technol. 96 (2005) 1889.

[22] Minami, E. and Saka, S., "Kinetics of hydrolysis and methyl esterification for biodiesel production in two-step supercritical methanol process" Fuel 85 (2006) 2479.

[23] Saka, S., Kusdiana, D., and Minami, E., "Non-catalytic Biodiesel Fuel Production with Supercritical Methanol Technologies" J. Sci. Ind. Res. 65 (2006) 420.

[24] Freedman, B., Butterfield, R.O., and Pryde, E.H., "Transesterification Kinetics of Soybean Oil" J. Am. Chem. Soc. 63 (1986) 1375.

[25] Zheng, S., Kates, M., Dube, M.A., and McLean, D.D., "Acid-catalyzed production of biodiesel from waste frying oil" Biomass & Bioenergy 30 (2006) 267. Ye-book.

16

Numerical Simulation for Achieving Optimum Dimensions of a Solar Chimney Power Plant

M. Ghalamchi[1,*], M. Ghalamchi[2], T. Ahanj[3]

[1]Department of Energy Engineering, Science and Research Campus, Islamic Azad University, Tehran, Iran
[2]Faculty of New Sciences &Technologies, University of Tehran, Tehran, Iran
[3]Faculty of Nuclear Engineering, University of Shahid Beheshti, Tehran, Iran
*Corresponding author: Mehrdad.Ghalamchi.mech@gmail.com

Abstract Renewable energies are playing a fundamental role in supplying energy, as these kinds of energies can be clean, low carbon and sustainable. Solar chimney power plant is a novel technology for electricity production from solar energy. A solar chimney power plant derives its mechanical power from the kinetic power of the hot air which rises through a tall chimney, the air being heated by solar energy through a transparent roof surrounding the chimney base. The performance evaluation of solar chimney power plant was done by FLUENT software by changing three parameters including collector slope, chimney diameter and entrance gap of collector. The results were validated with the solar chimney power plant which was constructed in Zanajn, Iran with 12 m height, 10 m collector radius and 10 degree Collector angle. By simulation and numerical optimization of many cases with dimensional variations, increasing 300 to 500 percent of chimney velocity and eventually increasing output power of system was observed in different cases.

Keywords: renewable energy, solar chimney, numerical simulation, Zanjan

1. Introduction

According to the shortage of the current energy resources and increasing the global energy demand and also, regarding to this reality that the energy obtained of the fossil fuels is environment damaging and nonrenewable, the demand of achieving a technology for exploitation of clean and renewable energies is sensed. Solar chimney power plant is one of the proper options for using the clean energy resources. The solar chimney technology is designed for preparing energy in large scales. In this type of power plant, the solar energy and subsequently the air movement are used. The air flow cause turbine rotation and the rotation is converted to electrical energy by a generator.

The conceptual design of solar chimney power plants was first propounded by professor Schaich in 1978 [1]. Before 1980, the system was built as an experimental sample in Manzanares, Spain [2]. The chimney diameter, collector radius, and height of the sample were 10 m, 122 m, and 194.6 m, respectively. The maximum output power of the system reached to 41 KW in September, 1982. Since then, many researchers became interested in this work and studied the related technologies for the high potential and vast applications of solar chimney power plants. Yan et al. [3] and Padki and Sherif [4] have done some of the Preliminary study on the thermo-fluid

analysis of a solar chimney power plant. In 1983, Krisst built a 10 W, 6 m collector diameter, 10 m height chimney in US [5]. In 1997, a chimney was built in Florida University and the plan was optimized for two times [6]. In aspect of small scales, a model with 0.14 W outlet power, 3.5 cm chimney radius, 2 m height, and 9 m² collector areas was built by Klink, Turkey [7].

Pretorius andKröger [8], Ninic [9], Onyango and Ochieng [10] investigated the influence of a developed convective heat transfer equation, more accurate turbine inlet loss coefficient, quality of collector roof glass, and various types of soil on the performance of a solar chimney power plant. Ming et al. [11] evaluated the temperature and pressure fields for air in solar chimney power plant. More investigation and simulations have been carried out by Lodhi [12], Bernardes et al. [13], Bernardes et al. [14], von Backstro ¨mand Gannon [15], Gannon and von Backstro¨m [16], Pastohr et al. [17], Schlaich et al. [18], Bilgen and Rheault [19].

The efficiency of solar chimney power plant was investigated according to the previous works. Since the practical measurements are much difficult, the simulation methods would be much easier for efficiency prediction of the different models of solar chimneys. Generally speaking, different combinations of chimney and collector dimensions can be built for purpose of electrical power production. Considering the cost reduction, that is much important that the optimized combination of chimney-collector dimensions would be known for the prediction.

In this paper, the mathematical model and experimental results of a solar chimney, which was built in University of Zanjan, are described.

2. Operation of Solar Chimney Power Plants

The solar chimney power plants include three main components: chimney, collector, and turbine generator. Chimney is a tall cylindrical structure which is equipped with turbine generator and installed on the middle of collector. The ground under the collector is covered by black bodies which are able to absorb the solar radiation energy. In general, collector is an enclosure which has one inlet and one exit, the collector inlet is the plant atmospheric air and the outlet is connected to the chimney base. The chimney can be installed vertically on the collector or may be laid as inclined on a mountain. This composition causes sunlight to convert to electrical energy and the conversion is divided to two stages. In the first stage, the conversion of solar energy to thermal energy which is done by using the thermal absorbers in the collector and the greenhouse effect increases the temperature. In the second stage, the chimney converts the thermal energy to kinetic energy and then is converted to electrical energy by a combination of wind turbine and generator. This energy conversion is done due to two reasons: the first reason is that the air temperature raising causes the air density decreasing, hence the air moves toward the chimney exit point, the second reason is the conic shape of the chimney which causes increasing the velocity of the air that move toward the center of the collector. The schematic of the chimney is shown in Figure 1.

The collectors, in the simplest state, can be made of glass or transparent plastics which are located horizontally with a space from ground. The transparent layer is able to transmit the radiation from sun position (short wave-length waves), but in return the radiations emitted by ground (long wave-length waves) are trapped. Therefore,

the ground under the collector is warmed up and causes warming the radial air flow. The collectors were made of polycarbonate sheets, while the ground was covered by thick polyethylene black films. The polyethylene films play the role of absorbent which are able to store the heat during the day and reject it at night. Hence, an annual-permanent power generation may be obtainable.

The chimney is similar to a pressure pipe with low friction loss. The upward movement force of the warm air is dependent on the collector air temperature, air volumetric rate, and chimney height. The volumetric rate and air velocity rising makes better movement of the turbine and finally more electrical power. The air flow which passes through the turbine, can be adjusted by the turbine blade angle. The produced mechanical energy is converted to electrical energy by coupling the turbine with a generator.

A solar chimney doesn't need the direct insulation, the energy can even be supplied in cloudy days.

3. Solar Chimney Set-up

For purpose of more studies regarding these kinds of power plants, an experimental sample was built in University of Zanjan, in 2010 [20]. The chimney height is 12 m and the collector has 10 m diameter. The collector angle must be designed in a way that the most possible heat could be absorbed, Zanjan city has the attitude of 36°, 68′ and longitude of 48°, 45′ [21]. So if we want to have the most absorbed heat by the collector, the collector output must have the height of 5 tanπ/6. The collector output was made with a height of 1m and the collector opening was 15 cm. The schematic and the dimensions of the built sample are shown in Figure 2 and Table 1.

Table 1. Dimensions of the built sample

Geometric parameter	Symbol	Amount (m)
Collector opening height	H_2	0.15
Collector output height	H_c-H_2	1
Chimney diameter	d_c	0.25
Collector diameter	d_2	10
Chimney height	H_c	12

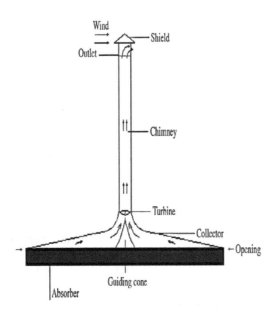

Figure 1. Schematic of the solar chimney

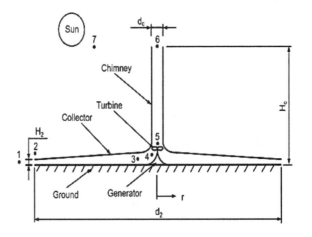

Figure 2. Schematic of the built model

Because of the UV resistance and proper cost of polyethylene, a 12 m length, 0.25 m diameter polyethylene pipe was used as the chimney. Due to the low diameter and height of the chimney, the friction inside the chimney is very low. The collector structure is made of 48 pieces of steel profiles, for strengthening the collector against wind and heavy snow a 4×4 steel profile was used and the collector bases was put into cement foundation. The greenhouse effect of the collector was supplied by the double-layer polycarbonate sheets. The grade of the material is UV resistant and the air between the sheet layers prevents heat loss. For temperature measuring in different zones of the power plant, SMT-160 sensors were utilized and a data-logger registered all the data round the clock. Totally, 12 temperature sensors were used for the measurements from which, 4 sensors were allocated for the chimney and 8 for the collector. The chimney sensors were positioned exactly at the middle with 3 m distances. Also, the collector sensors were put under the PC sheets with different height. Ambient temperature measuring was done using resistance sensors in 2 m height. An accurate anemometer, installed in 10 m height, was used for measuring the ambient wind velocity, and AVM-702 anemometer was utilized for measuring the air velocity inside the chimney at the opening. The photo of the power plant is shown in Figure 3.

Figure 3. Photo of the built solar chimney

In present study the performance evaluation of solar chimney power plant by changing three main parameters including collector slope, chimney diameter and entrance gap of collector in constant absorber of 78.5 m^2 investigated.

4. Governing Equation

Governing equations include the continuity equation, the Navier-Stokes equations, the energy equation and k-ϵ equations which are presented as followings:
Continuity equation:

$$\frac{\partial P}{\partial t}+\frac{1}{r}\frac{\partial}{\partial z}(r\rho u)+\frac{\partial}{\partial z}(\rho v)=0 \qquad (1)$$

Navier-Stokes equations:

$$\rho\frac{du}{dt}=-\frac{\partial P}{\partial r}+\frac{\partial}{\partial r}\left[2\mu\frac{\partial u}{\partial r}+\mu'\,\vec{\nabla}.\vec{v}\right]$$
$$+\frac{\partial}{\partial z}\left[\mu\left(\frac{\partial u}{\partial z}+\frac{\partial v}{\partial r}\right)\right]+\frac{2\mu}{r}\left(\frac{\partial u}{\partial r}-\frac{v}{r}\right) \qquad (2)$$

$$\rho\frac{dv}{dt}=-\frac{\partial P}{\partial z}+\rho g+\frac{\partial}{\partial z}\left[2\mu\frac{\partial v}{\partial z}+\mu'\vec{\nabla}.\vec{v}\right]$$
$$+\frac{1}{r}\frac{\partial}{\partial r}\left[\mu r\left(\frac{\partial u}{\partial z}+\frac{\partial v}{\partial r}\right)\right] \qquad (3)$$

Energy equation:

$$\rho c_p\left[\frac{\partial T}{\partial t}+\frac{1}{r}\frac{\partial}{\partial r}(rTu)+\frac{\partial}{\partial z}(Tv)\right]$$
$$=\frac{1}{r}\frac{\partial}{\partial r}\left(rw\frac{\partial T}{\partial r}\right)+\frac{\partial}{\partial z}\left(w\frac{\partial T}{\partial z}\right)+\frac{\partial P}{\partial t} \qquad (4)$$
$$+\frac{1}{r}\frac{\partial}{\partial r}(rPu)+\frac{\partial}{\partial z}(Pv)+\Phi$$

$k-\epsilon$ Equations:

$$\rho\left[\frac{1}{r}\frac{\partial}{\partial r}(rku)+\frac{\partial}{\partial z}(kv)\right]$$
$$=\frac{\partial}{\partial z}\left[\left(\mu+\frac{\mu_t}{\sigma_k}\right)\frac{\partial k}{\partial z}\right]+\frac{1}{r}\frac{\partial}{\partial r}\left[r\left(\mu+\frac{\mu_t}{\sigma_k}\right)\frac{\partial k}{\partial r}\right] \qquad (5)$$
$$+G_k+\beta g\frac{\mu_t}{\mathrm{Pr}_t}\frac{\partial T}{\partial z}-\rho\epsilon$$

$$\rho\left[\frac{1}{r}\frac{\partial}{\partial r}(r\epsilon u)+\frac{\partial}{\partial z}(\epsilon v)\right]$$
$$=\frac{\partial}{\partial z}\left[\left(\mu+\frac{\mu_t}{\sigma_\epsilon}\right)\frac{\partial \epsilon}{\partial z}\right]+\frac{1}{r}\frac{\partial}{\partial r}\left[r\left(\mu+\frac{\mu_t}{\sigma_\epsilon}\right)\frac{\partial \epsilon}{\partial r}\right] \qquad (6)$$
$$+C_{1\epsilon}G_k\frac{\epsilon}{k}-C_{2\epsilon}\rho\frac{\epsilon^2}{k}$$

Assuming one-dimensional, steady state flow, the following equation would be valid for mass flow rate:

$$\dot{m}=\rho_f A U_f \qquad (7)$$

Substituting flow cross section area in equation (1) yields in:

$$U_f=-\frac{\dot{m}}{2\pi r H_c\rho_f} \qquad (8)$$

The energy equation of the hot air inside the collector is:

$$\rho_a C_p U_f H_c\frac{\partial T_f}{\partial r}=h_c\left(T_f-T_c\right)+h_e\left(T_f-T_e\right) \qquad (9)$$

In this form of energy equation, the conduction heat transfer includes all the connecting surfaces. For finding the temperature of the surfaces, we need integration of equation (3). At first, a preliminary amount is assumed for the surfaces temperature and heat transfer coefficient. Assuming constant amount for the air density, substituting Uf of equation (2) in (3), and then integrating equation (3), the following simplified correlation is obtained:

$$T_f(r) = \frac{1}{2}\left[T_c + T_e + (2T_a - T_c - T_e)e^{\frac{2\pi h}{C_p \dot{m}}\left(r^2 - r_0^2\right)} \right] \quad (10)$$

The collector inlet air flow profile in a given mass flow rate is depicted in Figure 3. For proving the effect of the surface temperatures on the temperature profile of the fluid, three different amounts are selected for the ground and collector ceiling temperatures. Considering $\dot{m} = 10$ kg/s and h= 5 W/m2K, equation for is solved for constant wall temperature conditions. Figure 4 shows the temperature rising of the collector entered air. The temperature difference between the collector inlet and outlet shows the vertical efficiency of the system. The results demonstrate that the temperature between the inlet and outlet is increased by raising the temperatures of the ground and collector surface, so that, this temperature increase causes the system efficiency promotion.

Therefore, the energy equation of the collector air may be revised as:

$$\rho_a C_p u_f H_c \frac{\partial T_f}{\partial r} = h_c\left(T_f - T_c(r)\right) + h_e\left(T_f - T_e(r)\right) (11)$$

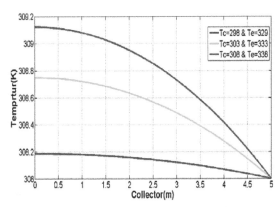

Figure 4. The temperature profile of the fluid flow along with collector $\dot{m} = 10 \mathrm{kgs}^{-1}$

5. Numerical Simulation of System

In this part of the study, the numerical simulation of the solar chimney power plant is presented. A physical model for a solar chimney power plant was built based on the geometrical dimensions of the prototype of Zanjan. The basic equations including the models discussed up to now were numerically solved with the help of the commercial simulation program FLUENT (6.3).The basic equations

were simplified to axisymmetric and steady state. Because a turbulence model is necessary for the description of the turbulent flow conditions, the standard k–e model and standard wall mode were selected to describe the fluid flow inside the collector and the chimney. The domain was discredited with 45,000 two-dimensional unstructured mesh elements. The grid was refined adaptively (hanging nodes) near walls (bend, glass, ground and chimney) and in the area of the turbine. This is required because large gradients appear near the walls. Boundary conditions of zanjan solar chimney power plant indicate at Figure 5 and Table 2.

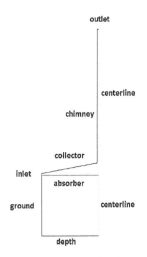

Figure 5. Boundaries locations

Table 2. Boundary Condition

Material	Boundary Condition		
	Boundary	Type	Condition
Fluid (Air)	Centerline	Axis	Symmetry axis
	Absorber	Wall	Convective heat transfer
	Collector	Wall	Convective heat transfer
	Chimney	Wall	Insulated surface
	Inlet	Pressure inlet	Atmospheric pressure
	Outlet	Pressure outlet	Atmospheric pressure
Solid (Soil)	Centerline	Axis	Symmetry axis
	Absorber	Wall	Convective heat transfer
	Ground	Wall	Insulated surface
	Depth	Wall	Constant temperature

All numerical calculations had to be carried out with the solver with double precision. The iteration error was at least 10^{-6} on all calculations, for the energy equation at least10^{-9}.Under these conditions, the solution converged in less than 2500 iterations. Figure 6 a. shows the collector temperature distribution and 6 b. velocity magnitude for $800(\mathrm{w/m}^2)$ with ambient temperature of 300 k.

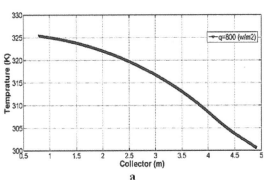

a

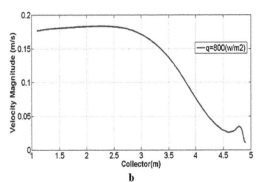

b

Figure 6. Collector temperature distribution

6. Results and Discussion

Figure 7 compares the experimental and numerical collector temperature distribution over 9 cm of absorber at 11:33 AM.

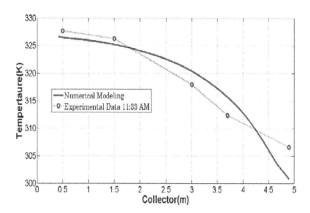

Figure 7. Experimental and numerical collector temperature distribution over 9 cm of absorber

Compare experimental and numerical results shows the good agreement. Figure 8 indicates the velocity magnitude distribution in collector with 0.15m inlet, 0.25m chimney diameter and with slope of 0,5,10 degrees over 9 cm of absorber.

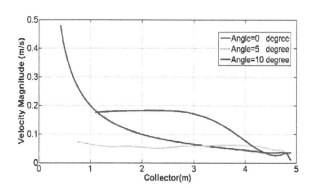

Figure 8. Velocity magnitude distribution of collector for different collector slope

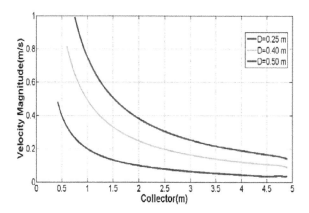

Figure 9. Velocity magnitude distribution of collector for different chimney diameter

Figure 8 appears the best angle of collector is zero. Because the air flow in the collector with collector slope of zero move in the straight stream lines and in the

turbulent regime of the air flow the eddy generation comes down. Figure 9 indicate the velocity magnitude distribution in collector with 0.15m inlet, zero collector slope and chimney diameter of 0.25,0.4 and 0.5 m over 9 cm of absorber.

Figure 9 shows the best diameter of chimney is 0.5 m for special constant of zanjan solar chimney absorber.

Figure 10 shows the best entrance height of collector is 0.1 m for special constant of Zanjan solar chimney absorber Figure 8, Figure 9 and Figure 10 indicates that the velocity magnitude of Zanjan Solar chimney power plant can improve 300 to 500 percent in different cases.

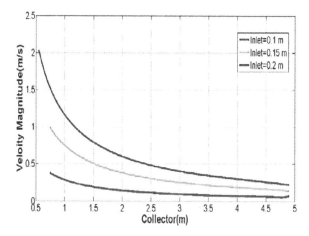

Figure 10. Velocity magnitude distribution of collector for different entrance height

7. Conclusion

In present study, the collector of Zanjan Solar chimney power plant mathematically modeled. The numerical model of collector validate with experimental data of temperature sensors. The performance evaluation of solar chimney power plant was done by FLUENT software by changing three parameters including collector slope, chimney diameter and entrance gap of collector. The results shows that the horizontal collector works better and the best chimney diameter and entrance gap of collector are 0.5 and 0.1 m respectively. By simulation and numerical optimization of many cases with dimensional variations, increasing 300 to 500 percent of chimney velocity and eventually increasing output power of system was observed in different cases.

Nomenclature

c_p Specific heat capacity (J/kg.K)

D Diameter (m)

h Convective heat transfer coefficient (W/m².K)

$\dot{m}$ Mass flow rate (kg/s)

P Pressure (Pa)

r Radial coordinate (m)

T Temperature (K)

u Velocity in the radial direction (m/s)

V Quantity of velocity vector (m/s)

v Velocity in the axial direction (m/s)

w Thermal conductivity (W/m.K)

z	Axial coordinate (m)
β	Thermal expansion coefficient (1/K)
ρ	Density (kg/m^3)

Subscripts

a	Ambient
e	Absorber
c	Collector
ch	Chimney
i	Inner
o	Out
r	Radial coordinate

References

[1] Bernardes, M.A., dos, S., Voss, A., Weinrebe, G. *Thermal and technical analyzes of solar chimneys.* Solar Energy 2003, 511-524.

[2] Haaf, W. Solar chimneys: part ii: preliminary test results from the Manzanares pilot plant. *Solar Energy* 14 (2), 141-161. May.1984.

[3] Yan, M.Q., Sherif, S.A., Kridli, G.T., Lee, S.S., Padki, M.M., 1991. Thermo-fluid analysis of solar chimneys. In: Morrow, T.B., Marshall, L.R., Sherif, S.A. (Eds.), Industrial Applications of Fluid Mechanics-1991, FED, vol 132. The American Society of Mechanical Engineers, New York, pp. 125-130.

[4] Padki, M.M., Sherif, S.A., 1999. On a simple analytical model for solar chimneys. *Energy Research* 23 (4), 345-349.

[5] Krisst, R.J.K., 1983. *Energy transfer system. Alternative Source Energy* 63, 8-11.

[6] Pasumarthi N, Sherif SA. *Experimental and theoretical performance of ademon- stration solar chimney model—Part II: experimental and theoretical results and economic analysis.* Int J Energy Res 1998; 22:443-61.

[7] Klink, H., 1985. *A prototype solar convection chimney operated under Izmit conditions. In: Proceedings of the 7th Miami* International Conference on Alternative Energy Sources, Veiroglu, TN, vol. 162.

[8] Pretorius JP, Kröger DG. Critical evaluation of solar chimney power plant per- formance. *Sol Energy* 2006; 80:535-44.

[9] Ninic N. Available energy of the air in solar chimneys and the possibility of its ground-level concentration. *Sol Energy* 2006; 80:804-11.

[10] Onyango FN, Ochieng RM. *The potential of solar chimney for application in rural areas of developing countries.* Fuel 2006; 85:2561-6.

[11] Ming, T., Liu, W., Xu, G., 2006. *Analytical and numerical investigation of the solar chimney power plant systems.* Int. J. Energy Res. 30 (11), 861-873.

[12] M.A.K. Lodhi, *Application of helio-aero-gravity concept in producing energy and suppressing pollution,* Energy Convers. Manage. 40(1999) 407-421.

[13] M.A. dos S Bernardes, A. Vob, G. Weinrebe, *Thermal and technical analyzes of solar chimneys,* Sol. Energy 75 (2003) 511-524.

[14] M.A. dos S Bernardes, R.M. Valle, M.F. Cortez, *Numerical analysis of natural laminar convection in a radial solar heater,* Int. J. Therm.Sci. 38 (1999) 42-50.

[15] T.W. Von Backstro¨m, A.J. Gannon, Compressible flow through solarpower plant chimneys, ASME J. Sol. *Energy Eng.* 122 (2000) 138-145.

[16] A.J. Gannon, T.W. von Backstro¨m, *Solar chimney turbine performance,* ASME J. Sol Energy Eng. 125 (1) (2003) 101-106.

[17] H. Pastohr, O. Kornadt, K. Gurlebeck, *Numerical and analytical calculations of the temperature and flow field in the upwind power plant,* Int. J. Energy Res. 28 (2004) 495-510.

[18] J. Schlaich, R. Bergermann, W. Schiel, G. Weinrebe, Design ofcommercial solar updraft tower systems-utilization of solar inducedconvective flows for power generation, ASME J. Sol. Energy Eng. 127(2005) 117-124.

[19] E. Bilgen, J. Rheault, *Solar chimney power plants for high latitudes,* Sol. Energy 79 (2005) 449-458.

[20] Kasaeian, A.B., Heidari, E., NasiriVatan, Sh., 2010. Experimental investigation of climatic effects on the efficiency of a solar chimney pilot power plant. *Renewable and Sustainable Energy Reviews* 20 (8). 5202-5206.

[21] Sabziparvar AA, Shetaee H. *Estimation of global solar radiation in arid and semi- arid climates of east and west Iran.* 2007, 32-55.

17

Optimization of Power Solar Dish-Stirling: Induced Effects of Heat Source Temperature and Working Fluid Temperature in Hot Side

Mohammad H. Ahmadi[1,*], Hosyen Sayyaadi[2]

[1]Renewable Energies and Environmental Department, Faculty of New Science and Technologies, University of Tehran, Tehran, Iran
[2]Faculty of Mechanical Engineering-Energy Division, K.N. Toosi University of Technology, Tehran, Iran
*Corresponding author: mohammadhosein.ahmadi@gmail.com

Abstract This paper presents an investigation on finite time thermodynamic evaluation and analysis of a Solar-dish Stirling heat engine. Finite time thermodynamics has been applied to determine the net power output and thermal efficiency of the Stirling system with finite-rate heat transfer, regenerative heat loss, conductive thermal bridging loss and finite regeneration process time. The model investigates the effects of the inlet temperature of the heat source and heat sink, the volumetric ratio of the engine, effectiveness of heat exchangers and heat capacitance rates on the net power output and thermal efficiency of the engine and entropy generation. The thermal efficiency of the cycle corresponding to the magnitude of the maximized power of the engine is evaluated. Finally, sensitivities of results in a variation of the thermal parameters of the engine are studied. The present analysis provides a good theoretical guideline for designing and operating of the Stirling heat engine systems.

Keywords: *stirling engine, thermal efficiency, entropy generation, solar dish, concentration ratio*

1. Introduction

The Stirling engine is a simple type of external-combustion engine that uses a compressible fluid as a working fluid. The Stirling engine can theoretically be a very efficient engine to convert heat into mechanical work at Carnot efficiency. The thermal limit for the operation of a Stirling engine depends on the material used for its construction. In most instances, the engines operate with a heater and cooler temperature of 923 and 338 K, respectively [1]. Engine efficiency ranges from about 30 to 40% resulting from a typical temperature range of 923–1073 K, and normal operating speed range from 2000 to 4000 rpm [2].

The classical analysis of the operation of real Stirling engines is that of Schmidt [3]. The theory provides for harmonic motion of the reciprocating elements, but retains the major assumptions of isothermal compression and expansion and perfect regeneration. It thus remains highly idealized, but is certainly more realistic than the ideal Stirling cycle [4].

Senft [5] developed an engine with 2°C temperature difference due to analysis on rotary mechanism. Senft [6] studied Crossley-Stirling engine that is described by two isochoric and two polytropic processes and presented an optimum compression ratio. Also he studied thermodynamic performance and physical restrictions while mechanical efficiency as function of volumetric ratio, temperature ratio and effectiveness of regenerator was developed [7]. Organ was studied effects of various parameters such as diameter, length and materials on regenerator performance, irreversibilities and temperature gradient in Stirling engine regenerator while regenerator was optimized [8,9]. Formosa and Despesse [10] modeled engine's output power and efficiency due to dead volume by implementing of the isotherm model.

Thombare and, Verma [11] gathered the available technologies and obtained achievements with regard to the analysis of Stirling engines and, at the end, presented some suggestions for their applications.

Tavakolpour et al. [12], experimentally investigated gamma type solar engine by implementing a flat plate solar collector which has 80C temperature difference and 900 W.m^{-2} density solar radiation can generate 1.2 W of output power in 30 rpm rotation speed. Ahmadi et al [13] Investigated of Solar Collector Design Parameters Effect onto Solar Stirling Engine Efficiency.

Shendage et. al [14] analyzed a rhombic feature for the design of a single cylinder, beta type Stirling engine of 1.5 kWe capacity for rural electrification. Eid [15] showed the performance of a beta-configuration heat engine having a regenerative displacer. The theoretical analysis of the engine is based mainly on Schmidt theory, the objective for the optimum looking dimensions. In comparison between the proposed engine which has a regenerative more power with more efficiency than the GPU-3 engine.

Podesser [16] investigated heated alpha type Stirling engine by applying flue gas in the outlet of biomass furnace. In their engine, the engine pressure was 33 bar and 600 rpm rotation speed which leads to produce 3.2 kW output power. Costae and Feidt [17] explored the effect of the variation of the overall heat transfer coefficient on the optimum state and on the optimum distribution of the heat transfer surface conductance in the area of the heat exchangers of the Stirling engine. Makhkamov [18] has formed mathematical and practical examination of working process and mechanical losses of a 1kW Stirling engine manufactured for solar operation. Also, He has also emphasized the necessity of simulation of the engine process for better understanding into the engine system.

A thermodynamic analysis of a gamma-type Stirling engine by using a quasi- steady flow model was implemented by Parlak, et al. [19].

Cinar et al. [20] constructed a beta-type Stirling engine which operate at atmospheric pressure. The tests performed on this engine have indicated that by increasing the heat source temperature, the engine speed, engine torque, and power output will be increased.

Minassians et al., constructed and studied the performance of Stirling engine with low temperature solar collector to generate power and heat [21]. A more complete theoretical model of a LTD Stirling engine developed by Robson et al. [22]. In this model, a full differential description of the major components of the engine, the behavior of the gas in the expansion and the compression spaces, the behavior of the gas in the regenerator, the dynamic behavior of the displacer, and the power piston/flywheel assembly was investigated.

Kongtragool [23] studied the influence of the regenerator efficiency and the dead volumes on the work as well as the efficiency of the machine. However, this study does not include the heat transfers through the temperature difference at the heat source and sink.

In 2005, Kongtragool and Wongwises [24] analyzed, theoretically, the power output of the gamma-configuration LTD Stirling engine. The former works on Stirling-engine power output calculation were considered and discussed. They indicated that the mean pressure power formula was the most appropriate for LTD Stirling-engine power output estimation. However, the hot-space and cold-space working fluid temperature was needed in the mean-pressure power formula. In 2005, Kongtragool and Wongwises [25] organized the optimum absorber temperature of a once-reflecting full-conical reflector for a LTD Stirling engine. A mathematical model for the overall efficiency of a solar-powered Stirling engine was developed. Both limiting conditions of maximum possible engine efficiency and power output were considered. Results disclosed that the optimum absorber temperatures obtained from both conditions were not significantly different and the overall efficiency in the case of the maximum possible engine power output was very close to that of the real engine of 55% Carnot efficiency.

The idea of coupling of solar concentrators to Stirling engines is a new technology which facilitates conversion of the solar energy into the electric power. In this regard a dish collector with the parabolic arrangement of its mirrors is used to concentrate the sun radiations in a focal point of the collector which the heat absorber of the engine is located i.e. the solar energy is collected and concentrated thanks to a parabola of mirrors.

For solar applications, Abdullah et al. [26] have presented the design considerations to be taken into account in designing a low-temperature differential double-acting Stirling engine.

On the basis of the conventional entropy techniques, for the studying of solar Stirling engine cycle performance, Costea et al. [27] included the effects of heat transfers, incomplete heat regeneration and irreversibilities of the cycle as conduction, pressure losses or mechanical friction between the moving parts. Timoumi et al. [28] developed an exact second order model counting all the losses at the same time. The method based on a lumped analysis approach leads to a numerical model and has been applied for the optimization of the General Motors GPU-3 [29,30].

The development of finite-time thermodynamics [31,32,33,34], a new discipline in modern thermo-dynamics, provides a powerful tool for performance analysis of practical engineering cycles. Several authors have studied the finite-time thermodynamic performance of the Stirling engine [35-42]. Ahmadi et al [35,36,37] developed an intelligent approach to figure power of solar Stirling heat engine by implementation of evolutionary algorithms Blank and Wu [38] studied the power output and thermal efficiency of a finite time, optimized, extra-terrestrial, solar-radiant Stirling heat engine, in which the heat source and sink were assumed to have infinite heat-capacity rates, obtaining expressions for optimum power and efficiency at optimum power.

Li et al developed a mathematical model for the overall thermal efficiency of solar powered high temperature differential dish Stirling engine with finite heat transfer and the irreversibility of regenerator and optimized the absorber temperature and corresponding thermal efficiency [39]. Tlili investigated effects of regenerating effectiveness and heat capacitance rate of external fluids in a heat source / sink at maximum power and efficiency [40]. Kaushik et al studied the effects of irreversibilities of regeneration and heat transfer of heat/sink sources [41,42].

In the present work by utilizing the finite time thermodynamic analysis method, the thermal efficiency of the system and the entropy generation correspond to output system power in order to discover the optimized design of the Stirling are analyzed.

The analysis investigates the sensitivity of the system operational variables against the maximum output power, Stirling thermal efficiency, entropy generation and thermal efficiency of the dish system.

2. System Description

In a solar - dish Stirling systems, mirrors of the parabolic shaped concentrator focuses the sunlight to the focal point of the concentrator where the hot end of the Stirling engine is located. Therefore, the solar energy with a relatively high temperature is transferred to the hot side heat exchanger of the Stirling engine. Figure 1 illustrates a schematic for a solar-dish Stirling engine connected to a solar dish concentrator. The solar-dish is equipped with a sun tracker which tracks the sun in order to have maximum solar energy transfer to the engine when the sun moves during the days. Hence, solar energy is absorbed and transferred to the working fluid in the hot space of an engine.

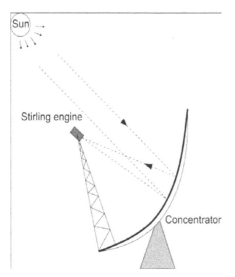

Figure 1. Schematic of a solar Stirling engine

Figure 2 is a schematic diagram of a Stirling heat engine cycle with finite-time heat transfer and regenerative heat losses. As shown in Figure 3, Stirling cycle consisted of four processes. Process 1-2 is an isothermal process, in which the working fluid after compressing at constant temperature, T_c and rejected heat to the heat sink at low temperature T_{L_1}, Therefore, the temperature of heat sink is increased to T_{L_2}. Then the working fluid crosses over the regenerator and warms up to T_h in an isochoric process 2-3. In process 3-4, the working fluid is expanded in a constant temperature T_h process and receives heat from the heat source in which its temperature is reduced from T_{H_1} into T_{H_2}. Last process (4-1), is an isochoric cooling process, where the regenerator absorbs heat from the working fluid.

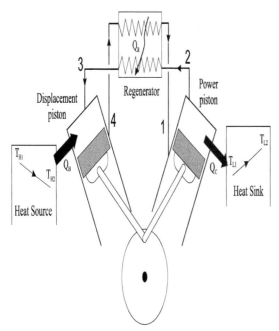

Figure 2. Schematic diagram of the Stirling heat engine cycle

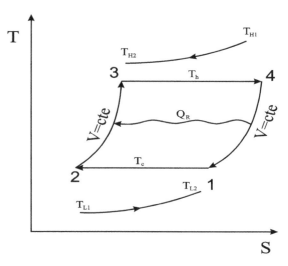

Figure 3. T-S diagram of a Stirling engine cycle

In a real cycle it is impractical to have an ideal heat transfer in the regenerator in which entire amount of absorbed heat (in the process 4-1) is transferred to the working fluid in the heating mode (process 2-3). Therefore a heat transfer loss denoted by ΔQ_R is occurred in the regenerator. In addition, a conduction heat transfer between the heat source and the heat sink namely as thermal bridge loss (Q_0) must be considered.

3. Thermodynamic Analysis of the System

The actual useful heat gain of the dish collector, considering conduction, convection and radiation losses is given by [25,39] as follows:

$$q_u = IA_{app}\eta_0 - A_{rec}[h(T_{H_{ave}} - T_0) + \varepsilon\delta(T_{H_{ave}}^4 - T_0^4)] \quad (1)$$

where I is the direct solar flux intensity, A_{app} is the collector aperture area, η_0 is the collector optical efficiency, A_{rec} is the absorber area, h is conduction/convection heat transfer coefficient, $T_{H_{ave}}$ is the average absorber temperature, T_0 is the ambient temperature, ε is an emissivity factor of the collector, δ is the Stefan's constant.

The thermal efficiency η_s of the dish collector is obtained as [25,39]:

$$\eta_s = \frac{q_u}{IA_{app}} = \eta_0 - \frac{1}{IC}[h(T_{H_{ave}} - T_0) + \varepsilon\delta(T_{H_{ave}}^4 - T_0^4)] \quad (2)$$

3.1. Regenerative Heat Losses in the Regenerator

It is important to mention that there also exists a finite heat transfer in the regenerative heat transfer (Q_R) is given by [39,40,41,42]:

$$Q_r = nC_v\varepsilon_R\left(T_h - T_c\right) \quad (3)$$

ΔQ_r is the heat loss during the two regenerative processes in the cycle. By the following relationship is obtained [39,40,41]:

$$\Delta Q_r = nC_v\left(1-\varepsilon_R\right)\left(T_h - T_c\right) \qquad (4)$$

n is the mass of the working fluid in mole, C_v is the specific heat capacity of the working fluid in the regenerative processes in terms of mole, ε_r is the effectiveness of regenerator, T_h and T_c are the working fluid temperatures in the hot space and cold space, respectively.

Owing to the influence of irreversibility of the finite-rate heat transfer, the time of the regenerative processes is not negligible in comparison to that of the two isothermal processes [37,38,39]. In order to calculate the time of the regenerative processes, one assumes that the temperature of the working fluid in the regenerative processes as a function of time is given by [37,38,39]:

$$\frac{dT}{dt} = \pm M_i \qquad (5)$$

where M is the proportionality constant which is independent of the temperature difference and dependent only on the property of the regenerative material, called regenerative time constant and the $\pm$ sign belong to the heating (i= 1) and cooling (i= 2) processes respectively [37,38,39].

$$t_3 = \frac{T_1 - T_2}{M_1} \qquad (6)$$

$$t_4 = \frac{T_1 - T_2}{M_2} \qquad (7)$$

3.2. The Amounts of Heat Released by the Heat Source and Absorbed by the Heat Sink

The heat released between heat source and working fluid (Q_h), the heat absorbed between the working fluid and the heat sink (Q_c) is obtained as follows:

$$Q_h = nRT_h Ln\lambda + nC_v\left(1-\varepsilon_R\right)\left(T_h - T_c\right) \qquad (8)$$

$$Q_c = nRT_c Ln\lambda + nC_v\left(1-\varepsilon_R\right)\left(T_h - T_c\right) \qquad (9)$$

On the other hand

$$Q_h = \left[C_H\varepsilon_H\left(T_{H_1} - T_h\right) + \xi C_H\varepsilon_H\left(T_{H_1}^4 - T_h^4\right)\right]t_h \quad (10)$$

$$Q_c = C_L\varepsilon_L\left(T_c - T_{L_1}\right)t_l \qquad (11)$$

where C_H and C_L are the heat capacitance rate of external fluids in the heat source and heat sink, respectively.

$$\varepsilon_H = 1 - e^{-N_H} \qquad (12)$$

$$\varepsilon_L = 1 - e^{-N_L} \qquad (13)$$

where ε_H and ε_L are the effectiveness's of the high and low temperature heat exchangers, respectively.

where $N_L = \dfrac{U_L A_L}{C_L}, N_H = \dfrac{U_H A_H}{C_H}$

The cyclic period
Using Eqs. (3)-(11), we get that the cyclic period t is:

$$
\begin{aligned}
t =\ & \frac{nRT_h Ln\lambda + nC_v\left(1-\varepsilon_R\right)\left(T_h - T_c\right)}{C_H\varepsilon_H\left(T_{H_1} - T_h\right) + \xi C_H\varepsilon_H\left(T_{H_1}^4 - T_h^4\right)} \\
& + \frac{nRT_c Ln\lambda + nC_v\left(1-\varepsilon_R\right)\left(T_h - T_c\right)}{C_L\varepsilon_L\left(T_c - T_{L_1}\right)} \\
& + \left(\frac{1}{M_1} + \frac{1}{M_2}\right)\left(T_h - T_c\right)
\end{aligned} \qquad (14)
$$

3.3. The Conductive Thermal Bridging Losses from Heat Source to the Heat Sink

This value is proportional to the average temperature difference of the heat source and heat sink and the cycle time it is obtained as follows:

$$Q_\circ = K_\circ\left(T_{H_{ave}} - T_{L_{ave}}\right)t_{cycle} \qquad (15)$$

$$T_{H_{ave}} = \frac{T_{H_1} + T_{H_2}}{2} \qquad (15a)$$

$$T_{L_{ave}} = \frac{T_{L_1} + T_{L_2}}{2} \qquad (15b)$$

We have [41,42]

$$T_{H_2} = \left(1-\varepsilon_H\right)T_{H_1} + \varepsilon_H T_h \qquad (15c)$$

$$T_{L_2} = \left(1-\varepsilon_L\right)T_{L_1} + \varepsilon_L T_c \qquad (15b)$$

Thus, using Eqs.(15a)-(15d), we have

$$Q_\circ = \frac{K_\circ}{2}\left[\begin{array}{c}\left(2-\varepsilon_H\right)T_{H_1} - \left(2-\varepsilon_L\right)T_{L_1} \\ + \left(\varepsilon_H T_h - \varepsilon_L T_c\right)\end{array}\right]t_{cycle} \qquad (16)$$

The net heat released from the heat source (Q_H) and the net heat absorbed by the heat sink (Q_L) are obtained as follows:

$$Q_H = Q_h + Q_\circ \qquad (17)$$
$$Q_L = Q_c + Q_\circ \qquad (18)$$

Considering the cyclic period of the Stirling engine, the output power, the thermal efficiency and entropy production of the engine are given by:

$$p = \frac{W}{t} = \frac{Q_H - Q_L}{t} \qquad (19)$$

$$\eta_t = \frac{Q_H - Q_L}{Q_H} \qquad (20)$$

$$\sigma = \frac{1}{t}\left(\frac{Q_L}{T_{L_{ave}}} - \frac{Q_H}{T_{H_{ave}}}\right) \qquad (21)$$

Substituting Eqs. (3)- (14) into Eqs. (19) and (20) we have,

$$P = \frac{nR(T_h - T_c)Ln\lambda}{\dfrac{nRT_h Ln\lambda + nC_v\left(1-\varepsilon_R\right)\left(T_h - T_c\right)}{C_H\varepsilon_H\left(T_{H_1} - T_h\right) + \xi C_H\varepsilon_H\left(T_{H_1}^4 - T_h^4\right)} + \dfrac{nRT_c Ln\lambda + nC_v\left(1-\varepsilon_R\right)\left(T_h - T_c\right)}{C_L\varepsilon_L\left(T_c - T_{L_1}\right)} + \left(\dfrac{1}{M_1} + \dfrac{1}{M_2}\right)\left(T_h - T_c\right)}, \qquad (22)$$

$$\eta_t = \frac{nR(T_h - T_c)Ln\lambda}{nRT_h Ln\lambda + nC_v(1-\varepsilon_R)(T_h - T_c)} \tag{23}$$
$$+ \frac{K_\circ}{2}\left[\begin{array}{c}(2-\varepsilon_H)T_{H_1} - (2-\varepsilon_L)T_{L_1} \\ + (\varepsilon_H T_h - \varepsilon_L T_c)\end{array}\right] t_{cycle}$$

With simplification and considering $M = \dfrac{C_V(1-\varepsilon_R)}{RLn\lambda}$

and $F = \dfrac{1}{nR\ln\lambda}\left(\dfrac{1}{M_1} + \dfrac{1}{M_2}\right)$ we obtain:

$$P = \frac{T_h - T_c}{\dfrac{T_h + M(T_h - T_c)}{C_H \varepsilon_H(T_{H_1} - T_h) + \xi C_H \varepsilon_H(T_{H_1}^4 - T_h^4)} + \dfrac{T_c + M(T_h - T_c)}{C_L \varepsilon_L(T_c - T_{L_1})} + F(T_h - T_c)} \tag{24}$$

$$\eta_t = \frac{T_h - T_c}{T_h + M(T_h - T_c) + \dfrac{K_\circ}{2}\left[\begin{array}{c}(2-\varepsilon_H)T_{H_1} \\ -(2-\varepsilon_L)T_{L_1} \\ +(\varepsilon_H T_h - \varepsilon_L T_c)\end{array}\right]} \cdot \left[\frac{T_h + M(T_h - T_c)}{C_H \varepsilon_H(T_{H_1} - T_h) + \xi C_H \varepsilon_H(T_{H_1}^4 - T_h^4)} + \frac{T_c + M(T_h - T_c)}{C_L \varepsilon_L(T_c - T_{L_1})} + F(T_h - T_c)\right] \tag{25}$$

For the sake of convenience, a new parameter $x = \dfrac{T_c}{T_h}$ is introduced into Eqs. (24) and (25), then we have:

$$P = \frac{(1-x)}{\left(\left(\dfrac{1+M(1-x)}{C_H\varepsilon_H(T_{H_1}-T_h)+\zeta C_H\varepsilon_H(T_{H_1}^4-T_h^4)}\right) + \left(\dfrac{x+M(1-x)}{C_L\varepsilon_L(xT_h-T_{L_1})}\right)+F(1-x)\right)} \tag{26}$$

$$\eta_t = \frac{(1-x)}{1+M(1-x)+\dfrac{K_o}{2}\left[\begin{array}{c}(2-\varepsilon_H)T_{H_1}-(2-\varepsilon_L)T_{L_1} \\ +T_h(\varepsilon_H-x\varepsilon_L)\end{array}\right]} \cdot \left[\left(\dfrac{1+M(1-x)}{C_H\varepsilon_H(T_{H_1}-T_h)+\zeta C_H\varepsilon_H(T_{H_1}^4-T_h^4)}\right) + \left(\dfrac{x+M(1-x)}{C_L\varepsilon_L(xT_h-T_{L_1})}\right)+F(1-x)\right] \tag{27}$$

For the optimization for maximum Power, Eq. (26) must be differentiated from T_h and put it equal to zero:

$$\frac{\partial P}{\partial T_h} = \circ \tag{28}$$

$$K_1 T_{h_{opt}}^8 + K_2 T_{h_{opt}}^5 + K_3 T_{h_{opt}}^4 + K_4 T_{h_{opt}}^3 \\ + K_5 T_{h_{opt}}^2 + K_6 T_{h_{opt}} + K_7 = 0 \tag{29}$$

where

$$K_1 = \zeta^2 C_H^2 \varepsilon_H^2 B_1 x \tag{29a}$$

$$K_2 = \zeta C_H \varepsilon_H x(2C_H \varepsilon_H B_1 - 3B_2 C_L \varepsilon_L x) \tag{29b}$$

$$K_3 = 2\zeta C_H \varepsilon_H x(3B_2 C_L \varepsilon_L T_L - B_1 B_3) \tag{29c}$$

$$K_4 = -3B_2 \zeta C_H \varepsilon_H C_L \varepsilon_L T_L^2 \tag{29d}$$

$$K_5 = C_H \varepsilon_H x(B_1 C_H \varepsilon_H - B_2 x C_L \varepsilon_L) \tag{29e}$$

$$K_6 = 2C_H \varepsilon_H x(B_2 C_L \varepsilon_L T_L - B_1 B_3) \tag{29f}$$

$$K_7 = B_1 x B_3^2 - B_2 C_H \varepsilon_H C_L \varepsilon_L T_L^2 \tag{29g}$$

$$B_1 = x + M(1-x) \tag{29h}$$

$$B_2 = 1 + M(1-x) \tag{29i}$$

$$B_3 = C_H \varepsilon_H T_H + \zeta C_H \varepsilon_H T_H^4 \tag{29j}$$

With Placement $T_{h_{opt}}$ in P_{max} and η, maximum power and its corresponding thermal efficiency are calculated as follows:

$$P = \frac{(1-x)}{\left(\left(\dfrac{1+M(1-x)}{C_H\varepsilon_H(T_{H_1}-T_{hopt})+\zeta C_H\varepsilon_H(T_{H_1}^4-T_{hopt}^4)}\right) + \left(\dfrac{x+M(1-x)}{C_L\varepsilon_L(xT_{hopt}-T_{L_1})}\right)+F(1-x)\right)} \tag{30}$$

$$\eta_t = \frac{(1-x)}{1+M(1-x)+\dfrac{K_o}{2}\left[\begin{array}{c}(2-\varepsilon_H)T_{H_1}-(2-\varepsilon_L)T_{L_1} \\ +T_{hopt}(\varepsilon_H-x\varepsilon_L)\end{array}\right]} \cdot \left[\left(\dfrac{1+M(1-x)}{C_H\varepsilon_H(T_{H_1}-T_{hopt})+\zeta C_H\varepsilon_H(T_{H_1}^4-T_{hopt}^4)}\right) + \left(\dfrac{x+M(1-x)}{C_L\varepsilon_L(xT_{hopt}-T_{L_1})}\right)+F(1-x)\right] \tag{31}$$

The maximum thermal efficiency of the entire solar-dish Stirling engine is a product of the thermal efficiency of the collector and the optimal thermal efficiency of the Stirling engine [39]. Namely:

$$\eta_m = \eta_s \eta_t \tag{32}$$

Therefore by substituting of Eqs. (2) And (31) into Eq. (32), we have the following expression for thermal efficiency of the entire solar-dish Stirling engine:

$$\eta_m = \left\{\eta_0 - \frac{1}{IC}[h(T_{H_1} - T_0) + \varepsilon\delta(T_{H_1}^4 - T_0^4)]\right\}$$

$$\frac{(1-x)}{1+M(1-x)+\dfrac{K_o}{2}\begin{bmatrix}(2-\varepsilon_H)T_{H_1}-(2-\varepsilon_L)T_{L_1}\\+T_{hopt}(\varepsilon_H - x\varepsilon_L)\end{bmatrix}}$$

$$\begin{bmatrix}\left(\dfrac{1+M(1-x)}{C_H\varepsilon_H\left(T_{H_1}-T_{hopt}\right)+\zeta C_H\varepsilon_H\left(T_{H_1}^4-T_{hopt}^4\right)}\right)\\+\left(\dfrac{x+M(1-x)}{C_L\varepsilon_L\left(xT_{hopt}-T_{L_1}\right)}\right)+F(1-x)\end{bmatrix} \qquad (33)$$

4. Numerical Results and Discussion

In order to evaluate the effect of the heat source working fluid inlet temperature (T_{H_1}), heat capacitance rates (C_L, C_H), the effectiveness of the regenerator (ε_R), effectiveness of the heat exchangers ($\varepsilon_H, \varepsilon_L$) and the heat

leak coefficient (k_0) on the powered Stirling heat engine system, all the other parameters will be kept constant as $x = 0.5$, $C_H = C_L = 150 WK^{-1}$, $n = 1mol$, $\lambda = 2$, $R = 4.3 Jmol^{-1}K^{-1}$, $C_v = 15 Jmol^{-1}K^{-1}$ $\varepsilon_L = 0.7$, $\varepsilon_H = 0.7$, $\varepsilon_R = 0.9$, $T_{L_1} = 288K$, $T_{H_1} = 1300K$, $T_0 = 288K$, $C = 1300$, $\delta = 5.67\times10^{-8} W.m^{-2}.K^{-4}$, $h = 20W.m^{-2}.K^{-1}, I = 1000W.m^{-2}, \varepsilon = 0.9, \eta_0 = 0.9$.

In the Figure 4 different system parameters against the heat source working fluid inlet temperature in the different heat capacities rate of the heat source are plotted. And in these figures by increasing the heat source working fluid temperature, maximum output power, Stirling thermal efficiency and entropy generation will be raised. Also the overall thermal efficiency of the dish system reaches a peak at about 900K and after that decreases. In Addition, at the given heat source working fluid inlet temperature all parameters (the maximum output power, Stirling thermal efficiency, entropy generation and thermal efficiency of the dish system) have increased with increasing of the heat capacities rate of the heat source.

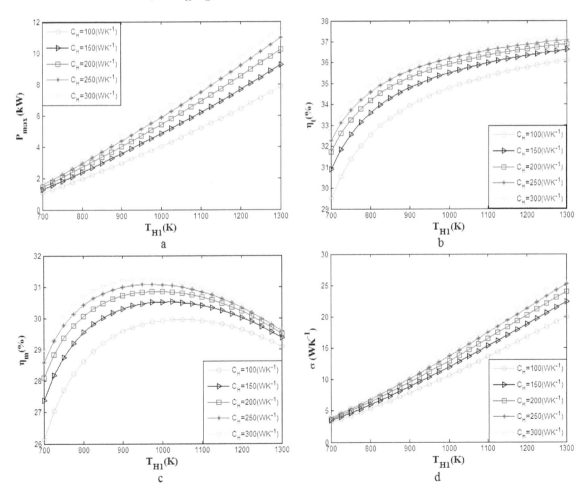

Figure 4. Variation of (a) maximum output power, (b) Stirling thermal efficiency, (c) thermal efficiency of the dish system and (d) entropy generation with heat source working fluid inlet temperature at a different heat capacities rate of heat source

In the Figure 5, all discussed parameter in Figure 4 are illustrated for the heat capacities rate of the heat sink. Surprisingly all parameters show the same trend against the heat source working fluid inlet temperature.

Furthermore at given heat source working fluid inlet temperature, all mentioned parameters increase with increasing the of the heat capacities rate of the heat sink.

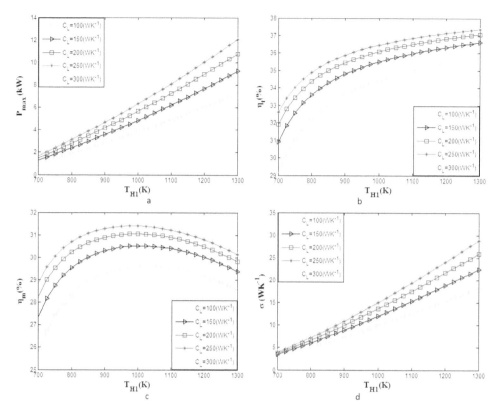

Figure 5. Variation of maximum (a) output power, (b) Stirling thermal efficiency, (c) thermal efficiency of the dish system and (d) entropy generation with heat source working fluid inlet temperature in the different heat capacities rate of heat sink

From Figure 6 it can be seen that, the maximum output power, the thermal efficiency and entropy generation increase considerably with increasing of the heat source working fluid inlet temperature at various values of the hot side heat exchanger effectiveness (0.4-0.6). But the thermal efficiency of the Stirling dish system in the temperature of about 1000K reaches a maximum and then decreases.

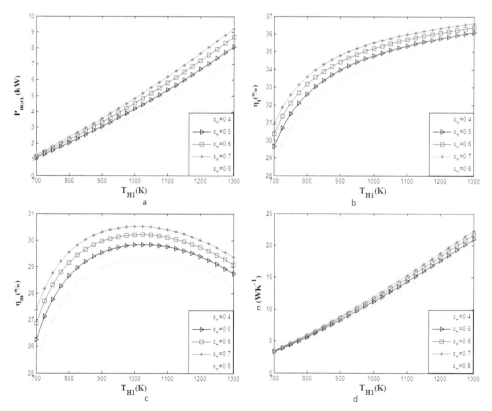

Figure 6. Effects of heat source working fluid inlet temperature and effectiveness of the hot side heat exchanger on the maximum (a) output power, (b) Stirling thermal efficiency,(c)thermal efficiency of the dish system and (d) entropy generation

Also in the Figure 7 the similar demonstrated parameters in the Figure 6 are plotted against the heat source working fluid inlet temperature at the different effectiveness of the cold side heat exchanger. As it can be seen overall trends are the same as Figure 6.

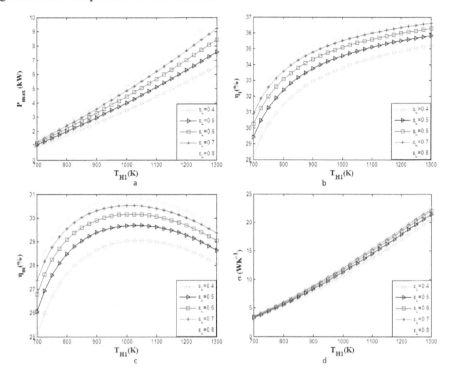

Figure 7. Effects of heat source working fluid inlet temperature and effectiveness of the cold side heat exchanger on the (a) maximum output power, (b) Stirling thermal efficiency, (c) thermal efficiency of the dish system and (d) entropy generation

In the Figure 8 different system parameters against the heat sink working fluid inlet temperature in the different heat capacities rate of the heat source are shown. As can be seen, in all graphs with increasing the heat sink working fluid temperature, maximum output power, Stirling thermal efficiency, dish system efficiency and entropy generation decrease. In Addition, at the given heat sink working fluid inlet temperature all four discussed parameters are increased with increasing in the heat capacities rate of the heat source.

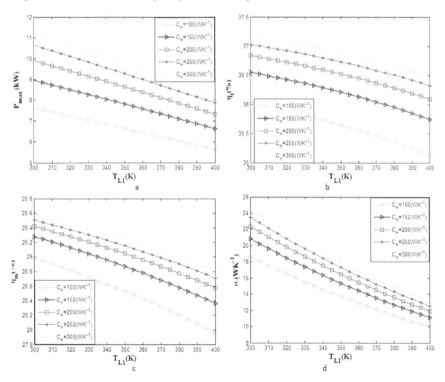

Figure 8. Variation of (a) maximum output power, (b) Stirling thermal efficiency, (c) thermal efficiency of the dish system and (d) entropy generation and heat sink working fluid inlet temperature at a different heat capacities rate of heat source

From Figure 9 all parameters in Figure 8 are illustrated for the heat capacities rate of the heat sink. Similarly all parameters express same manner versus the heat sink working fluid inlet temperature. Also at given heat sink working fluid inlet temperature, all parameters increase with increasing the in the heat capacities rate of the heat sink.

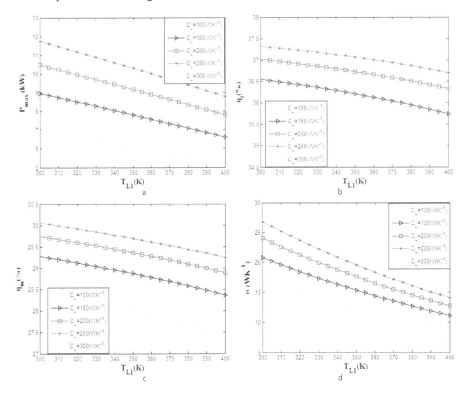

Figure 9. Variation of (a) maximum output power, (b) Stirling thermal efficiency, (c) thermal efficiency of the dish system and (d) entropy generation and heat sink working fluid inlet temperature at different heat capacities rate of heat sink

From Figure 10 it can be observed that, the maximum output power, the thermal efficiency, the dish system efficiency and entropy generation decrease with increasing of the heat sink working fluid inlet temperature at different values of the hot side heat exchanger effectiveness. Moreover, at given heat sink working fluid inlet temperature, all parameters increase by increasing of the hot side heat exchanger effectiveness from 0.4 to 0.8.

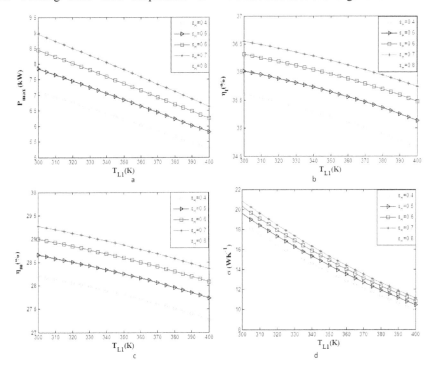

Figure 10. Variation of (a) maximum output power, (b) Stirling thermal efficiency, (c) thermal efficiency of the dish system and (d) entropy generation with heat sink working fluid inlet temperature at the different effectiveness of the hot side heat exchanger

In the Figure 11 the similar considered parameters in the Figure 10 are scheming against the heat sink working fluid inlet temperature at the different effectiveness of the cold side heat exchanger. As it can be pictured overall trends are the same as Figure 10.

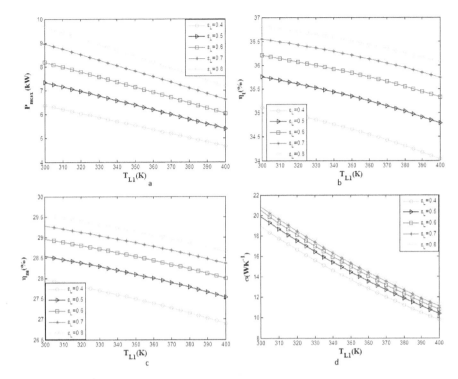

Figure 11. Variation of (a) output power, (b) Stirling thermal efficiency, (c) thermal efficiency of the dish system and (d) entropy generation with heat sink working fluid inlet temperature at the different effectiveness of the cold side heat exchanger

5. Conclusions

FTT modeling of the solar-dish Stirling engine was performed. In the model non-ideal performance of the regenerator, finite time heat transfer in the heat exchangers, conductive/radiative heat transfers in the absorber (heat source), conductive heat transfer in the heat sink and conductive thermal bridge loss between the heat source and heat sink were taken into account. Finite time thermodynamic analysis has been accepted a straightforward approach for optimization of thermodynamic systems. In the present work sensitivity of different engine's parameters in order to investigate the behavior of the various Stirling system parameters are analyzed. This analysis it also can be applied in determining the function of other related variables such as the volume ratio and the temperature ratio.

The results show parameters such as the heat capacity rate of heat source and heat sink and fluid temperature of the heat source and heat sink play an important role in system design. Also, the presented thermodynamic model is valuable for the design and analysis of the dish-Stirling.

References

[1] G Walker. Stirling engines. Oxford: Clarendon Press; 1980 p. 24-5, see also pages 50, 52, 73.

[2] WB Stine. Stirling engines. In: Kreith F, editor. The CRC handbook of mechanical engineers. Boca Raton: CRC Press; 1998. p. 8-67 see also pages 8-76.

[3] Schmidt, Theorie der geschlossenen calorischen Maschine von Laubroyund Schwartzkopff in Berlin, Z. Ver. Ing., 1861, 79p.

[4] G. Walker, Stirling-cycle machines, Clarendon Press, Oxford, 1973, 156 p.

[5] JR. Senft, An ultra-low temperature differential Stirling engine, Proceeding of the fifth international Stirling engine conference, Paper ISEC 91032, Dubrovnik, May 1991.

[6] JR. Senft, Mechanical Efficiency of Heat Engines, Cambridge University Press, 2007.

[7] JR. Senft, Theoretical Limits on the Performance of Stirling Engines, International Journal of Energy Research Vol. (22), 1998, P. 9 91-1000.

[8] AJ. Organ, The Regenerator and the Stirling Engine, Mechanical Engineering Publications Limited, London, 1997.

[9] AJ. Organ, Stirling air engine thermodynamic appreciation, J. Mechanical Engineering Science: Part C, 214, 2000, P. 511-536.

[10] Formosa, F., G. Despesse,. Analytical model for Stirling cycle machine designs. Energy Conversion and Management, 51, 2010, P. 1855-1863.

[11] Thombare, D.G, S.K. Verma,. Technological development in the Stirling cycle engines. Renewable and sustainable Energy Reviews, 12, 2008, P. 1-38.

[12] AR. Tavakolpour, A. Zomorodian, AA. Golneshan. Simulation construction and testing of a two-cylinder solar Stirling engine powered by a flat-plate solar collector without regenerator. Renewable Energy, 33, 2008, P. 77-87.

[13] M.H. Ahmadi, H. Hosseinzade, Investigation of Solar Collector Design Parameters Effect onto Solar Stirling Engine Efficiency, Applied Mechanical Engineering, 1, 2012, 1-4.

[14] D.J. Shendage, S.B. Kedare, S.L. Bapat, An analysis of beta type Stirling engine with rhombic drive mechanism, Renewable Energy, 36 (1), 2011, 289-297.

[15] E. Eid, Performance of a beta-configuration heat engine having a regenerative displacer, Renewable Energy, vol. 34 (11), (2009), 2404-2413.

[16] E. Podesser, "Electricity Production in Rural Villages with Biomass Stirling Engine", Renewable Energy, 16 (1-4), 1999, 1049- 1052.

[17] M Costea, M Feidt. the effect of the overall heat transfer coefficient variation on the optimal distribution of the heat transfer

surface conductance or area in a Stirling engine. Energ Convers Manage 39, 1998, 1753-63.

[18]   K. Makhkamov and D. B. Ingham, "Analysis of the working process and mechanical losses in a Stirling engine for a solar power unit," ASME J. Sol. Energy Eng. 122 (2000), 208.

[19]   N. Parlak. Thermodynamic analysis of a gamma type Stirling engine in non-ideal adiabatic conditions. Renewable Energy 34 (1), (2009), 266-73.

[20]   C Cinar, S Yucesu, T Topgul, M Okur. Beta-type Stirling engine operating at atmospheric pressure. Appl Energy 81, (2005), 351-7.

[21]   A. Minassians, SR. Sanders, Stirling engine for Distributed low-Cost Solar-Thermal-Electric Power Generation, Journal of Solar Energy Engineering: ASME, 133, 2011, 011015-2.

[22]   A Robson, T Grassie, J Kubie. Modelling of a low temperature differential Stirling engine. Proceedings of the Institution of Mechanical Engineers, Part C: Journal of Mechanical Engineering Science 221, 2007, 927-943.

[23]   B Kongtragool, S Wongwises. Thermodynamic analysis of a Stirling engine including dead volumes of hot space, cold space and regenerator. Renew Energy 31, (2006), 345-59.

[24]   B Kongtragool, S Wongwises. Investigation on power output of the gamma-configuration low temperature differential Stirling engines. Renewable Energy 30, (2005), 465-76.

[25]   B Kongtragool, S Wongwises. Optimum absorber temperature of a once-reflecting full conical concentrator of a low-temperature differential Stirling engine. Renewable Energy 31, (2006), 345-59.

[26]   S Abdullah, BF Yousif, K Sopian. Design consideration of low temperature differential double-acting Stirling engine for solar application. Renew Energy 30, (2005), 1923-41.

[27]   M Costa, S Petrescu, C Harman. The effect of irreversibilities on solar Stirling engine cycle performance. Energy Convers Manage 40, (1999), 1723-31.

[28]   Y Timoumi, I Tlili, S Ben Nasrallah. Design and performance optimization of GPU-3 Stirling engines. Energy 33 (7), (2008), 1100-14.

[29]   WR Martini. Stirling engine design manual. NASA CR-168088; 1983.

[30]   Percival WH. Historical review of Stirling engine development in the United States from 1960 to 1970. NASA CR-121097; 1974.

[31]   B. Andresen, RS. Berry, A Nitzan and P Salamon, Thermodynamics in finite time. I. The step Carnot cycle, Phys Rev A, 15, (1977), pp. 2086-93.

[32]   Chen L, Wu C, Sun F. Finite time thermodynamics optimization or entropy generation minimization of energy systems. J Non-Equilibrium Thermodyn 1999; 24: 327.

[33]   S. Petrescu, M. Costea, G. Stanescu,, Optimization of a cavity type receiver for a solar Stirling engine taking into account the influence of the pressure losses, finite speed losses, friction losses and convective heat transfer, ENSEC' 93, Cracow, Poland, 1993.

[34]   HG Ladas, OM Ibrahim, Finite-time view of the Stirling engine, Energy, 19 (8), (1994), pp. 837-43.

[35]   Ahmadi MH, GhareAghaj SS, Nazeri A. Prediction of power in solar stirling heat engine by using neural network based on hybrid genetic algorithm and particle swarm optimization. Neural Comput & Applic 2013; 22: 1141-50.

[36]   Ahmadi MH, Sayyaadi H, Dehghani S, Hosseinzade H. Designing a solar powered Stirling heat engine based on multiple criteria:

maximized thermal efficiency and power. Energy Convers Manage 2013; 75: 282-91.

[37]   Ahmadi MH, Mohammadi AH, Dehghani S, Barranco-Jiménez Marco A. Multiobjective thermodynamic-based optimization of output power of solar dish- Stirling engine by implementing an evolutionary algorithm. Energy Convers Manage 2013; 75: 438-45.

[38]   DA Blank, C Wu. Power optimization of an extraterrestrial solar-radiant Stirling heat engine. Energy 20 (6), (1995), 523-30.

[39]   L. Yaqi and et al, Optimization of solar-powered Stirling heat engine with finite-time thermodynamics, Renewable Energy, 36 (2011), pp. 421-427.

[40]   I Tlili, "Finite time thermodynamic evaluation of endoreversible Stirling heat engine at maximum power conditions", Renew & Sustain Energy Review, 16 (4), 2012, 2234-2241.

[41]   SC Kaushik, S Kumar, Finite time thermodynamic evaluation of irreversible Ericsson and Stirling heat engines, Energy Convers Manage, 42 (2001), pp. 295-312.

[42]   SC Kaushik, S Kumar, Finite time thermodynamic analysis of endoreversible Stirling heat engine with regenerative losses, Energy, 25 (2000), pp. 989-1003.

Nomenclature

A	Area, m2
C	Heat capacitance rate, WK-1
C_v	Specific heat capacity, Jmol-1K-1
K_0	Heat leak coefficient, WK-1
n	Number of mole
N	Number of heat transfer units
P	Power output, W
Q	Heat, J
R	The gas constant, Jmol-1K-1
S	Entropy, JK-1
T	Temperature, K
t	Time, s
U	Overall heat transfer coefficient, WK-1m-2
V	Volume of the working fluid, m3
W	Output work, j
Subscripts	
L	Cold side/Heat sink
H	Heat source
h	Hot side
1	Inlet
2	Outlet
$1, 2, 3, 4$	State points
R	Regenerator
Greek	
λ	Ratio of volume during the regenerative processes
ε	Effectiveness and emissivity factor
η	Thermal efficiency
σ	entropy production

18

Electricity from Waste –Bibliographic Survey

Anuradha Tomar*, **Anushree Shrivastav, Saurav Vats, Manuja, Shrey Vishnoi**

Department of Electrical & Electronics Engineering, Northern India Engineering College, New Delhi, India
*Corresponding author: tomar.anuradha19@gmail.com

Abstract Presented here is a bibliographic survey, which covers the work done in the period ranging from 1913 to 2013; towards realization of feasible methods of electricity generation using waste materials. This paper is the outcome of thorough analysis of various literatures available from the earlier research. This paper would be a great source of literature summarized in one paper. This paper would be helpful as a one stop guide, it will introduce the subject to the reviewer, give him an idea of all the previous work done in this field, the chronological research done by some scientist as well as the development in technology/methodology all along the period.

Keywords: bio energy, thermal energy, energy from waste

1. Introduction

The feasting habits of modern consumer routines are causing a huge worldwide waste problem. Having crowded local landfill volumes, many first world countries are now transferring their refuse to third world countries. This is having a disturbing impact on ecosystems and cultures throughout the world. Some substitute energy companies are emerging new ways to recycle waste by generating electricity from landfill wastes, waste heat from industries and nuclear power plants, bio-energy and many other miscellaneous sources. In count to wind and solar energy, the purported bio-fuels are becoming progressively common. Breeding energy through burning, vaporizing, or fermenting biomass such as waste plant material, vegetable waste, and manure are well-founded methods. Microbial fuel cell is also a new concept which is accomplished of directly generating energy from materials such as waste water.

The bibliography has been divided into the following sections.
1. Electricity from Bio-Energy.
2. Electricity from Thermal-Energy.
3. Electricity from other Miscellaneous Sources of Energy.

2. Electricity from Bio-Energy

Peters [1] mentioned that millions of tons of municipal solid waste (MSW) and disposing it are serious problem in many countries. So an alternative to this has found by recovering energy from waste and thus generating 1.5 quad million Btu of electricity in United States. Disposal problems are reduced and amount of municipal solid waste in landfills are also less toxic now.

Porteous [2], reported that based on the non fossil fuel obligation (NFFO) allocations, energy generation from

MSW and landfill gas combustion have been inspected. Emphasis has been given on energy recycling, waste management. Further he re-examined MSW waste for extracting energy by incineration process and concluded increment in energy generating capacity and reduction in volume of disposed waste.

Zahedi [3], presented the paper on waste to energy technology in Australia. House hold waste which is organic has been examined in five different cities of Australia with accurate amount. So estimation of electricity production is also given in accurate units. Waste to energy conversion has been done by incineration process which involves burning of waste, this adding a benefit of waste disposal in less quantity.

Maunder et. al. [4], described the extraction of bio fuel from biomass and waste material. Former one included agricultural residues and forestry operations. Later one includes Municipal solid waste, human waste, and industrial waste. Bio fuel leads to production of electricity by transformation of all waste. Research was going through advanced technologies for conversion of biomass and waste into better efficient electrical energy.

Walker [5], presented the establishment of thames water services as provider of water and sewerage services in london. Covering the wide perimeter among different cities, the anaerobic absorption of sewage sludge produces methane which acts as energy source producing 8 m of electricity. The spark ignited engines were implemented which improved the power output capacity.

Chakravarthi [6], emphasized on the treatment of cattle waste and thus generated energy and biogas can be utilized in respective ways. Energy from waste using anaerobic digestion process reduces environmental pollution thus generating a low budget energy source for farms. Similarly, biogas can be used for indoor usage like lightening, water heating, and grain drying. Overall cattle waste is utilized thus providing suitable benefits with generation of fertilizer for crops and plants and cleaning of lands.

Kishinevsky and Zelingher [7], emphasized the use of first fuel cells that harnessed anaerobic digester gas energy by producing medium electricity and heat thereby absorbing the harmful pollutants that can cause green house effect. Anaerobic digester process is used in many wastewater treatment plants where anaerobic digester gas (ADG) is generated as by product and is a mixture of methane and carbon dioxide.

Tomberlin & Moorman [8], represented the benefits of energy generation from municipal solid waste. A thorough power output details are discussed in this paper. The harms and benefits of using combustion of waste have been elaborated.

Justin, et. al. [9], design of bio fuel cells could extract electricity from Escherichia Coli and human white blood cells. This bio fuel cell can be used as an energy source for implantable electronic device. The current obtained from both of these experiments have been discussed along with appropriate proportions taken.

Varadi et al [10], drawed the attention towards accumulation of Municipal solid waste into enormous heaps. This problem was dealt by setting up waste to energy plant at high temperature which generates steam and thus generates electricity. Heat can also be used in distillation of portable water and industrial hot water requirements. 144 MWh of electricity was generated with new waste to energy systems in addition to more electricity from recovered heat from furnace flue gases.

Monier et. al. [11], presented microbial Fuel Cell (MFC) technology for electricity generation from the organic matter. The metabolic activity of bacteria on the biodegradable substrates converts the chemical energy stored in substrates to the electrical energy. The operation and current developments in MFC are presented in detail.

Varadi and Takacs [12], described a new plant module supporting combustion at 1150 degree Celsius temperature, 180 metric tons of municipal solid waste and generating a minimum of 144 MWh of electricity, as well as 912,000 litres of potable water a day. The other specifications of plant were also discussed.

Zhi et al. [13], analyzed the effect of suspended sludge on power production to improve the designs of Microbial Fuel Cells (MFCs).Two parallel single-chambered MFCs with or without suspended sludge, operated in fed-batch mode was investigated to study the effect of suspended sludge on electricity generation.

Kayes and Tehzeeb [14], discussed Waste-to-Energy(WTE) technologies in detail. WTE involves converting various elements of municipal solid waste such as paper, plastics and food wastes to generate energy by either thermo chemical or biochemical processes. The thermo chemical techniques like combustion, gasification, thermal de-polymerization, plasma arc gasification or pyrolysis and the biochemical processes including anaerobic digestion, hydrolysis, Mechanical Biological Treatment (MBT) were illustrated.

Amin et. al. [15], suggested the use of waste materials for the generation of electricity instead of conventional fuels, in developing countries like Bangladesh. A brief account of waste-to-energy projects in some parts of the world was presented. The future of waste-to-energy systems in Bangladesh was outlined.

Sarker et. al. [16], proposed alternative fuel created from domestic sources on account of rising prices of fossil fuels and the increasing environmental and health problems. The outlines contained descriptions of a new alternative hydrocarbon fuel which is produced from abundant waste plastic materials and discovered by Natural State Research Inc. (NSR). NSR fuel has the potential to generate electricity as well.

Curry and Pillay [17], emphasized on the application of anaerobic digestion to the organic wastes produced in urban environments. A case study from the downtown campus of Concordia University in Montreal, Canada is also included. Thus, anaerobic digestion is viewed as a critical solution to growing garbage problems and simultaneously providing valuable energy in urban areas.

Baidoo et. al. [18], focused on three major problems of pending energy crisis, the environmental degradation due to waste and the environmental degradation due to greenhouse emissions. Efficient plasma gasification technology can be the solution of the above three problems. A closed loop renewable power generation system was being developed that can be integrated into local communities. A model of energy and economic analysis of the system was presented using visual basic in the form of graphical user interface (GUI). Incineration and micro-turbine models were also discussed.

Baidoo et. al. [19], presented the energy analysis of a plasma gasification system. Plasma furnace modelling was also outlined on the basis of data obtained from the Technical University of Lodz, Poland.

Nikolaeva et. al. [20], concentrated on the adequate management and treatment of piggery and dairy wastes. The authors figured out anaerobic technology ('AFBRs') as a solution to the environmental problems derived from intensive animal breeding. The successful application of anaerobic technology (AFBRs) to treat the waste water of dairy milk with the utilization of biogas in order to produce electricity for the dairy equipment in the milk farm located in the province of Cartago, Costa Rica was also outlined.

Tang et. al. [21], described a RF (radio frequency) Plasma Pyrolysis reactor. The outlines revealed that the reactor could be utilized for the pyrolysis treatment of biomass at different input powers and operating pressures.

Kallimanni, et. al. [22], presented the waste food can be used in biogas generation by anerobic digestion. At temperature 4° Celsius, waste stored. auto-mixing anerobic reactor technology (AART) was improved digestor with rapid rate of biomethanation compared to conventional digester. It would act as boon responsible for producing cooking gas manure and renewable energy. Wilfried Ngyz Mbav et. al. [23], presented the population energy harnessing from gases emitted from land fill. This source of energy is viable and clean, mitigate green house gas emission and manage the rampant waste production proved as remedy for power shortage.

Cuijie et. al. [24], presented MSWI bottom ash used or road base construction with reduction in resource consumption. Compared to the conventional road saved electricity 51% and 41% diesel. Due to heavy metal as content enhances toxicity via water acts as disadvantage and need for research.

Bardi and Astoefi [25], presented conversation from waste – bed temp. to specified referenced value for maximize steam production, reference temperature has chosen. Waste incineration process described with control methodologies and new approach.

Nan et al. [26], presented an energy recyclable burn in technology for electronic ballast for HPS lamps. It simulated transitive characteristic of HID lamps worked at high frequency, processed frequency ballast output power and recycle power back to grid discussed working principle and mathematical analysis.

Khan [27], studied and explored the potential of thermoelectric power from baggage by sugar mills. Eight mills are used as a model and took interview of officials. Total production of baggage in recent years will provide 478 mW/hr when averaged from models. It would be boon for power crisis area. It acts as a fuel comes under biomass energy source.

Paurali [28], discussed basics of plasma gasification technology, reviews challenges and opportunities for implementation of it. He showed it is one of the best methods to get rid from landfills, emission of green house gases and save underground water.

Rafi et al. [29], presented a real study of Faridabad area on the basis of power generation by IRES (Integrated Renewable Energy Source). For reducing burden on land fill, MSD is converted to biomass which could be used as bio fertilizers have cost of Rs. 3-5 per kg. Power generation by IRES considered three sources solar, wind and biomass. Potential of solar, wind and biomass energy calculated organic fractions of MSW conversion reduced emission of green house gases.

Kramer et. al. [30], discussed how to regulate the organic waste such as food, animal and human waste converted by anaerobic process with production of hydrogen. Hydrogen production influenced because of some advantages like reduce green house emission, fuel cells and in reciprocating engine. Solar thermal system and its process overviewed and showed produced electricity, heated building and obtain portable water. H production graph showed relation between temperature vs intial pH and substrate concentration Vs pH.

Bin Jobli et. al., [31] presented a main objective to analyze the potential of designed heat pump system when banana peels act as source of energy by recovered heat from it. For make this experiment more fruitful, heat pump applied to heat exchanger. Maintain temperature difference and amount of energy extracted, 3-4 watt heat produced1 kg dry matter. Data record from data logger and monitored during experimental setup. Evaluation of moisture content and heat recovery from decomposition system was done. A graph showed heat extraction analysis with power output and temperature profile.

Curry and Pillay [33], presented the MSD or any kind of waste has become major global issue which will soon be uncontrollable. Pie chart represented annual MSW composition in US. Anaerobic digestion and incineration methods for reducing waste. But generation of green house gas due to land fill after incineration proved it would not suitable technology. Small scale anerobic digestion and plasma gasification justified as best method for disposal. Plasma gasification alone produced 1.5-1 MWh energy.

Seyed Kamran foadMarashi and Karimina [34], discussed the application of petrochemical wastewater in generation of electricity by using a membrane -less microbial fuel cell. It is renewable energy produced from organic and inorganic material using microorganisms. Fuel and micro-organism engaged are PTA waste water and sludge of anerobic contact filter.

Ahiduzzam and Islam [35], discussed the use of surplus rise husk in production of green power. A survey was done on four major rise processing zones in Bangladesh to estimate potential husk available for generation electricity.

Ahsan and Chowdhury [36], identified the application of house hold produced biomass as a source of green electricity using a biogas digester. It is economically and technically viable for operation of appliances such as biogas light bulb, stove and dual fuel hybrid generator when compared with appliances using conventional energy sources.

Khan and Chowdhury [37], presented treatment of tannery waste fulfills to aim i.e. waste treatment and harness energy from it. Pie chart showed percentage of animal hides and skin in tanneries. Tannery water treated in both manners by gasification and anaerobic digestion. Annual energy yield from both mesophillic and thermophilic methods was compared Government regulatory framework and tanner's owner's support in mitigating pollution and harness electricity from tannery waste.

Lohani et. al. [38], presented two methods incineration and anaerobic digestion compared which proved better results from anaerobic digestion technology by installing a 200 litres capacity of 60 US$ digesters in homes. By product of digester used as bio fuel and bio-fertilizer. At homes waste can treated and useful tool for energy harness.

Belonio et al. [39], discussed about the rise husk as a fuel for cooking and rural electrification. Stove operation based on rise husk fuel during test some parameters considered such as fuel consumption thermal efficiency percentage char produce specific gasification rate with some design tips. Water temperature profile graph showed performance in three different runs. Light has thrown upon some aspects like operating cost analysis and environmental aspect with some trouble shooting guide.

Verma et. al. [40], wrote a paper that revealed that tremendous amount of biogass resources are available on earth. Now it is evident that biogas can be substituted instead of fossil fuels and by investigation all possible generation and utilization are yet to be explored.

Maier and Street [41], focused more on environmental, economic, and social areas somewhat than the perception of a public body's requirements and the economic realities of solid waste management. They guarantee investors with satisfied returns and revenues from treatment of stated amounts of solid waste. Finally the findings from the case study provide interesting insights into the conditions under which state of the art MSW incineration plants with energy recovery using the Clean Development Mechanism are competitive with low cost landfills.

Namuli [42], wrote a paper in which rural farms have installed biomass waste to energy conversion systems to solve their manure disposal problems. These systems can however be sources of revenue to the farms, through sale of electricity to the grid. Three biomass wastes to energy conversion systems were optimized with the objective of obtaining maximum revenues. Two of the systems had the right installed generation capacity required to maximize revenue. These were Emerling farm and Sunny Knoll farm. Emerling farm's system was however not being operated in a way that would maximize revenue. With an improved operational strategy, Emerling farm can increase its cost savings by 18%.

Ushimaru [43], proposed that power plants will deliver much needed power provision increased agricultural productions, help in the access to higher technology and it will also reduce waste for society and local community at a vast scale. This project is actually based on the sound business model so once the operation is underway the project will have many revenue sources from sales of sales of chickens, processing of poultry and meat processing waste from other large farms in the province, sales of electricity to the local utility company, sales of ashes to fertilizer producers, sales of food flavoring pastes to food processors, and ales of feedstock to bio-fuel processors. With the help of external funding this project will proceed much speedy and with a higher probability of confidence for success. The revenue system acquired from this operation will be used to pay back the initial funding investments.

Khelidj et al. [44], carried out a study of viability of a biogas manufacture project and they investigated to prove the important possible of biogas. They started this research by making inventories on different sources of wastes by making a very essential potential of more than 173 billion of cubic meter of biogas in Algeria.

Sulistyo et. al. [45], concluded that the C/N ratio has affected the biogass production. The satisfying results were found when C/N ratio came out to be 20 for white mustard and cow manure, green mustard, rice straw and cow manure and the rate of high production of biogas occurred at 21st day of incubation period. It was observed that methogenic micro organisms have noticeable effect for manure formation.

3. Electricity from Thermal Energy

Hobson [46], performed in depth study of the various forms of waste energy available and the general conditions governing their employment for the purpose of electrical generation. The considerations involved in the disposal of the power generated were also outlined. The "waste heat" was interpreted in its broader sense as including both waste steam and waste gases.

Crawford [47], discussed the usefulness of the waste heat in both thermal and nuclear power stations in order to improve their thermal efficiency. The author suggested district heating as a way of doing this. The concept of total energy was widely developed.

Palmer [48], represented the loss of significant energy in exhaust systems, cooling water and similar working systems. These losses are recovered to reduce the cost that helps in achieving economic stability. The process of cogeneration has been encouraged for considering the lost energy. In this paper, new technological advancements are described by four systems set up in four different cities in United States.

Haidar and Ghojel [49], the non renewable energy resources must be reduced because of scarcity with more development of renewable sources. They elaborated the recovery of heat from a diesel engine using thermo electric generators. Waste heat recovery system is mounted on an exhaust pipe and detailed analysis is discussed.

Solbrekken et al. [50], presented the implementation of shunt attach generation (thermoelectric technology) for changing waste heat from microprocessor into electrical energy which is used for driving a cooling fan for cooling the chip. Low voltage fan works in conjunction with heat driven source. Remaining heat is used to keep the temperature of a chip below a critical value.

Othman et al. [51], provided a detailed analysis of the research conducted on electronic plastic waste to determine its potential as a source of energy. Proximate analysis, ultimate analysis and the heavy metal content analysis of the plastic waste sample were covered. The average heat value for an electronic waste was found to be 30,872.42 kJ/kg or 7,375 kcal/kg. Thus, electronic waste can be used as a source of energy in the future.

Deshpande and Pillai [52], give a detailed analysis on green air conditioning technologies in automobiles. The research work supports the use of silica gel-water adsorption system for air conditioners of vehicles. The above technology has been tested using a four stroke diesel engine from Mahindra and results are presented in the paper.

Bornnert and Burki [53] presented a technical solution to convert low and medium temperature waste heat into electricity in highly energy intensive industries like cement. Heat conversion into electricity with ORC (Organic Rankine Cycle) power plant was covered in detail. ORC power plants can boost electrical energy efficiency in cement plants by upto 20%, reduce indirect CO_2 emissions considerably and save water at the same time. The outlines revealed the fact that high and rising energy costs and the requirement to reduce CO_2 emissions are the main drivers to invest in heat recovery systems.

Xie et al. [54], discussed the various aspects of CMOS Micro-electromechanical systems-based thermoelectric power generators (TPGs) for conversion of waste heat into electrical power. The design, modeling, characterization and fabrication steps of TPGs were covered in detail.

Razak et al. [55], high-lighted harnessing electricity from heat energy harvesting using thermo couple concept. Bismuth Telleuride (Bi2Te3) material produced power and convert heat energy to electrical energy with having temperature differences between two sides.

Megha Tak [56], concluded that the Thermoelectric Generator or TEG waste heat recovery system could potentially offer significant fuel economy improvements. If this is achieved successfully on large scale applications such as automotive, a noticeable saving in fuel consumption can be achieved by using it in automobile sector. They have proved that it is possible to use thermoelectric convertors to light up the car headlights for large SUV cars using 80 Amp-hr on higher batteries. There is also potential to increase the conversion rate from heat to electrical energy, by using materials with better Seebeck coefficient difference and increasing efficiency of TEG's. This application, on a real scale would help in prevention of large amount of heat, preventing the environment also from damage.

Derakhshandeh et al. [57], elaborated a new generation method coordinated with PEV's charging based on DOPF for an IMG consisting of 12 factories with CHP systems,

PV generation systems coupled with PV storages and 6 types of PEVs. Both the network security and factories constraints are added in DOPF formulation. It decreases the cost of IMG by optimizing the hourly heat and electricity generation schedules for individual factories. The problem of optimization is subjected to both electric and thermal needs keeping in mind the possibilities of heat transfer between relative factories. It manages the factories such that part of required electricity is purchased from the upstream network when the price of electricity is lower than the generation cost. Otherwise, the IMG will sell electricity to increase the overall profit. It considers PEVs with time and energy related constraints as coordinated loads and optimizes their charging rates in order to minimize the cost associated with vehicle charging and to maintain the voltage profile within the acceptable limits. Finally, based on the analyses of this paper, introduction of PV generation systems coupled with PV storages in IMGs could have positive effects on their scheduling solution and minimizing the overall cost since the peak of most industrial electricity loads occur during daytime and usually coincide with the maximum output of PV generation.

Vineetha V. and Shibu K. [58], studied on single chambered less microbial fuel cell and the effect of using two types of anode on diary wastewater treatment and electricity production which resulted in maximum electricity and voltage production on day 4th using iron coated carbon anode compared to plain carbon anode. Carole-Jean Wu [59], discussed about harvesting of heat energy of modern computing systems using COTS TEGs. They showed that with a single TEG, they can recover wasted heat from the CPU to significant electrical energy on a real-system.

4. Electricity from Miscellaneous Sources of Energy

Charles P. Steinmetz [60], demonstrated that the efficient utilization of the America's energy supply requires generating electric power where on earth hydraulic or fuel energy is available, and collecting the power electrically, just as we distribute it electrically. Richard L. Nailen [62], illustrates the use of induction generators for the process industries as a key to energy conservation. Emphasis is given to the induction generators producing "free" electric power from process energy that would otherwise be wasted. The operating principles, design and performance using performance curves, protection and control measures, along with application considerations of induction generators in waste heat applications are outlined. Examples of industrial use and the utility rates are also covered in brief.

Hammons and Geddes [63], illustrated the implication of alternative energy sources for energy generation which can replace conventional energy sources in United kingdom. Several energy sources are examined in this paper which includes, tidal power, wind power, small scale hydro turbine, use of natural gas coming out from land, wave power, municipal solid waste, biogas conversion, geothermal power which could generate an ample amount of water.

Palanichamy et al. [64] illustrated the support of government of India in saving non-renewable resources by developing more efficient energy sources. Here new technological advancements about extraction of energy from waste are discussed and comparisons of new developments have been done with old technologies.

Leung and Hui [65] reviewed the adoption of renewable energy developments in HONGKONG. Technological trends are reviewed in two stages. First stage included use of solar power, wind power and energy from waste. Second stage was to implement a building integrated photovoltaic system which gives output of 55 KW.

Tompros [66], focused on reduction of energy cost in cement industry by taking benefits of energy conservation devices. Exhaust gases and clinker coolers are utilized for generation of electricity. The comparison of heat source in cement industry with heat sources in kalina cycle power plant is shown. Thus purchase cost of power has been cutoff to some extent.

Inamdar et. al. [67], present an energy management architecture that is applicable on domestic appliances. By the effective use of latest information and communication technology, the proposed architecture is concerned about three main functions: real-time estimation of the energy consumption of the home environment; control of domestic appliances energy use; and autonomous identification and management of standby devices.

H.P and R.P. [68], explained IT based Energy Auditing for effective energy management and conservation. Computers and advanced metering technologies can be employed in industrial sector. The IT based Energy Auditing components like data measurement, data analysis program, advanced data management, network communication are also covered in detail. A case study supporting the use of Information Technology in energy management is also mentioned.

Thokair and Mansi [69], discussed application of waste hydrogen in desalination and power plant. Hydrogen calculated from a process and then used in different areas like cooling generators, generating electricity using fuel cell in UPS and source of electrolysis process. Hydrogen enhanced ½ to 1 full load efficiency of generator. Use of recovered hydrogen leads to save money. Jae-Do Park [71], described about an efficient Microbial fuel cells also known as MFC, energy harvesting system using DC/DC converters has been presented. The proposed energy harvesters capture the energy from multiple MFCs at individually controlled operating points and at the same time forms the energy into a usable shape. The proposed parallel operation system consists of multiple harvesting converters for each MFC and a single voltage boost converter. The proposed control scheme has been validated experimentally and a successful result has been shown.

Zhang and Wang [72], told that high-concentration photovoltaic (HCPV) is a highly promising technology to directly convert plentiful solar energy to electricity. However, even for the most advanced HCPVs, about 60% of the concentrated solar energy is rejected as waste heat; therefore, it is desirable to utilize the massive waste heat from HCPV modules. Considering the nature of low-grade waste thermal energy, a micro scale organic cycle (MaRC) offers a promising solution. In a subcritical MaRC, sub cooled refrigerant is usually pumped into a micro channel heat sink of each multi-junction photovoltaic cell. In this paper, a complete micro channel flow boiling model is developed based on distributed mass, energy and momentum conservation laws. Detailed MaRC thermal-fluid analysis is conducted to evaluate the effects of working fluid, inlet sub cooling, axial fluid/cell temperature distribution and critical heat flux on cogeneration efficiency. The performance analysis indicates that the HCPV/MORC system can achieve a net 8.8% increase of power generation efficiency in comparison to liquid-cooled HCPV at ambient temperature. The proposed HCPV /MaRC configuration shows great promise in large-scale applications of HCPV solar power generation.

Habib et al. [73], promoted closed cycle standalone than other conventional power plants due to its higher efficiency and smaller size. MHD generators also don't consist of any rotating parts. Therefore they don't require much maintenance. But the advantage of MHD are offset by difficulties of high temperature requirement (2500-3000 K) and high magnetic flux densities (5 to 6 Tesla) involving costly superconducting magnet technologies. Swapna kumari B. Patil et. al. [74], dedicated research on nonconventional waste material harvesting model taking into account economy, environmental effects and power generation related to harvesting of various waste samples. In this study several waste management scenes are compared along with the various energy and environmental factors among alternate waste management strategies; the effect of waste diversion through reuse and effective conservation of energy to restore electrical power on environmental releases and economy.

5. Conclusion

The alarming rate of increase in the waste materials around us can no more be acceptable. Hazardous wastes poison our planet and negatively affect the health of millions of people worldwide. Also, conventional sources of energy like fossil fuels are depleting at a faster rate which will lead us to energy crisis. The concept of 'Electricity from wastes' gives us a way to manage energy as well as waste. Conducted literature survey included several research papers and categorized them in 3 areas that are Bio-energy, Thermal Energy and Miscellaneous sources of energy. In bio energy we encountered many methods like incineration, anaerobic digestion, plasma gasification etc. useful in generating electricity by mitigating the bio waste. While in Thermal energy we focused on saving waste heat energy and converting into electricity. Miscellaneous sources of energy consists many waste sources like electronic waste, plastic waste, waste hydrogen in desalination etc. Every energy source has its own indispensability in its area. Hence, we conclude from our work that by effective waste management we can keep earth green forever and save ourselves from energy crisis. Even though a conclusion may review the main results or contributions of the paper, do not duplicate the abstract or the introduction. For a conclusion, you might elaborate on the importance of the work or suggest the potential applications and extensions.

References

[1] Stanton M. Peters, "Cogeneration Fueled by Solid Waste Utilizing a new technology", IEEE Transactions on Power Apparatus and Systems, Vol. PAS-101, No. 10, October, 1982, pp. 3951-3956.

[2] A. Porteous, "Developments in, and environmental impacts of, electricity generation from municipal solid waste and landfill gas combustion", IEE Proceedings-A, Vol. 140, No. I, January 1993, pp. 86-93.

[3] Ahmad Zahedi, "Investigation of feasibility of establishing waste to energy facility in Australia", IEEE, 1994.

[4] D.H. Maunder, K.A. Brown and K.M. Richards, "Generating electricity from biomass and waste", Power Engineering Journal, August 1995.

[5] S. Walker, "Energy From waste in the sewage treatment process", IEEE Conference Publication No. 419, 1996.

[6] Janani Chakravarthi, "Biogas and energy production from cattle waste", IEEE 1997.

[7] Y. Kishinevsky and S. Zelingher, "A 200 Kw Onsi Fuel Cell On Anaerobic Digester Gas", IEEE, 1999.

[8] Gregg Tomberlin& Brad Moorman, "Energy generation through the combustion of municipal solid waste", 2004.

[9] Gusphyl A. Justin, Yingze Zhang, Mingui Sun, and Robert Sclabassi, "Biofuel Cells: A possible power source for implantable electronic devices", Proceedings of the 26th Annual International Conference of the IEEE, September 2004.

[10] Sz. Váradi, L. Strand, and J. Takács, "Clean Electrical Power Generation from Municipal Solid Waste", IEEE, 2007.

[11] J. M. Monier, L. Niard, N. Haddour, B. Allard and F. Buret, "Microbial Fuel Cells: from Biomass (waste) to Electricity", IEEE, 2008, page 663-668.

[12] A. Sz. Varadi, and J. Takacs, "Electricity Generation from Solid Waste by Pilot Projects", SPEEDAM 2008, International Symposium on Power Electronics, Electrical Drives, Automation and Motion, page 826-831.

[13] Yinfang Zhi, Hong Liu, and Lu Yao, "The effect of Suspended Sludge on Electricity Generation in Microbial Fuel Cells", IEEE, 2008, page 2923-2927.

[14] Imrul Kayes and A. H. Tehzeeb, "Waste to energy: A Lucrative alternative", IEEE, 2009.

[15] Md.Shahedul Amin et. al, "The Potential of Generating Energy from Solid Waste materials in Bangladesh", 2009.

[16] MoinuddinSarker et. al., "New alternative Energy from Solid Waste Plastics", 2009.

[17] Nathan Curry and Dr. Pragasen Pillay, "Converting food waste to usable energy in the Urban Environment through anaerobic digestion", IEEE Electrical Power and Energy Conference, 2009, page 14.

[18] Ransford R. Baidoo, F. Yeboah and H. Singh, "Energy and Economic analysis of closed loop Plasma Waste-to-Power Generation Model and in comparison with Incineration and Micro-Turbine Models", IEEE Electrical Power and Energy Conference, 2009, page 1-7.

[19] Ransford R. Baidoo, F. Ferguson and F. Yeboah, "Energy and Energy analysis of Plasma Waste-to-Power Generation Model", IEEE Electrical Power and Energy Conference, 2009, page 1-5.

[20] Svetlana Nikolaeva et. al., "A Sustainable Management of Treatment Plant for Dairy Wastes with the use of its by-products", PICMET 2009 Proceedings, Portland, Oregon USA, August 2-6, page 1745-1750.

[21] LanTang et. al., "Plasma Pyrolysis of Biomass for production of gaseous fuel to generate electricity", IEEE, 2010.

[22] Vish. Kallimanni et. al., "Design and development of a compact high rate digester for rapid bio-methanation from a kitchen waste for Energy generation", IEEE ICSET 2010, Kandy, Sri Lanka, 6-9 Dec 2010.

[23] Geng Cuijie, Chen Dezhen, Sun Wenzhore and Liu Pu, "Life Cycle Assessment for Road base Construction using Bottom Ash from Municipal Solid Waste Incineration in Shanghai", IEEE, 2010.

[24] Silvia Bardi and Acessandro Astoefi, "Modeling and Control of a Waste-to-Energy Plant Waste-Bed Temperature Regulation", IEEE, 2010.

[25] CHEN Nan and Henry Shu-hung CHUNG, "An Energy-Recyclable Burn-in Technology for Electronic Ballasts for HID Lamps", IEEE, 2010.

[26] Mohammad Rafiq Khan, "Potential of Thermoelectric Power from Bagasse by Sugar Mills of Pakistan", IEEE, 2010.

[27] Masoud Paurali, "Application of Plasma Gasification Technology in Waste to Energy—Challenges and Opportunities", IEEE Transactions on Sustainable Energy, Vol. 1, No. 3, October 2010.

[28] K M Rafi, Majid Jamil and A Mubeen, "Energy Potential & Generation through IRES in city of Faridabad-India, IEEE Global Humanitarian Technology Conference, 2011.

[29] Robert Kramer, Libbie Pelter, Kraig Kmoitck, Ralph Branch, Alexandru Colta, Bodgan Popa and everting and John Petterson, "Modular Waste/Renewable Energy System for Production of Electricity, Heat and Portable Water in Remote Location", IEEE Global Humanitarian Technology Conference, 2011.

[30] Mohamad Iskandar bin Jobli, Diana Kertinibinti Monis and Khee Kian Peng, "Analysis of Waste Thermal Energy from Banana Peels Using Decomposition Process for Heat and Generation", IEEE First Conference on Clean Technology CET, 2011.

[31] Emmanuel P.Leano and Sandhya Babel, "Electricity Generation from Anaerobic Sludge and Cassava Wastewater Subjected to Pretreatment Methods using Microbial Fuel cell", IEEE First Conference on Clean Technology CET, 2011.

[32] Nathur Curry and Dr. Pragasen Pillay, "Waste-to-Energy Solutions for the Urban Environment", IEEE, 2011.

[33] Seyed Kamran, Foad Marashi, and Hamid Reza Karimina, "Electricity generation from Petrochemical waste water using a membrane-less single chamber microbial fuel cell", Second Iranian Conference on Renewable Energy and Distributed Generation, 2012.

[34] Md. Ahiduzzam, and A. K. M. Sadrul Islam, "Assessment of Rice Husk Energy Use for Green Electricity Generation in Bangladesh", IEEE, 2012.

[35] Asif Ahsan, and Shahriar Ahmed Chowdhury, "Feasibility Study of Utilizing Biogas from Urban waste", IEEE, 2012.

[36] Ajmiri Sabrina Khan and Shahriar Ahmed Chowdhury, "Potential of Energy from Tannery Waste in Bangladesh", IEEE, 2012.

[37] S. P. Lohani, A. Satyal, S. Timilsina, S. Parajuli, and P. Dhilai, "Energy Recovery Potential from Solid Waste in Kathmandu Valley", IEEE, 2012.

[38] Alexis T. Belonio, Md. Aktaruzzam and Bhuiyan, "Design of a Continuous -Type Rice Husk Gasifier Stove and Power Generation Device for Bangladesh Household", IEEE, 2012.

[39] Amit Verma, Rahul Singh, Ranjeet Singh Yadav, Neeraj Kumar, Priti Srivastava, "Investigations on Potentials of Energy from Sewage Gas and their Use As Stand Alone System", IEEE, 2012.

[40] Sebastian Maier and Alexandre Street, "Model for the economic feasibility of energy recovery from municipal solid waste in Brazil", 2012 IEEE.

[41] R. Namuli, "Maximization of Revenue from Biomass Waste to Energy Conversion Systems on Rural Farms", 2012 IEEE.

[42] Kenji Ushimaru, "Sustainable Green Energy Production from Agricultural and Poultry Operations", 2012 IEEE.

[43] B. Khelidj, B. Abderezzak and A. Kellaci, "Biogas Production Potential in Algeria: Waste to Energy Opportunities", IEEE, 2012.

[44] Hary Sulistyo, Yogyakarta Surakarta, "Biogas Production from Traditional Market Waste to Generate Renewable Energy", IEEE, 2013.

[45] H.Hobson, "The Utilisation of Waste Heat for the generation of Electrical Energy", Newcastle Students' section, 25 January, 1915, page 844-848.

[46] J.J. Crawford, "Total Energy a realistic answer to fuel conservation", Electronics and Power, 31 May, 1973, page 210-212.

[47] James D. Palmer, "Cogeneration From Waste Cogeneration From Waste Energy Streams", IEEE Transactions on Power Apparatus and Systems, Vol. PAS-100, No. 6, June, 1981, pp. 2831-2836.

[48] Jihad G. Haidar and Jamil I. Ghojel, "Waste heat recovery from the exhaust of low-power diesel engine using thermoelectric genetators", 20th International Conference on Thermoelectrics, 2001.

[49] Gary L. Solbrekken, Member, IEEE, Kazuaki Yazawa, Member, IEEE, and Avram Bar-Cohen, Fellow, IEEE, "Heat Driven Cooling Of Portable Electronics Using Thermoelectric Technology", IEEE, 2008.

[50] N. Othman et. al., "Electronic Plastic Waste Management in Malaysia: The Potential of Waste to Energy Conversion", Proceedings of ICEE 2009, 3rd International Conference on Energy and Environment, Malacca, Malaysia, 7-8 December 2009, page 337-342.

[51] A. C. Deshpande and RM Pillai, "Adsorption Air-Conditioning (AdAC) for Automobiles using Waste Heat recovered from Exhaust Gases", Second International Conference on Emerging Trends in Engineering and Technology, ICETET-09, page 19-24.

[52] Thomas Bormert and Thomas Burki, "Waste heat conversion into Electricity", IEEE, 2010.

[53] Jin Xie, Chengkuo Lee and Hanhua Feng, "Design, Fabrication and Characterization of CMOS MEMS-based Thermoelectric Power Generators", Journal of Micro electromechanical Systems, Vol. 19, No. 2, April 2010, page 317-324.

[54] Ahmad Nazri, AbdRazak and Nursyarizalmohd. Norr and Taib Ibrahim, "Heat Energy Harvesting for Portable Power Supply", The 5th International Power Engineering and Optimization Conference (PEOCO2011), Shah Alam, Selangor, Malaysia, 6-7 June, 2011.

[55] MeghaTak, "Converting Waste Heat from Automobiles to Electrical Energy", IEEE, 2012.

[56] S. Y. Derakhshandeh,, Amir S. Masoum, Sara Deilami, Mohammad A. S. Masoumand M. E. Hamedani Golshan", Coordination of Generation Scheduling with PEVs Charging in Industrial Microgrids", IEEE, 2013.

[57] Vineetha V. Shibu K, "Electricity Production Coupled With W Astew after Treatment using Microbial Fuel Cell", IEEE, 2013.

[58] Carole-Jean Wu, "Architectural Thermal Energy Harvesting Opportunities for Sustainable Computing", IEEE, 2013.

[59] Charles P. Steinmetz, "America's Energy Supply", 1918 IEEE.

[60] John E. Heer, Jr. and D. Joseph Hagerty, "Energy: Refuse turns resource: Diverted from landfills to hammer mills, municipal waste becomes an economic energy resource," 1974 IEEE.

[61] Richard L. Nailen, "Watts from Waste Heat -Induction Generators for the Process Industries", IEEE Transactions on Industry Applications, Vol. 1 A-19, No.3, May/June, 1983, page 470-475.

[62] T J Hammons and A G Geddes, "Assessment of Alternative energy sources for generation of electricity in the UK following privatization of the electric supply industry", IEEE Transactions on Energy Conversion, Vol. 5, No. 4, December 1990, pp. 609-615.

[63] Panote Wllaipon et. al, "Study on the potential of corn cob engineer-generator for electricity generation in Thailand", IEEE, 2002.

[64] K.M. Leung and Jimmy W.W. Hui, "Renewable Energy Development in Hong Kong", IEEE International Conference on Electric Utility Deregulation, Restructuring and Power Technologies, April 2004.

[65] Mark D. Mirolli, "The Kalina Cycle for cement kiln waste heat recovery power plants", IEEE, 2005.

[66] Spyridon Tompros et. al., "Enabling Applicability of Energy saving applications on the appliances of the Home Environment", IEEE Network, November/December, 2009, page 8-16.

[67] Dr. Inamdar H.P. and Hasabe R.P., "IT based Energy Management through Demand Side in the Industrial Sector".

[68] N.AL- Thokair and S.H. Mansi, "A Study of Using Waste Hydrogen in Desalination and Power Plant as an Energy Source", 2011 IEEE

[69] Ehsaneh Shahhaidar, Olga Boric - Lubecke, Reza Ghorbani and Michael Wolfe, "Electromagnetic Generator as Respiratory Effort Energy Harvester, IEEE, 2011.

[70] Jae-Do Park and Zhiyong Ren, "Hysteresis-Controller-Based Energy Harvesting Scheme for Microbial Fuel Cells With Parallel Operation Capability", IEEE, 2012.

[71] TieJun Zhang and Evelyn N. Wang, "Design of a Microscale Organic Rankin Cycle for High-Concentration Photovoltaic Waste Thermal Power Generation", IEEE, 2012.

[72] Salman Habib, Ariful Haque and Jubeyer Rahman, "Production of MHD Power from Municipal Waste & Algal Biodiesel", IEEE, 2012.

[73] Swapnakumari B.Patil et. al., "Green Energy Revolution in Economic Power Generation-Composite MFC", IEEE, 2012.

[74] Satish Kumar, R and Rama Chandra, T.V., "Solid waste Management System Using Spatial Analysis Tools", 2000 National conference.

[75] Sonelgaz Group Company, "Renewable energy and energy efficiency program", Ministry of Energy and Mines", http://www.mem-algeria.org (2011), Accessed March 2011.

[76] Stambouli AB, "Algerian renewable energy assessment: the challenge of sustainability", Energy Policy 2011, 39:4507-4519.

[77] Stambouli AB, "Promotion of renewable energies in Algeria: strategies and perspectives. Ren. Sust." Energy Reviews 2011, 15:1169-1181.

[78] Population Reference Bureau (PRB), "World population data sheet. 2011.", http://www.prb.org (2011). Accessed 2011.

[79] BP Statistical Review of World Energy, "Statistical review London", http://www.bp.com/statisticalreview (2011). Accessed June 2011.

[80] Boudries R, Dizene R, "Potentialities of hydrogen production in Algeria. Int. J. Hydrogen Energy", 2008, 33: 4476-4487.

[81] Himri Y et. al., "Review and use of the Algerian renewable energy for sustainable development Ren. Sust.", Energy Reviews 2009, 13: 1584-1591.

[82] Gourine L, "Country report on the solid waste management: Algeria. The regional solid waste exchange of information and expertise network in Mashreq and Maghreb countries", http://www.sweep-net.org/content/algeria (2010), Accessed July 2010.

[83] Bendjoudi Z, Taleb F, Abdelmalek F, Addou A, "Healthcare waste management in Algeria and Mostaganem department", Waste Manage 2009, 29: 1383-1387.

[84] Sefouhi L, Kalla M, Aouragh L, "Health care waste management in the hospital of Batna City (Algeria)",Paper presented at the Singapore International Conference on Environment and Bio Science, Singapore; 2011.

[85] Alamgir M, Ahsan A, "Characterization of MSW and nutrient contents of organic component in Bangladesh", EJEAFCHE 2007, 6 (4): 1945-1956.

[86] Mc Kendry P, "Energy production from biomass (part 1): overview of biomass", Bioresour Technol 2002, 83: 37-46.

[87] Guermoud N et. al., "Municipal solid waste in Mostaganem City (Western Algeria)", Waste Manage 2009, 29: 896-902.

[88] McKendry P, "Energy production from biomass (part 2), conversion technologies", BIO RESOURE TECHNOLY 2002, 83: 47-54.

[89] Münster M, Lund H, "Use of waste for heat, electricity and transport-Challenges when performing energy system analysis", Energy 2009, 34: 636-644.

19

Performance Analysis of a Water Savonius Rotor: Effect of the Internal Overlap

Ibrahim Mabrouki, Zied Driss*, Mohamed Salah Abid

Laboratory of Electro-Mechanic Systems (LASEM), National School of Engineers of Sfax (ENIS), University of Sfax, B.P. 1173, Road Soukra km 3.5, 3038 Sfax, TUNISIA
*Corresponding author: zied.driss@enis.rnu.tn

Abstract The water Savonius rotor is classified as a vertical axis water rotor like the Darrieus, Gyromill or H-rotor. The advancing blade with concave side facing the water flow would experience more drag force than the returning blade, thus forcing the rotor to rotate. In this work, we are interested on the study of the of the internal overlap effect of a water Savonius rotor. The experimental results is developed using a hydraulic test bench. The test bench consists on an intake, a control gate, a penstock, a canalization, a test section, an outflow and a pump. A detailed description of the global characteristics is presented such as power, dynamic torque, power and its coefficients.

Keywords: Savonius rotor, internal overlap, hydraulic test bench, power coefficient, torque coefficient

1. Introduction

Savonius rotor is a unique fluid-mechanical device that has been studied by numerous investigators since 1920s. Applications for the Savonius rotor have included pumping water, driving an electrical generator, providing ventilation, and agitating water to keep stock ponds ice-free during the winter [1,2,3,4]. Savonius rotor has a high starting torque and a reasonable peak power output per given rotor size, weight and cost, thereby making it less efficient. From the point of aerodynamic efficiency, it cannot compete with high-speed propeller and the hydro-kinetic turbine electricity generation is mainly aimed for rural use at sites remote from existing electricity grids. It is a useful tool for improving the quality of life of people in these locations and for stimulating local economies. These rotors also can be considered for the wide variety of application like tides, marine currents, channel flows and water flows from industrial processes. Different designs of water current rotor are available for the extraction of energy from the river water or canals. Based on the alignment of the rotor axis with respect to water flow, two generic classes exist. They are horizontal axis turbine (axial turbines) and vertical axis turbine (cross flow turbines). Horizontal axis turbines are mainly used for extraction of the ocean energy. These turbines are expensive for small power applications. Vertical turbines generally used for small scale power generation and these are less expensive and required less maintenance compared to horizontal axis water turbines. Savonius rotor, helical turbine, Darrieus turbine and H-shaped Darrieus

are commonly used vertical axis turbines. Various types of water current turbines are being installed and tested worldwide for various ranges of powers. GCK technology limited (USA), installed a Gorlov helical water turbine (diameter of 1 m and height of 2.5 m) in the Uldolomok Strait off the coast of Korea. Similarly Verdant Power Ltd. (USA) installed a three bladed horizontal axis water turbine as free flow turbine in east river New York [5]. Alternative Hydro Solutions Ltd. in Ontario has developed vertical axis turbines specifically meant to harness the water energy from river [6]. Literature suggests that there is a gaining of popularity for water turbines [7,8,9]. Horizontal axis turbines are common in tidal energy converters and majority of marine current turbines are horizontal axis turbine [10]. They are very similar to modern day wind turbines from design and structural point of view. In the vertical axis turbines domain, the Darrieus turbines, Savonius turbine and Gorlov helical water turbines are generally used. The Gorlov turbine has the blades of helical structure. Gorlov [11] proposed a new helical turbine to convert kinetic energy of flowing water into electrical or mechanical energy. Many researchers have adopted various techniques to maximize the performance and improve the starting torque characteristics of Savonius turbine with wind as working medium. These include use of guide vanes, V-plate deflector [12], deflector plate [13,14] and blade with flat and circular shielding [15]. Some of these techniques require change in design of blade and other involves supplementary devices addition to the system. For example, Mohamed et al. [16] carried out a numerical analysis for identifying the optimum shielding of the returning blade of a Savonius wind rotor. Two dimensional numerical investigations were carried out using OPAL (Optimizing Algorithms) along with commercial CFD package FLUENT at a fluid velocity of 10 m/s and a tip speed ratio of 0.7. Investigation on the modified Savonius rotor (with shaft) reported by Modi and Fernando [17] was an effort in the direction of improvement of performance of Savonius wind turbine by changing the shape of the blade. Modified Savonius rotor with shaft was reported to have a maximum coefficient of power around 0.32. However, these tests were based on closed wind tunnel testing and coefficient of power was obtained by extrapolation. Kamoji et al. [18] investigated the performance of modified forms of conventional rotors with and without central shaft between the end plates.

According to these anterior studies, it has been noted a paucity on the study of the water Savonius rotor performance [19,20,21,22]. For thus, We are interested in this paper on the experimental characterization of a vertical axis water Savonius rotor type using a hydraulic test bench equipped with a specific instrumentation. In particular, a detailed description of the internal overlap effect on the water Savonius rotor was developed.

2. Materials and Method

Figure 1. Hydrodynamic test bench

The used hydrodynamic test bench consists on an intake, a control gate, a penstock, a canalization, a test section, an outflow and a pump. The collector is a parallelepiped where the water flow inside a square tank located above the test section shape on a closed circuit (Figure 1). In this section, we are interested on the study of the internal overlap effect of a water Savonius rotor. This rotor consists of two buckets of diameter d=100 mm and height H=200 mm. It is assembled on a common axis and secured with a screw nut at an angle of 180° (Figure 2). Particularly, we have considered the internal overlap equals to (e-e')/d=0, (e-e')/d=0.2 and (e-e')/d=0.3 (Figure 3). In this work, the experimental study involved the rotation of the Savonius water rotor. Experimental tests for the determination of global characteristics such as the power, the dynamic torque, and its coefficients required the use of the hydrodynamic test bench. Otherwise, the test bench should be equipped with a specific instrumentation for the development of various

experimental tests necessary in the laboratory scale. To achieve this goal, we use a permanent magnet dynamo creating a constant magnetic flux through the coil of the rotor, driven by the turbine rotation. The flux variation undergoes by rotating the coil creates a voltage proportional to the rotational speed. Why the higher water flow increases, the turbine rotates faster, and the generated current increases. By connecting the resistor with the multi-meter in the output of the generator, the generator is rotated by the dynamometer. For a same rotational speed imposed by the dynamometer, we change the electrical resistance and we measure the current supplied by the generator and the rotational speed of the Savonius rotor. Indeed, the dynamic torque and the power can be deduced.

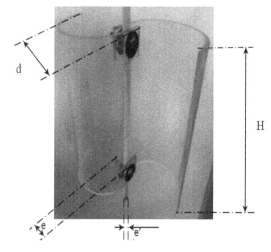

Figure 2. Geometrical arrangement

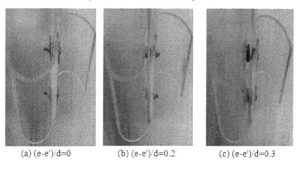

(a) (e-e')/d=0   (b) (e-e')/d=0.2   (c) (e-e')/d=0.3

Figure 3. Internal overlap cases

3. Experimental Results

3.1. Power

Figure 4 shows the variation of the power in function of the rotation speed for different internal overlaps equals to (e-e')/d=0, (e-e')/d=0.2 and (e-e')/d=0.3. These curves are superimposed on the same scale. These results are obtained at a speed water equal to V=2.45 m.s^{-1} corresponding to a Reynolds number equal to Re=588300. According to these results, these curves have a parabolic shape. Moreover, these results show that the internal overlap has a direct effect on the presentation of these curves. In fact, we find that the power reaches the most important values for the internal overlap (e-e')/d=0.3. With the decrease of the internal overlap, a gradual decrease of the power is then reported. In fact, it has been noted that the maximum value of the power is equal to P=19.28 W. It is obtained in the case of an internal overlap equal to (e-e')/d=0.3 for a rotation speed equal to Ω=737 rpm. With the decrease of the internal overlap, we find that the extremum characteristic decreases on value. For example, for the internal overlap (e-e')/d=0 the maximum value of the power is equal to P=15.05 W for a rotation speed Ω=685 rpm. The increase of the power value is due to the decrease of the Savonius rotor diameter. This implies that when the diameter decreases the Savonius rotor power decreases also.

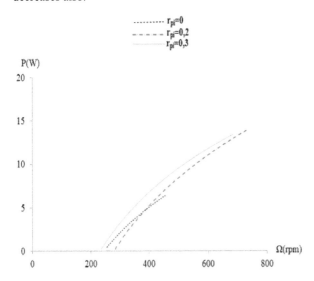

Figure 4. Variation of the power P for different internal overlap

3.2. Dynamic Torque

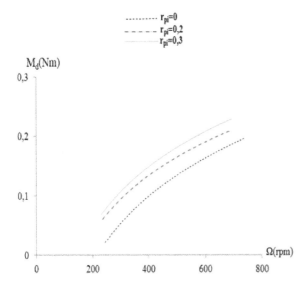

Figure 5. Variation of the dynamic torque M_d for different internal overlaps

Figure 5 shows the variation of the dynamic torque M_d in function of the speed of rotation for different internal overlaps equals to (e-e')/d=0, (e-e')/d=0.2 and (e-e')/d=0.3. These curves are superimposed on the same scale. These results are obtained at a speed water equal to V=2.45 m.s^{-1} corresponding to a Reynolds number equal to Re=588300. According to these results, these curves have a parabolic shape. Moreover, these results show that the internal

overlap has a direct effect on the presentation of these curves. In fact, we find that the dynamic torque reaches the most important values for an overlap (e-e')/d=0.3. With the internal overlap decrease, a gradual decrease of the dynamic torque M_d is then reported. In fact, it has been noted that the maximum value of the dynamic torque is equal to M_d=0.25 N.m. It is obtained in the case of an internal overlap (e-e')/d=0.3 for a rotation speed equal to Ω=737 rpm. With the decrease of the internal overlap, we find that the extremum characteristic decreases of value. For the internal overlap (e-e ')/d=0, the maximum value of the dynamic torque is equal to M_d=0.21 N.m for a rotation speed Ω=685 rpm. The increase of the dynamic torque value M_d is due to the decrease of the Savonius rotor diameter. This implies that when the diameter decreases, the dynamic torque of the Savonius rotor decreases also.

3.3. Power Coefficient

Figure 6 shows the variation of the power coefficient C_p in function of the specific speed for different internal overlaps equals to (e-e')/d=0, (e-e')/d=0.2 and (e-e')/d=0.3. These curves are superimposed on the same scale. These results are obtained at a water speed equal to V=2.45 m.s^{-1}. In these conditions, the Reynolds number is equal to Re=588300.

According to these results, these curves have a parabolic shape. Moreover, these results show that the internal overlap has a direct effect on the presentation of these curves. In fact, we find that the power coefficient reaches the most important value for the overlap equal to (e-e')/d=0.3. With the decrease of the internal overlap, a gradual decrease of the power coefficients values C_p is then reported. In fact, it has been noted that the maximum value of the power coefficient is equal to C_p=0.327. It is obtained in the case of an internal overlap equal to (e-e')/d=0.3 for a specific speed equal to λ=2.51. With the decrease of the internal overlap, we find that the extremum characteristic decreases of value. For the internal overlap (e-e')/d=0, the maximum value of the power coefficient is equal to C_p=0.215 for a specific speed λ=3.027. The increase of the power coefficient value C_p is due to the decrease of the Savonius rotor diameter. This implies that when the diameter increases, the Savonius rotor efficiency increases also.

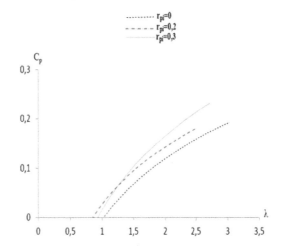

Figure 6. Variation of the power coefficients C_p for different internal overlap.

3.4. Dynamic Torque Coefficient

Figure 7 shows the variation of dynamic torque coefficient in function of the specific speed for different internal overlap equal to (e-e')/d=0, (e-e')/d=0.2 and (e-e')/d=0.3. These curves are superimposed on the same scale. These results are obtained at a speed water equal to v=2.45 m.s^{-1} corresponding to a Reynolds number equal to Re=588300.

According to these results, these curves have a parabolic shape. Moreover, these results show that the internal overlap has a direct effect on the presentation of these curves. In fact, we find that the dynamic torque coefficient reaches the most important values for an overlap (e-e')/d=0.3. With the decrease of the internal overlap, a gradual decrease of the of dynamic torque coefficient C_{Md} is then reported. In fact, it has been noted that the maximum value of the dynamic torque coefficient is equal to C_{Md}=0.26. It is obtained in the case of an internal overlap (e-e')/d=0.3 for a specific speed equal to λ=2.51. With the decrease of the internal overlap, we find that the extremum characteristic decreases of value. For the internal overlap (e-e')/d=0, the maximum value of the dynamic torque coefficient is equal to C_{Md}=0.155 for a specific speed λ= 3.02. The increase of the dynamic torque coefficient value C_{Md} is due to the decrease of the Savonius rotor diameter. This fact implies that when the diameter decreases the dynamic torque coefficient Savonius rotor decreases also.

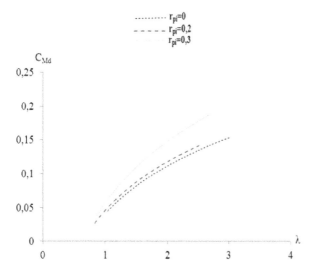

Figure 7. Variation of the dynamic torque coefficient C_{Md} for different internal overlaps

4. Conclusion

In this paper, we focalise our attention on the study of the overlap effect on the global characteristics of the water Savonius rotor. Particularly, we have studied the variation of the power, the dynamic torque, and its coefficients depending on the rotational and the specific speed. In this work, we confirm that the global characteristics of the Savonius rotor increases in the used test section with the increase of the overlap.

In the future, we suggest the deflector addition to improve the rotors performance.

Acknowledgement

The authors would like to thank the Laboratory of Electro Mechanic Systems (LASEM) members for the financial assistance.

Nomenclature

C_{Md} dynamic torque coefficient
C_p power coefficients
d rotor diameter [mm]
H Rotor height [mm]
M_d Dynamic torque [N.m]
P power [W]
V Water speed [m/s]
ρ Density of the water [kg.m^{-3}]
μ Dynamic viscosity [m.s^{-2}]
λ specific speed
Ω Rotational frequency [rad.s^{-1}]

Acknowledgement

The authors would like to thank the Laboratory of Electro-Mechanic Systems (LASEM) members for the financial assistance.

References

[1]   Modi VJ, Roth NJ, Pittalwala A. Blade configuration of the Sovonius rotor with application to irrigation system in Indonesia. In: Proceedings of 16th intersociety energy conversion engineering conference, Atlanta, GA, USA, 1981.

[2]   Clark RN, Nelson V, Barieau RE, Gilmore E. Wind turbines for irrigation pumping. Journal of Energy 1981; 5: 104-8.

[3]   Modi VJ, Fernando MSUK, Yokomizo T. An integrated approach to design of a wind energy operated integrated system. ASME/AIAA Paper No. 98-0041, 1998.

[4]   Vishwakarma R. Savonius rotor wind turbine for water pumping—an alternate energy source for rural sites.Journal of Institution of Engineers (India) 1999; 79: 32-4.

[5]   Rourke FO, Boyle F, Reynolds A. Renewable energy resources and technologies applicable to Ireland. Renew Sust Energy Rev 2009; 13: 1975-84.

[6]   Khan MJ, Iqbal MT, Quaicoe JE. River current energy conversion systems: progress, prospects and challenges. Renew Sust Energy Rev 2008; 12: 2177-93.

[7]   Khan MJ, Bhuyan G, Iqbal MT, Quaicoe JE. Hydrokinetic energy conversion systems and assessment of horizontal and vertical axis turbines for river andtidal applications: A technology status review. Appl Energy 2009; 86: 1823-35.

[8]   Anyi M, Kirke B. Evaluation of small axial flow hydrokinetic turbines for remote communities. Energy Sust Dev 2010; 14: 110-6.

[9]   Guney MS, Kaygusuz K. Hydrokinetic energy conversion systems: a technology status review. Renew Sust Energy Rev 2010; 14:2 996-3004.

[10] Rourke FO, Boyle F, Reynolds A. Marine current energy devices: current status and possible future applications in Ireland. Renew Sust Energy Rev 2010; 14: 1026-36.

[11] Gorlov AM. Helical turbines for the Gulf stream: conceptual approach to design of a large scale floating power farm. Marine Technol 1998; 35: 175-82.

[12] Shaughnessy BM, Probert SD. Partially-blocked Savonius rotor. Appl Energy 1992; 43: 239-49.

[13] Huda MD, Selim MA, Sadrul Islam AKM, Islam MQ. The performance of an Sshaped Savonius rotor with a deflecting plate. RERIC Int Energy J 1992; 14 (1): 25-32.

[14]   Ogawa T, Yoshida H, Yokota Y. Development of rotational speed control systems for a Savonius-type wind turbine. J Fluids Eng 1989; 111: 53-8.

[15]   Alexander AJ, Holownia BP. Wind tunnel tests on a Savonius rotor. J Ind Aerodynam 1978; 3: 343-51.

[16]   Mohamed M, Janiga G, Pap E, Thevenin D. Optimisation of Savonius turbinesusing an obstacle shielding the returning blade. Renew Energy 2010; 35: 2618-26.

[17]   Modi VJ, Fernando MSUK. On the Performance of the Savonius Wind Turbine. JSolar Energy Eng 1989; 111: 71-81.

[18]   Kamoji MA, Kedare SB, Prabhu SV. Experimental investigations on single stage modified Savonius rotor. Appl Energy 2009; 86 (7-8): 1064-73.

[19]   I. Mabrouki, Z. Driss and MS. Abid, "Influence of the hight on Characteristics of Savonius Hydraulic Turbine," International Conference on Mechanics and Energy ICME'2014, March 18-20, 2014, Monastir, TUNISIA.

[20]   I. Mabrouki, Z. Driss and MS. Abid, "Computer modeling of 3D turbulent free surface flow in a water channel with and without obstacle," International Conference on Mechanics and Energy ICME'2014, March 18-20, 2014, Monastir, TUNISIA.

[21]   I. Mabrouki, Z. Driss and MS. Abid, "Hydrodynamic test bench design for the study of the water turbines," International Symposium on Computational and Experimental Investigations on Fluid Dynamics CEFD'2013, March 18-20, 2013, Sfax, TUNISIA.

[22]   I. Mabrouki, Z. Driss and MS. Abid, "Experimental investigation of the height effect of water Savonius rotors," International journal of mechanics and application, 2014: 4: 8-12.

20

Sulfonic Acid Group Functionalized Ionic Liquid Catalyzed Hydrolysis of Cellulose in Water: Structure Activity Relationships

Ananda S. Amarasekara[*], **Bernard Wiredu**

Department of Chemistry, Prairie View A&M University, Prairie View, Texas, USA
*Corresponding author: asamarasekara@pvamu.edu

Abstract Catalytic activities of eight sulfonic acid group functionalized ionic liquids in water were compared for the hydrolysis of Sigmacell cellulose (DP ~ 450) in the 150-180°C temperature range by measuring total reducing sugar (TRS) and glucose produced. The catalytic activity of acidic ionic liquids with different cation types decreases in the order imidazolium > pyridinium > triethanol ammonium cation. Among the sulfonic acid group functionalized imidazolium ionic liquids, the catalysts which contain a single imidazolium ion and a flexible linker between sulfonic acid group and the imidazolium ionic liquid core structure are the most active catalysts.

Keywords: cellulose, hydrolysis, ionic liquid catalyst, structure-activity relationship

1. Introduction

Efficient hydrolysis of polysaccharides in lignocellulosic biomass to fermentable monosaccharides is a challenging step and the primary obstacle for the large scale production of cellulosic ethanol from abundant lignocellulosic biomass [1,2,3,4]. Enzyme technologies are currently being tested in more than a dozen of cellulosic ethanol pilot plants in the US and other parts of the world; yet these operations are confronting major challenges in bringing the production cost competitive with gasoline [5]. This is due to factors like high enzyme cost of current cellulase preparations, inability to recycle the enzyme, and energy costs associated with high pressure, high temperature pretreatment step [6,7,8]. The alternative method of saccharification using dilute aqueous sulfuric acid at high temperature and pressure is the older method used in the cellulosic ethanol plants in the 1940's. In fact acid saccharification was replaced by enzyme methods developed in the last two decades due to certain disadvantages of acid hydrolysis. Some of the main disadvantages include poor sugar yields, resulting in low ethanol yield, formation of fermentation inhibitors, and the high energy cost associated with operating at temperatures above 250°C at high pressures [9,10]. Even though the direct aqueous acid saccharification gives comparatively low sugar yields, several research groups have taken a renewed interest in recent times in reviewing this classical method due to its simplicity and lower cost compared to enzymatic saccharification, which nevertheless requires an energy intense pretreatment [11,10,12].

Ionic liquids are well known [13,14] for their ability to dissolve cellulose and our interest in the search for efficient catalytic methods for saccharification of cellulose has led us to develop -SO₃H group functionalized Brönsted acidic ionic liquids as solvents as well as catalysts for the degradation of cellulose and cellulosic biomass [15,16,17]. Later we found that these sulfuric acid derivatives with ionic liquid characteristics can be used as catalysts in aqueous phase as well [18]. For example, a dilute aq. solution of acidic ionic liquid 1-(1-propylsulfonic)-3-methylimidazolium chloride was shown to be a better catalyst than aq. sulfuric acid of the same H^+ ion concentration for the degradation of cellulose at moderate temperatures and pressures [18]. In the cellulose hydrolysis experiments using aq. solutions of 1-(1-propylsulfonic)-3-methylimidazolium chloride and sulfuric acid of the same acid strength, these catalysts have been shown to produce total reducing sugar (TRS) yields of 29.5 and 22.0 % respectively [18].

This enhanced catalytic activity of the 1-(1-propylsulfonic)-3-methylimidazolium chloride when compared to sulfuric acid can be explained as a result of an interaction or binding of the ionic liquid catalyst on the cellulose surface, which facilitates the approach of -SO₃H group for the hydrolysis of the glycosidic link. Furthermore, this observation can be seen as an important lead for the development of an ionic liquid based cellulase mimic type catalyst for depolymerization of cellulose. We hypothesized that this binding ability may depend on the nature of ionic liquid core structure of the catalyst. As far as we are aware the effects of structural variations of any type of ionic liquid core structures, anions, and the separation between -SO₃H group and cationic core in the

Brönsted acidic catalysts in the hydrolysis of cellulose in aqueous phase is not known. Therefore, in an attempt to develop a recyclable, simple enzyme mimic type acid catalyst and as an extension of our earlier work [15,16,17,18,19] on sulfonic acid substituted imidazolium ionic liquid catalysts, we have studied a series of sulfonic acid group functionalized Brönsted acidic ionic liquid catalysts shown in Figure 1 for the hydrolysis of cellulose in water at moderate temperatures and pressures.

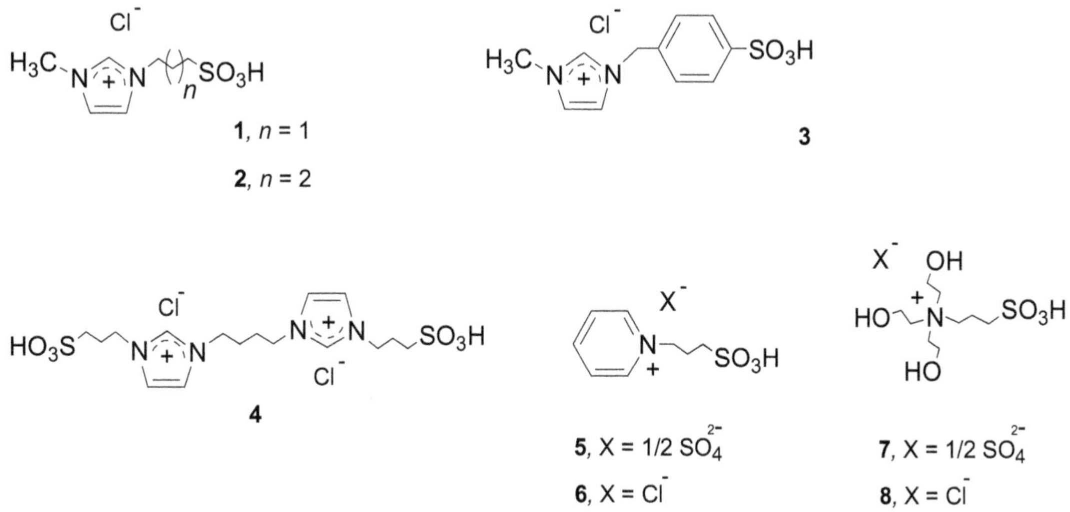

Figure 1. Sulfonic acid group functionalized Brönsted acidic ionic liquid catalysts (**1-8**)

2. Experimental

2.1. Materials and Instrumentation

Sigmacell cellulose - type 101 (DP ~ 450, from cotton linters), chemicals for the synthesis of sulfonic acid group functionalized ionic liquid catalysts (**1-8**) were purchased from Aldrich Chemical Co. USA. ^{1}H NMR Spectra were recorded in DMSO-d_6 on a Varian Mercury plus spectrometer operating at 400 MHz and chemical shifts are given in δ (ppm) downfield from TMS (δ = 0.00). ^{13}C NMR were recorded in the same spectrometer operating at 100 MHz; chemical shifts were measured relative to δ $(CD_3)_2SO$ and converted to δ (TMS) using δ $(CD_3)_2SO$ = 39.52. Cellulose hydrolysis experiments were carried out in 25 mL stainless steel solvothermal reaction kettles with Teflon inner sleeves, purchased from Lonsino Medical Products Co. Ltd., Jingsu, China. These reaction kettles were heated in a preheated Cole-Palmer WU-52402-91 microprocessor controlled convention oven with ±1°C accuracy. Total reducing sugars (TRS, total of glucose and glucose oligomers with reducing groups) and glucose concentrations in aqueous solutions were determined using a Carey 50 UV-Vis spectrophotometer and 1 cm quartz cells.

2.2. Synthesis of Sulfonic Acid Group Functionalized Ionic Liquid Catalysts (1-8)

Synthesis of Ionic Liquid Catalysts 1, 2, 3, 5, and 6

Catalysts **1**, **2**, **3**, **5**, and **6** were prepared following literature procedures [15,20]. The structures and purity of the samples were confirmed by ^{1}H, ^{13}C NMR and acid group determination by titration with standardized aqueous sodium hydroxide, using phenolphthalein as the indicator.

Synthesis of Ionic Liquid Catalyst 4

A mixture of 1,1'-(1,4-butanediyl)-*bis*-imidazole (3.80 g, 20.0 mmol) [21] and 1,3-propanesultone (4.89 g, 40.0 mmol) was heated in a closed round bottom flask at 110°C for 14 h in an oil bath to give imidazolium salt as a white solid mass. The product was cooled in an ice bath at 0°C, cold concentrated hydrochloric acid (4.06 g, 40.0 mmol) was slowly added, closed reaction flask was allowed to warm to room temperature in 2 h, and then heated at 90°C for 14 h in an oil bath. The resulting liquid product was washed with *t*-butyl methyl ether (2x10mL) and dried under vacuum overnight to give sulfonic acid group functionalized ionic liquid catalyst **4** as a colorless viscous oil, 9.64, 96% yield. Titration of a sample with standardized 0.05 M aqueous sodium hydroxide using phenolphthalein as the indicator showed that product is > 99% pure.

^{1}H NMR δ 1.76 (4H, m), 2.07 (4H, m), 2.47 (4H, m), 4.20 (4H, m), 4.27 (4H, t, J = 6.8Hz), 4.76 (2H, bs), 7.79 (4H, s), 9.26 (2H, s)

^{13}C NMR δ 26.3, 28.8, 47.9, 48.2, 60.4, 122.9, 136.8

Synthesis of Ionic Liquid Catalyst 7

A mixture of triethanolamine (2.98 g, 20.0 mmol) and 1,3-propanesultone (2.44 g, 20.0 mmol) was heated in a closed round bottom flask at 110°C for 14 h in an oil bath to give triethanolammonium salt as a colorless solid mass. The product was cooled in an ice bath to 0°C, cold concentrated sulfuric acid (1.00 g, 10.0 mmol) was slowly added, closed reaction flask was allowed to warm to room temperature in 2 h, and then heated at 90°C for 14 h in an oil bath. The resulting product was washed with *t*-butyl methyl ether (2x10mL) and dried under vacuum overnight to give catalyst **7** as a pale yellow viscous oil, 5.81 g, 91% yield. Titration of a sample with standardized 0.05 M aqueous sodium hydroxide using phenolphthalein as the indicator showed that product is > 99% pure.

^{1}H NMR δ 2.58 (2H, m), 3.27-3.35 (10H, m), 3.63 (6H, m,), 4.78(3H, bs), 7.75(1H, bs)

^{13}C NMR δ 22.1, 28.8, 46.9, 52.6, 60.4

Synthesis of Ionic Liquid Catalyst 8

A mixture of triethanolamine (2.98 g, 20.0 mmol) and 1,3-propanesultone (2.44 g, 20.0 mmol) was heated in a closed round bottom flask at 110°C for 14 h in an oil bath to give triethanolammonium salt as a colorless solid mass. The product was cooled in an ice bath to 0°C, cold concentrated hydrochloric acid (2.03 g, 20.0 mmol) was slowly added, closed reaction flask was allowed to warm to room temperature in 2 h, and then heated at 90°C for 14 h in an oil bath. The resulting liquid product was washed with *t*-butyl methyl ether (2x10mL) and dried under vacuum overnight to give catalyst **8** as a pale yellow viscous oil, 5.66 g, 92% yield. Titration of a sample with standardized 0.05 M aqueous sodium hydroxide using phenolphthalein as the indicator showed that product is > 99% pure.

^{1}H NMR δ 2.50 (2H, m), 3.27-3.35 (10H, m), 3.73 (6H, m,), 4.43(3H, bs), 9.00(1H, bs)

^{13}C NMR δ 22.9, 31.6, 45.7, 50.8, 60.9

2.3. General Experimental Procedure for Hydrolysis of Cellulose Samples in Aqueous Brönsted Acidic Ionic Liquid Catalyst Solutions

Stock solutions of the Brönsted acidic ionic liquid catalyst solutions were prepared by dissolving appropriate amounts of these acids in deionized water to give acid concentration of 0.0321 mol H^{+}/L in each solution. The accuracy of the concentration was checked by titration with standardized aq. NaOH solution using phenolphthalein as the indicator. Sigmacell cellulose-type 101 (DP ~ 450) (0.030 g, 0.185 mmol of glucose unit of cellulose) was suspended in 2.00 mL of aqueous acid solution in a 25 mL high pressure stainless steel reaction kettle with Teflon inner sleeve. The reaction kettle was firmly closed and heated in a thermostated oven maintained at the desired temperature for 3.0 h. Then reaction kettle was removed from the oven and immediately cooled under running cold water to quench the reaction. The contents were transferred into a centrifuge tube and diluted to 10.0 mL with deionized water, neutralized by drop wise addition of 0.5 M aq. NaOH, and centrifuged at 3500 rpm for 6 min. to precipitate the solids before total reducing sugar (TRS) determination using 3,4-dinitrosalicylic acid (DNS) method [22]. The glucose formed was measured using glucose oxidase/peroxidase enzymatic assay [23]. A series of experiments were carried out in duplicate in the 150-180°C temperature range to study the variations in TRS and glucose yields at different temperatures. The plots of changes in percent yields of total reducing sugar (TRS) and glucose produced at different temperatures are shown in Figure 2 and Figure 3 respectively.

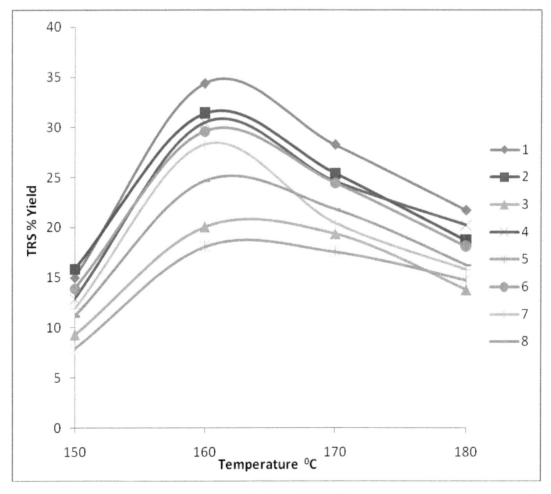

Figure 2. The changes in % yields of total reducing sugar (TRS) produced during the hydrolysis of Sigmacell cellulose (DP ~ 450) in aqueous solutions of sulfonic acid group functionalized Brönsted acidic ionic liquid catalysts 1-8 at different temperatures. All acid solutions are 0.0321 mol H^{+}/L, reaction time: 3.0 h, 0.030 g of Sigmacell cellulose in 2.00 mL of aq. acid was used in all experiments. Averages of duplicate experiments

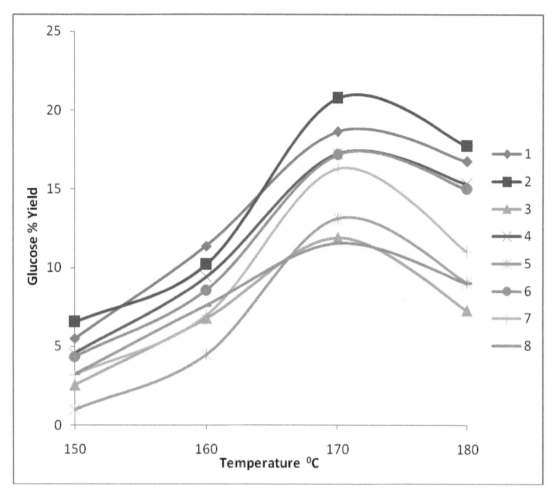

Figure 3. The changes in % yields of glucose produced during the hydrolysis of Sigmacell cellulose (DP ~ 450) in aqueous solutions of sulfonic acid group functionalized Brönsted acidic ionic liquid catalysts 1-8 at different temperatures. All acid solutions are 0.0321 mol H+/L, reaction time: 3.0 h, 0.030 g of Sigmacell cellulose in 2.00 mL of aq. acid was used in all experiments. Averages of duplicate experiments

2.4. Analysis of Hydrolyzate

TRS Assay

A 1.00 mL portion of the clear hydrolyzate solution from the centrifuge tube was transferred into a vial and 2.50 mL of deionized water was added. To this, was added 0.50 mL of DNS reagent [22] and the mixture was incubated in a water bath maintained at $90°C$ for 5 min. The reagent blank sample was prepared with 3.50 mL of deionized water and 0.50 mL of DNS reagent and heated similar to the samples. Then the absorbance was measured at 540 nm, against the reagent blank, and TRS concentrations in solutions were calculated by employing a standard curve prepared using glucose.

Glucose Assay

A 0.20 mL portion of the clear hydrolyzate solution from the centrifuge tube was transferred into a vial, and diluted with 1.80 mL deionized water. At zero time, reaction was started by adding 2.00 ml of glucose oxidase-peroxidase assay reagent [23] to the vial and mixing thoroughly, and the vial was incubated in a water bath at 37°C for 30 min. Then reaction was quenched by adding 2.00 mL of 6 M HCl to give a pink solution. The reagent blank was prepared by mixing 2.00 mL of deionized water and 2.00 mL of assay reagent, and was treated similarly. Then the absorbance was immediately measured at 540 nm against the reagent blank and glucose concentration in

the solution was calculated by employing a standard curve prepared using glucose.

3. Results and Discussion

3.1. Synthesis and Characterization of Brönsted Acidic Ionic Liquid Catalysts (1-8)

Ionic liquid catalysts **1**, **2**, **3**, **5**, and **6** (Figure 1) were prepared following literature procedures [15,20]. The synthesized samples showed ^{1}H, ^{13}C NMR spectroscopy data in agreement with the published data of these compounds [15,20]. The synthesized samples were confirmed as > 99% pure by titration with standardized aqueous sodium hydroxide, using phenolphthalein as the indicator.

The dicationic ionic liquid catalyst **4** was prepared by condensation of the known 1,1'-(1,4-butanediyl)-*bis*-imidazole [21] with two equivalents of 1,3-propanesultone and then acidification of the resulting salt with an equivalent amount of concentrated hydrochloric acid. The product **4** showed two low field peaks in the ^{1}H NMR spectrum at δ 7.79 (4H, s) and 9.26 (2H, s) for imidazolium ring protons, characteristic of the imidazolium chloride moieties [21]. The ^{13}C NMR spectrum of **4** showed seven peaks indicating the symmetry of the molecule and the two low field peaks at δ

122.9, 136.8 could be assigned to C-4,5 and C-2 in the imidazolium rings [21]. The triethanolammonium ionic liquid catalysts **7** and **8** were prepared by condensation of triethanolamine and 1,3-propanesultone to give triethanolammonium salt and then acidification with equivalent amounts of conc. H_2SO_4 and HCl respectively. The ^{13}C NMR spectra of compounds **7** and **8** showed five peaks each and the purity of the samples were further confirmed by titration with standardized sodium hydroxide solution.

3.2. Comparison of the Activities of Brönsted Acidic Ionic Liquid Catalysts (1-8)

In this study, eight Brönsted acidic ionic liquid catalysts (**1-8**) shown in Figure 1 were compared for the hydrolysis of Sigmacell cellulose type 101 (DP ~ 450) samples. The acidic ionic liquid catalysts used are thermally stable in the temperature range used in this study [24]. The average TRS and glucose yields produced in a series of experiments conducted in eight acid media at 150-180°C temperature range are shown in Figure 2 and Figure 3 respectively. These results show that catalytic activities of sulfonic acid group functionalized acidic ionic liquids depends on the ionic liquid core structure and the anion in the catalyst. For all the Brönsted acidic ionic liquid catalysts studied the highest TRS yields are observed for samples heated at 160°C for 3h, whereas the highest glucose yields are seen for samples heated at 170°C for 3h. The catalysts 1-(1-propylsulfonic)-3-methylimidazolium chloride (**1**) and 1-(1-butylsulfonic)-3-methylimidazolium chloride (**2**), produced the highest acid TRS yields of 34.4 and 31.4% respectively. The same two catalysts **1** and **2** produced the highest glucose yields as well; 18.6 and 20.8 % respectively at 160°C. The lowest TRS yield is produced from pyridinium ionic liquid catalyst N-propylsulfonic pyridinium sulfate (**5**).

In the comparison of monocationic ionic liquid catalysts with Cl⁻ counter ion (**1**, **2**, **6**, and **8**), the activity of the Brönsted acidic ionic liquid catalysts generally decreases in the order of imidazolium > pyridinium > triethanolammonium. The dicationic imidazolium ionic liquid catalyst **4** showed the third highest TRS and glucose yields of 30.5 and 17.2% respectively. Interestingly two binding domains on the catalyst failed to enhance the catalytic activity, when compared to catalysts **1**, and **2**. Even though imidazolium cation containing ionic liquids **1** and **2** are the most active ones, another imidazolium ionic liquid catalyst **3** with arylsufonic acid function showed very poor catalytic activity. This catalyst (**3**) showed the highest TRS and glucose yields of 20.0 and 11.9 % respectively. This observation suggests the importance of a flexible linker to connect the -SO₃H group to the imidazolium cation for a favorable approach towards the glycosidic links in cellulose.

4. Conclusion

We have shown that catalytic activity of the sulfonic acid group functionalized acidic ionic liquids depends on the ionic liquid core structure and the anion in the catalyst. From the series of Brönsted acidic ionic liquid catalysts studied catalysts **1** and **2** produced the highest TRS yields

of 34.4 and 31.4% respectively at 160°C, and the same two catalysts produced the highest glucose yields of 18.6 and 20.8 % respectively at 170°C. The results indicate that structure - activity studies can be used as a tool in designing of an efficient small molecule catalyst system that can hydrolyze cellulose in the aqueous phase at moderate temperatures and pressure.

Acknowledgments

We thank American Chemical Society-PRF grant UR1-49436, NSF grants CBET-0929970, HRD-1036593, and USDA grant CBG-2010-38821-21569 for financial support.

Statement of Competing Interests

The authors have no competing interests (financial or others) that may have influenced this study or the conclusions drawn from this study.

References

[1]   Geddes, C.C., Nieves, I.U., and Ingram, L.O., "Advances in ethanol production". *Current Opinion Biotechnol.*, 22(3), 312-319, 2011.

[2]   Huang, R., Su, R., Qi, W., and He, Z., "Bioconversion of Lignocellulose into Bioethanol: Process Intensification and Mechanism Research". *Bioenerg. Res.*, 1, 1-21, 2011.

[3]   Zhu, J.Y. and Pan, X.J., "Woody biomass pretreatment for cellulosic ethanol production: Technology and energy consumption evaluation". *Bioresource Technol.*, 101(13), 4992-5002, 2010.

[4]   Brethauer, S. and Wyman, C.E., "Review: Continuous hydrolysis and fermentation for cellulosic ethanol production". *Bioresource Technol.*, 101(13), 4862-4874, 2010.

[5]   Zhang, P.F., Zhang, Q., Pei, Z.J., and Wang, D.H., "Cost estimates of cellulosic ethanol production: A review". *J. Manufact. Sci. Eng. Transact. ASME*, 135(2), article: 12005, 2013.

[6]   Sukumaran, R.K., Singhania, R.R., Mathew, G.M., and Pandey, A., "Cellulase production using biomass feed stock and its application in lignocellulose saccharification for bio-ethanol production". *Renewable Energ.*, 34(2), 421-424, 2009.

[7]   Alvira, P., Tomás-Pejó, E., Ballesteros, M., and Negro, M.J., "Pretreatment technologies for an efficient bioethanol production process based on enzymatic hydrolysis: A review". *Bioresource Technol.*, 101(13), 4851-4861, 2010.

[8]   Zhu, J.Y., Pan, X., and Zalesny Jr, R.S., "Pretreatment of woody biomass for biofuel production: Energy efficiency, technologies, and recalcitrance". *Appl. Microbiol. Biotechnol.*, 87(3), 847-857, 2010.

[9]   Hu, F. and Ragauskas, A., "Pretreatment and Lignocellulosic Chemistry". *Bioenerg. Res.*, 5(4), 1043-1066, 2012.

[10] Lenihan, P., Orozco, A., O'Neill, E. Ahmad, M.N.M., Rooney, D.W., and Walker, G.M., "Dilute acid hydrolysis of lignocellulosic biomass". *Chem. Eng. J.*, 156(2), 395-403, 2010.

[11] Gurgel, L.V.A., Marabezi, K., Zanbom, M.D., and Curvelo, A.A.D.S., "Dilute acid hydrolysis of sugar cane bagasse at high temperatures: A Kinetic study of cellulose saccharification and glucose decomposition. Part I: Sulfuric acid as the catalyst". *Ind. Eng. Chem. Res.*, 51(3), 1173-1185, 2012.

[12] Taherzadeh, M.J., and Karimi, K., "Acid-based hydrolysis processes for ethanol from lignocellulosic materials: A review". *BioResources*, 2(3), 472-499, 2007.

[13] Wang, H., Gurau, G., and Rogers, R.D., "Ionic liquid processing of cellulose". *Chem. Soc. Rev.*, 41(4), 1519-1537, 2012.

[14] Mäki-Arvela, P., Anugwom I., Virtanen, P., Sjöholm, R., and Mikkola J.P., "Dissolution of lignocellulosic materials and its constituents using ionic liquids-A review". *Indust. Crop. Product.*, 32(3), 175-201, 2010.

[15] Amarasekara, A.S., and Owereh, O.S., "Hydrolysis and decomposition of cellulose in brönsted acidic ionic liquids under mild conditions". *Ind. Eng. Chem. Res.*, 48(22), 10152-10155, 2009.

[16] Amarasekara, A.S., and Owereh, O.S., "Synthesis of a sulfonic acid functionalized acidic ionic liquid modified silica catalyst and applications in the hydrolysis of cellulose". *Catal. Commun.*, 11(13), 1072-1075, 2010.

[17] Amarasekara, A.S., and Shanbhag, P., "Degradation of Untreated Switchgrass Biomass into Reducing Sugars in 1-(Alkylsulfonic)-3-Methylimidazolium Brönsted Acidic Ionic Liquid Medium Under Mild Conditions". *Bioenerg. Res.*, 6(2), 719-724, 2013.

[18] Amarasekara, A.S., and Wiredu, B., "Degradation of cellulose in dilute aqueous solutions of acidic ionic liquid 1-(1-propylsulfonic)-3-methylimidazolium chloride, and p-toluenesulfonic acid at moderate temperatures and pressures". *Ind. Eng. Chem. Res.*, 50(21), 12276-12280, 2011.

[19] Amarasekara, A.S., and Wiredu B., "Brönsted Acidic Ionic Liquid 1-(1-Propylsulfonic)-3-methylimidazolium-Chloride Catalyzed Hydrolysis of D-Cellobiose in Aqueous Medium". *Int. J. Carbohyd. Chem.*, 2012, 6-9, 2012.

[20] Amarasekara, A.S., and Wiredu, B., "Single reactor conversion of corn stover biomass to C5–C20 furanic biocrude oil using sulfonic acid functionalized Brönsted acidic ionic liquid catalysts". *Biomass Conversion Biorefin.*,1, 1-7, 2013.

[21] Amarasekara, A.S., and Shanbhag, P., "Synthesis and characterization of polymeric ionic liquid poly(imidazolium chloride-1,3-diylbutane-1,4-diyl)". *Polym. Bull.*, 67(4), 623-629, 2011.

[22] Breuil, C., and Saddler, J.N., "Comparison of the 3,5-dinitrosalicylic acid and Nelson-Somogyi methods of assaying for reducing sugars and determining cellulase activity". *Enzym. Microb. Technol.*, 7(7), 327-332, 1985.

[23] Bergmeyer, H.U., Bernt, E., ed. Methods of Enzymatic Analysis. ed. H.U. Bergmeyer, Academic Press: NewYork. pp 1205-1212, 1974.

[24] Amarasekara, A.S., and Owereh, O.S., "Thermal properties of sulfonic acid group functionalized Brönsted acidic ionic liquids". *J. Therm. Anal. Calorim.*, 103(3), 1027-1030, 2011.

21

Wake Interaction of NREL Wind Turbines Using a Lattice Boltzmann Method

Jun Xu[*]

Department of Engineering Technology, Tarleton State University, Stephenville, USA
*Corresponding author: junxu@tarleton.edu

Abstract Wind turbines installed in arrays in a wind farm tend to experience reduced power production and increased fatigue load on the blades which can prematurely wear down turbine hardware. In this paper, the aerodynamic characteristics of a single wind turbine was studied numerically first. Then the aerodynamic impacts of three in-line wind turbines were investigated. Turbine wake simulations were performed on NREL unsteady aerodynamics experiment phase VI two-bladed wind turbines. The Lattice Boltzmann Method (LBM) was used to model the wakes behind turbines. The ability of LBM to capture wake evolution and detailed flow characteristics were explored for a single and three in-line turbines. The model results provide an insight on the turbine wake interactions and demonstrate that the LBM can simulate the complexity of the wake interactions efficiently.

Keywords: *CFD, NREL Phase VI, wind turbine, wake interaction, Lattice Boltzmann Method*

1. Introduction

With increasing interest in energy independence and sustainability, there is an urgent need to assess the performance of current and future power generation technologies. As one of the promising energy sources, wind energy is becoming more important due to its availability and relatively low impact on the surrounding environment. The U.S. is currently on a path to generate 20% of its electricity from wind by 2030, which is a 10-fold increase compared to the current value of roughly 2% (DOE report 2008). In recent years, the cost of wind energy has also declined steadily as a result of dramatic improvements to turbines, blades, and gearboxes, and increases in the rotor diameter and the height of turbine towers. With increased height and size, wind turbines installed in a wind farm tend to experience increased mechanical stresses and fatigue loads on the blades, gearbox, and tower, due to increased wind speed and wake interferences among turbines [1]. Structural failure of wind turbine blades, which leads to high maintenance costs and intermittent operation, is partly attributed to unsteady aerodynamic forces that originate from the vorticity contained in shed wake as shown in Figure 1 (right). The optimal placement of wind turbines in a wind farm is influenced by these interactions. Similarly, wind shear, time variations of the rotor, and the tower's effect on the rotor of an upwind turbine significantly influence the turbine wake. All these flow complications result in overall increases in the turbulence fluxes. An in-depth understanding of the wake interference of wind turbines is necessary in order to improve the design of the blades so as to reduce their impact on the surrounding environment,

and to provide more accurate parameterizations of turbulent fluxes in various weather conditions. Challenges for the alleviation of the undesirable effects fall in the category of flow control/modification. Three distinctly identifiable problems related to the wake interference include: 1) suppression of vortex shedding to minimize vibrations and reduce fatigue loads, 2) modification of wake for drag reduction, and 3) reduction of noise generated by multiple scales of turbulence in the separated shear layers. In resolving these issues, Computational Fluid Dynamics (CFD) has proven to be an indispensable tool to shed light on the performance of wind energy systems both at the individual turbine level and wind farm level [2,3].

In general, a fluid can be described at the molecular level where fluid molecules collide with each other (Molecular Dynamics simulation). This scale is microscale. On the other hand, a fluid can also be studied at the macroscopic level using the continuum approximation where conservation of energy, mass, and momentum can be achieved [5]. The gap between the microscale and macroscale can be bridged by Lattice Boltzmann Method (LBM), which is in mesoscale.

At the macroscale level, the traditional mesh-based CFD approaches are based on the numerical solution of Navier-Stokes (N-S) equations. In order to solve N-S equations numerically, this continuum domain must be meshed. This discretization procedure requires that the flow domain of interest be meshed into small control volumes. This meshing process becomes increasingly costly and tricky when the domain is a complex shape, e.g. the wake flow field generated by wind turbines. Typically, a large number of mesh points needs to be used to be able to accurately capture the detailed flow physics near the turbine blades.

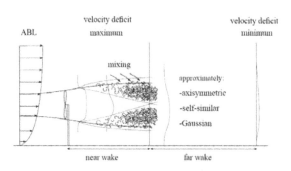

Figure 1. Aerodynamics of the wake of a wind turbine (left) and vortical wake visualization behind a horizontal axis wind turbine (right) in the wind tunnel at NASA Ames Research Center [4]

There exists various mesh-based CFD approaches for the computation of turbulent flows [3,6]. For example, the Direct Numerical Simulation (DNS) approach provides all the details of turbulence as it is capable of resolving the smallest scales. However, it requires extensive computational resources, thereby making it impractical at least presently for solving realistic problems such as wind turbine aerodynamics [7]. The more often used time-filtered Reynolds Averaged Navier-Stokes (RANS) modeling requires validated turbulence models. The advantage of RANS is that a fully resolved computation can be carried out with a few million mesh points, which makes it possible to reach a full 3D solution [2]. There have been some attempts to simulate wind turbine aerodynamics by the RANS models such as k-ω SST model [8,9,10,11]. However, these models encounter difficulties in predicting flows in massively separated regions with a reasonable accuracy. In addition, due to the dissipative nature of RANS turbulence models, the tip vortices in the wake of the blades dissolve right after the flow separation. This dissipative nature of RANS models was also observed by Hedges et al. [12] for a simplified landing gear. In addition, for stall controlled wind turbines, all RANS closure models show lack of accuracy to simulate the stalled flow regime at high wind speeds of about 10m/s or above. The third approach to deal with turbulence is the space-filtered Large Eddy Simulation (LES) where large eddies (scales) of turbulence are numerically computed, while the small scales are modeled in terms of large scales through subgrid-scale modeling [13]. LES is a good compromise between the RANS and the DNS approach, because it is more versatile than RANS and less costly than DNS. In fact, LES has shown some promising results, and efforts are continuing to explore various prospects of wind turbine applications [14,15,16,17,18,19]. For a typical wind turbine, a mesh consisting of around 2 million cells for a standard RANS rotor computation is sufficient. On the other hand, pure LES requires a computational grid of 300 million points to compute the initial transient of the development of a tip vortex [20]. In addition, the requirement of time accurate calculations makes LES significantly more expensive than steady-state computations. Typically, numerical simulation of a real wind farm requires development of a model comprised of multiple wind turbines. Hence, it can be deduced that the amount of computing time to carry out LES of a typical wind farm facility would be exceptionally high and computationally intensive even with the available state-of-the-art computational resources today.

As an alternative approach to the traditional mesh-based CFD methods as mentioned above, LBM simulates flow problems using a fundamentally different approach at the kinetic level where the property of the collection of particles is represented by a distribution functions [21,22]. LBM has many attractive features. First, by using kinetic theory, the underlying physics is simpler since it is restricted to capturing the kinetic behavior of particles or collections of particles as opposed to attempting to solve non-linear PDEs, which is very difficult. It is also easy to apply for complex domains, easy to treat multi-phase and multi-species flows without a need to trace the interfaces between different phases.

Wind farms have gradually increased in size over the years. Spacing between the turbines becomes an important issue as power losses due to wake blocking effects from wind turbine wake interference which can be dramatic in tightly spaced wind farms (spacing is about 4 – 8D, D is the rotor diameter). On the other hand, spacing beyond 8-10D is considered expensive due to the high cost of installing cables to the wind farm [23,24]. Hence, there are substantial benefits to be gained from accurate modeling of wind turbine wake interferences in wind farm design to minimize power losses, reduce fatigue loads, and minimize operational and maintenance costs. Although many strides have been made in the understanding and modelling of wind turbine aerodynamics, current mesh-based CFD models suffer from a number of limitations in terms of the capability of handling complex geometries, computation efficiency, and accuracy. The current work is an attempt to explore an alternate framework to overcome/alleviate these limitations using a mesh-free LBM approach. Therefore, it can provide a valuable design/analysis tool to the wind energy community to optimize wind farm layouts and render the power outputs more predictable for planning and financing purposes.

As a first step, a three-turbine system was investigated in this paper and the insights gained from this simple system can be extended to study large wind farms and their interaction with Atmospheric Boundary Layer (ABL) using Weather Research and Forecasting data in the future. In this paper, the problem to be considered is the aerodynamic interference between three in-line wind turbines and the wake blocking effect of the upwind turbine on the downwind turbines. The purpose of this study is to evaluate the feasibility of LBM in simulating the wind aerodynamics. Because the interference between the wakes of turbines in a farm reduces the total power production (i.e. wake blocking effect), compared with an equal number of standalone turbines, we investigated the wake blocking effect of an array of three in-line wind turbines. This interference also increases turbulence intensity because of mixing from surrounding wakes, resulting in a marked increase in dynamic loadings of turbine blades. Therefore, it is important to investigate the effect of this dynamic loading on the blades. We investigated the effects of wind shear, and transient motion of rotor and tower on the wake produced by one wind turbine; and we also investigated the dynamic interferences (wake blocking) between the wakes of three in-line turbines. The work presented in this paper aims to understand the fundamental nature of wake interactions of multiple turbines.

Recently, mesh-free methods such as LBM have been applied to a flow field around a wind turbine. It should be noted that the technique is still in early stages as far as application to wind turbine simulations is concerned. For example, researchers have started to use mesh-free LBM to simulate wind turbine aerodynamics using Power FLOW [25]. The authors simulated and compared the 3D flow field around a two-bladed wind turbine against the test data in terms of the normal force coefficient, the mean torque, and the pressure coefficient by the National Renewable Energy Laboratory (NREL Phase VI turbine). However, wake interference between turbines and aero-elastic coupling between wind field and turbine blades were not further investigated.

2. Computational Model and Methods

The CFD modeling effort evaluates and identifies efficient computational strategies for the study of the various aspects associated with wind turbine aerodynamics. These include individual blade characteristics, coupled blade-tower flow fields, and multiple wind turbines. In this paper, the overall approach is to apply LBM to resolve the complex aerodynamics of wind turbines. Though relatively new, the LBM is a promising method that has been intensively studied in recent years due to its natural affiliation to parallel computing and being mesh free [26]. It is important to point out that typically meshing and establishing the topology of the computational domain is the largest load on the computational resources.

The time-dependent Boltzmann's transport equation is

$$\frac{\partial f}{\partial t} + v \cdot \nabla f = \Omega \qquad (1)$$

Where Ω is the collision operator. The set of particles that has a given discrete velocity at a given spatial location is called the particle distribution at that grid point. The particle distribution is the main computational element of this method. The probability distribution function is

$$f = f(x, v, t). \qquad (2)$$

All fluid properties and their evolution can be derived from probability distribution [22]. In fact, macroscopic variables are statistical moments of the particle distribution function . Density can be obtained by

$$\rho(x, t) = \int f(x, v, t) \delta v. \qquad (3)$$

Linear momentum can be obtained by

$$\rho(x, t) u(x, t) = \int f(x, v, t) v \delta v. \qquad (4)$$

It can be shown that above LBM approaches the following macroscopic Navier-Stokes equations if the density variation is small enough

$$\frac{\partial \rho}{\partial t} + \nabla \cdot \rho u = 0 \qquad (5)$$

$$\frac{\partial u}{\partial t} + (u \nabla) u = -\nabla P + \nu \nabla^2 u. \qquad (6)$$

The above N-S equations describe the flow of incompressible fluids. Where u is the flow velocity, P is the pressure, and ν is the kinematic viscosity.

The wind turbine model used in the current study is the Phase VI wind turbine unsteady state experimental setup by the National Renewable Energy Laboratory (NREL) as shown in Figure 2 [27]. This experiment tested a two-bladed, twisted and tapered, 10-m diameter, stall-regulated wind turbine operating at 72 rpm in the NASA Ames Research Center 80 ft x 120 ft (24.4m x 36.3m) Wind Tunnel and provided a definitive set of turbine air-loads and performance measurements over an extensive matrix of well-controlled operating conditions. Figure 2 (right) shows the reconstructed model turbine used in the current study.

Two cases were simulated. The first case deals with a single turbine, tested during the phase VI of the NREL unsteady aerodynamics experiment. The same wind tunnel dimensions were reconstructed in the simulation. Figure 3 shows the rotor blade that has a radius of 5.029 m. The blade is made of S809 airfoils, which is designed specifically for the use in horizontal axis turbines. The experimental conditions and parameters are described in [28] and need not be repeated in detail here.

Figure 2. Unsteady aerodynamics experimental turbine in the NASA Ames 80 ft by 120 ft wind tunnel (left) and Model turbine used in current study (right)

Figure 3. Three-dimensional solid model of NREL turbine blade

The second case focuses on the wake interaction of three identical NREL turbines which are in-line and separated by 7D as shown in Figure 4. D is the turbine rotor diameter. In all the simulations, the inlet velocity of the virtual wind tunel was set at 5m/s at the hub height, pitch angle 3 degree, and rotational speed is 72 rpm. All the simulations were implemented in a LBM solver FLOW. To fully resolve the wake structures, the number of lattice elements was dynamically adapted to the wakes as shown in Figure 5. The smallest length scale resolved is set to 0.1m in all the simulations.

Figure 4. Computational layout for three in-line turbines

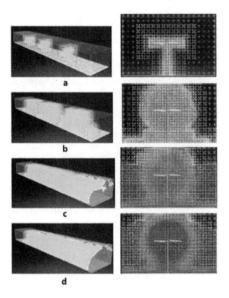

Figure 5. (a) Number of lattice elements 309965 at t=0 s. (b) Number of lattice elements 14826509 at t=5 s. (c) Number of lattice elements 30321611 at t=10 s. (d) Number of lattice elements 36199938 at t=15s

3. Results and Discussion

3.1. Results for Single Wind Turbine

In the first case, we simulated the wind turbine aerodynamics of a single turbine using a general LBM solver. The turbine system of our choice is the NREL Phase VI two-bladed turbine. This turbine system has been extensively examined both experimentally and numerically in the past [19,27,29,30,31,32,33,33]. We feel that this system is ideal for modeling purposes since a rich set of experimental data is available which provides a sufficient matrix of data points to evaluate and constrain our computational framework.

First, we investigated the aerodynamics of a single turbine. The vortical structures are shown in Figure 6 with varying time. The inlet velocity of the virtual wind tunel was set at 5m/s at the hub height, pitch angle 3 degree, and rotational speed is 72 rpm. Clearly, the LBM was able to

capture the time evolution of the vortical structure, which closely resembles the experimental observation shown in Figure 1. Figure 7 illustrates blade surface velocity, pressure, vorticity, and turbulence intensity at t=10s. In addition, the impact of the tower is clearly shown as the wake is shedding off the tower and interacts with the wake from the blades.

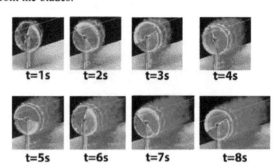

Figure 6. Single Turbine Results showing vortical structure around the turbine at varying times t=1, 2, 3, 4, 5, 6, 7, 8 s, respectively. Inlet wind velocity was set to 5m/s at hub height. The NREL Phase VI two-bladed turbine was used in the simulation

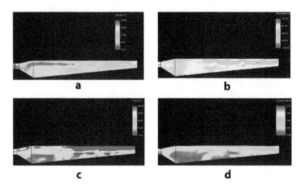

Figure 7. surface contours at 10s. (a) velocity. (b) pressure. (c) vorticity. (d) turbulence intensity

3.2. Results for Three in-line Wind Turbines

Next, we simulated three in-line wind turbines to investigate the wake interaction amondg turbines. The first turbine is 5D from inlet, where D is the turbine rotor diameter. The spacing between turbines is 7D. Figure 8 shows vortical structure around the turbine at varying times t=1, 3, 5, 15 s, respectively. Inlet wind velocity was set to 5m/s at hub height. The NREL Phase VI two-bladed turbine was used in the simulation. The yaw angle is 0 degree and pitch angle is 3 degree for each turbine. The free stream at the inlet was modeled with a sheared velocity to make the results more relevant to realistic configurations. In Figure 8, we can see the structure of the wake behind each turbine and how the downstream turbine's wakes are interfered by the upstream turbine, a hallmark of wind blocking effect. Figure 9 shows the variation of the wake profile with increasing time in vertical cutting planes. We can see the structure of the wake, including the acceleration at the tip vortices and the detached flow behind the hub. It is interesting to note how the velocity field impinging on the second turbine and second on the third depends strongly on the position of the turbine from free stream inlet, for this close positioned turbines. It is evident that downwind turbines are markedly influenced by the upwind turbines.

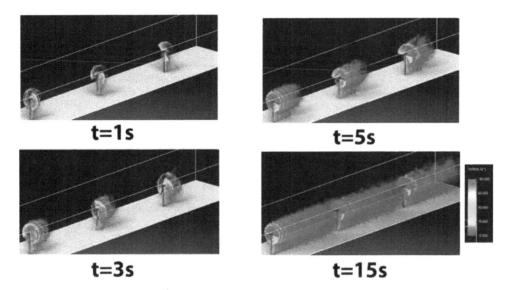

Figure 8. Wake Interaction among three inline turbines. Turbine Results showing vortical structure around the turbine at varying times t=1, 3, 5, 15 s, respectively. Inlet wind velocity was set to 5m/s at hub height. The NREL Phase VI two-bladed turbine was used in the simulation

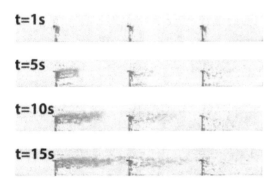

Figure 9. Three in-line turbine results showing instaneous vortical structure around the turbine at varying times t=1, 5, 10, 15 s, respectively. Inlet wind velocity was set to 5m/s at hub height. The NREL Phase VI two-bladed turbines were used in the simulation. The vorticity magnitude contours are plotted with the use of identical scales

The instantaneous velocity, vorticity, preesure and turbulence intensity contours on the rotor blades are shown in Figure 10. The contours show considerable spanwise variations in addition to chordwise variations from upwind turbine to downwind turbines. Please note that, for illustration purpose, blades from three in-line turbines are superposed in Figure 10.

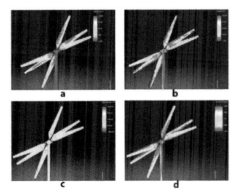

Figure 10. Surface coutours of three in-line turbines (superposed) at 14.84 s. (a) velocity. (b) vorticity. (c) pressure. and (d) turbulence intensity

4. Conclusion

The current work demonstrates the use of transient CFD simulation using LBM to capture the fidelity of turbine wake modeling. Two cases of LBM simulations of the NREL Phase VI wind turbines with 0° yaw angle, and 3° pitch angle were carried out in this study. The results demonstrate that LBM is able to capture the basic flow physics of turbine wakes, including trends in wake structure that are comparable with those seen in other works, including wake evolution and interaction. In addition, three in-line turbines were simulated, and the effect of the wake from upwind turbines was analyzed in terms of vorticity, velocity, and pressure profiles. This study presents a technique to implement a multiple turbine wake simulation with the ability to study wake interaction in wind farms in the future. Developing a high fidelity/efficiency CFD tool of wind turbine aerodynamics is a challenging task due to the multiphysical and multiscale nature of the problem as shown in previous studies. This paper demonstrates a simulation-based design tool to study wake interferences of wind turbines. It is expected that the use of a mesh-free approach to better resolve the wind turbine geometry and motions, coupled with rigorous verification and validation, will significantly enhance the understanding of wind turbine flow physics and improve the accuracy of predicting aero-elastic coupling between the flow field and the turbine blades. This simulation-based design tool then can be used by the wind turbine community to increase the efficiency of wind turbines through the development of optimal control strategies to minimize the blocking effects and maximize the power outputs.

Acknowledgement

The author wishes to gratefully acknowledge the financial support provided by the Tarleton State University Organized Research Grant (ORG).

References

[1] R. Gomez-Elvira, A. Crespo, E. Migoya, F. Manuel, J. Hernandez, Anisotropy of turbulence in wind turbine wakes, Journal of Wind Engineering and Industrial Aerodynamics 93 (2005) 797-814.

[2] J.N. Sorensen, Aerodynamic Aspects of Wind Energy Conversion, Annual Review of Fluid Mechanics, Vol 43 43 (2011) 427-448.

[3] A.C. Hansen, C.P. Butterfield, Aerodynamics of Horizontal-Axis Wind Turbines, Annual Review of Fluid Mechanics 25 (1993) 115-149.

[4] S. Schreck, The NREL Full-Scale Wind Tunnel Experiment, Wind Energy (2002) 77-84.

[5] U. Frisch, Lattice Gas Automata for the Navier-Stokes Equations - a New Approach to Hydrodynamics and Turbulence, Physica Scripta 40 (1989) 423-423.

[6] B. Sanderse, S.P. van der Pijl, B. Koren, Review of computational fluid dynamics for wind turbine wake aerodynamics, Wind Energy 14 (2011) 799-819.

[7] J. Peinke, Oberlack, M., Talamelli, A., Progress in Turbulence III, Editoin Edition, Springer, 2010.

[8] J. Johansen, H.A. Madsen, M. Gaunaa, C. Bak, N.N. Sorensen, Design of a Wind Turbine Rotor for Maximum Aerodynamic Efficiency, Wind Energy 12 (2009) 261-273.

[9] J.N. Sorensen, W.Z. Shen, Numerical modeling of wind turbine wakes, Journal of Fluids Engineering-Transactions of the Asme 124 (2002) 393-399.

[10] A. Bechmann, N.N. Sorensen, F. Zahle, CFD simulations of the MEXICO rotor, Wind Energy 14 (2011) 677-689.

[11] N.N. Sorensen, A. Bechmann, P.E. Rethore, F. Zahle, Near wake Reynolds-averaged Navier-Stokes predictions of the wake behind the MEXICO rotor in axial and yawed flow conditions, Wind Energy 17 (2014) 75-86.

[12] L.S. Hedges, A.K. Travin, P.R. Spalart, Detached-Eddy Simulations over a simplified landing gear, Journal of Fluids Engineering-Transactions of the Asme 124 (2002) 413-423.

[13] P. Sagaut, Large Eddy Simulation for Incompressible Flows, Editoin Edition, Springer, 2006.

[14] P. Chatelain, S. Backaert, G. Winckelmans, S. Kern, Large Eddy Simulation of Wind Turbine Wakes, Flow Turbulence and Combustion 91 (2013) 587-605.

[15] H. Lu, F. Porte-Agel, Large-eddy simulation of a very large wind farm in a stable atmospheric boundary layer, Physics of Fluids 23 (2011).

[16] M. Calaf, C. Meneveau, J. Meyers, Large eddy simulation study of fully developed wind-turbine array boundary layers, Physics of Fluids 22 (2010).

[17] J. Meyers, C. Meneveau, Optimal turbine spacing in fully developed wind farm boundary layers, Wind Energy 15 (2012) 305-317.

[18] M.J. Churchfield, S. Lee, J. Michalakes, P.J. Moriarty, A numerical study of the effects of atmospheric and wake turbulence on wind turbine dynamics, Journal of Turbulence 13 (2012) 1-32.

[19] J.O. Mo, A. Choudhry, M. Arjomandi, Y.H. Lee, Large eddy simulation of the wind turbine wake characteristics in the numerical wind tunnel model, Journal of Wind Engineering and Industrial Aerodynamics 112 (2013) 11-24.

[20] O. Fleig, M. Lida, C. Arakawa, Wind turbine blade tip flow and noise prediction by large-eddy simulation, Journal of Solar Energy Engineering-Transactions of the Asme 126 (2004) 1017-1024.

[21] S. Succi, The Lattice Boltzmann Equation for Fluid Dynamics and Beyond, Editoin Edition, Oxford University Press, 2001.

[22] S. Chen, G.D. Doolen, Lattice Boltzmann method for fluid flows, Annual Review of Fluid Mechanics 30 (1998) 329-364.

[23] A.R. Henderson, C. Morgan, B. Smith, H.C. Sorensen, R.J. Barthelmie, B. Boesmans, Offshore wind energy in Europe - A review of the state-of-the-art, Wind Energy 6 (2003) 35-52.

[24] R.J. Barthelmie, L. Folkerts, G.C. Larsen, K. Rados, S.C. Pryor, S.T. Frandsen, B. Lange, G. Schepers, Comparison of wake model simulations with offshore wind turbine wake profiles measured by sodar, Journal of Atmospheric and Oceanic Technology 23 (2006) 888-901.

[25] M.-S.K. Franck Perot, and Mohammed Meskine, NREL wind turbine aerodynamics validation and noise predictions using a Lattice Boltzmann Method, 18th AIAA/CEAS Aeroacoustics Conference (33rd AIAA Aeroacoustics Conference), 2012.

[26] D.Z. Yu, R.W. Mei, W. Shyy, A multi-block lattice Boltzmann method for viscous fluid flows, International Journal for Numerical Methods in Fluids 39 (2002) 99-120.

[27] M.M. Hand, Simms, D.A., Fingersh, L.J., Jager, D.W., Cotrell, J.R., Schreck, S., and Larwood, S.M., Unsteady Aerodynamics Experiment Phase VI: Wind Tunnel Test Configurations and Available Data Campaigns, NREL, 2001.

[28] J. Jonkman, S. Butterfield, W. Musial, G. Scott, Definition of a 5-MW reference wind turbine for offshore system development, National Renewable Energy Laboratory, 2009, pp. 1-14.

[29] D.O. Yu, J.Y. You, O.J. Kwon, Numerical investigation of unsteady aerodynamics of a Horizontal-axis wind turbine under yawed flow conditions, Wind Energy 16 (2013) 711-727.

[30] R. Lanzafame, S. Mauro, M. Messina, Wind turbine CFD modeling using a correlation-based transitional model, Renewable Energy 52 (2013) 31-39.

[31] Y. Li, K.-J. Paik, T. Xing, P.M. Carrica, Dynamic overset CFD simulations of wind turbine aerodynamics, Renewable Energy 37 (2012) 285-298.

[32] N. Sezer-Uzol, O. Uzol, Effect of steady and transient wind shear on the wake structure and performance of a horizontal axis wind turbine rotor, Wind Energy 16 (2013) 1-17.

[33] N.N. Sorensen, S. Schreck, Computation of the National Renewable Energy Laboratory Phase-VI rotor in pitch motion during standstill, Wind Energy 15 (2012) 425-442.

22

Designing of a Standalone Photovoltaic System for a Residential Building in Gurgaon, India

Abhik Milan Pal[1,*], Subhra Das[1], N.B.Raju[2]

[1]Renewable Energy Department, Amity School of Applied Science, Amity University, Gurgaon, India
[2]National Institute of Solar Energy, Gurgaon, India
*Corresponding author: mailtoabhik@yahoo.in

Abstract Photovoltaic power system, through direct conversion of solar irradiance into electricity, can be used as electrical power source for home to meet its daily energy requirement. In this paper detailed design of a standalone photovoltaic power system for uninterrupted power supply of a residential building in a typical urban area is presented. The process of acquiring photovoltaic power involves designing, selecting and determining specifications of different components that are used in the system conforming the load estimation. Accomplishment of this process depends on a variety of factors, such as geographical location, weather condition, solar irradiance, and load consumption. This paper outlines in detail the procedure for specifying each component of the standalone photovoltaic power system and as a case study, a residence in Gurgaon, India with typical energy consumption is selected. Detailed cost analysis including installation and maintenance of a solar PV system during its life span have been carried out also. The analysis shows though the initial investment is high, still, within few years it not only returns this amount but also gain substantial dividend during the system life span.

Keywords: photovoltaic array, inverter, charge controller, battery, module orientation, payback period

1. Introduction

Energy plays a fundamental role in our daily activities. The degree of development and civilization of a country is determined by the amount of energy utilized by its human beings. Energy demand is increasing day by day due to increase in population, urbanization and industrialization. The world's fossil fuel supply viz. coal, petroleum and natural gas, main source of energy until now, will thus be exhausted in a few hundred years [1]. On one hand, the rate of energy consumption increasing, on the other hand, fuel supply depleting - it will lead to energy crisis one day. It will also results in inflation, poverty and global warming [2]. Hence alternative or clean renewable energy sources have to be explored and developed to meet future energy requirement. Solar energy, wind energy, etc. are clean, inexhaustible and environment- friendly resources among the renewable energy options.

Utilizing solar energy we can fulfil our daily energy needs during sunshine hours. But neither solar nor wind energy system can provide a continuous supply of energy demand throughout a day, like in the night time or in other conditions when sunshine or wind power is not there or not enough to fulfil the demand. At that off-time extra energy storage devices are required to meet the demand. In this context, standalone solar power systems are now being contemplated.

Different places on the globe experience different climatic conditions. Total solar irradiance that reaches the surface of earth varies with time of day, season, location and weather conditions. Therefore, design of a standalone solar system cannot have only one standard. Location is a major aspect that will affect photovoltaic power system design and it varies from place to place [3]. India is blessed with enough sun shine which can meet our energy demand without any compromise and it is also pollution free. Standalone PV system is a popular concept in rural areas of India where national electricity grid connection facility is not available. But in urban areas where grid connection system is easily available, it is not a common practice to use solar power. There is a general impression that grid energy from conventional sources is much less costly compared to solar and other alternate energy sources.

One of the objective of this paper is to estimate the potential of solar photovoltaic power system in urban areas taking for example, Gurgaon area inthe state of Haryana in India. For this purpose, a typical residential building in Haryana is taken up for designing and developing a system based on its daily load requirement. Equipment specifications are provided based on availability of the best components in market. In addition to the design considerations, we have done a detailed cost analysis of the system in this paper. As expected initial cost of solar power plant installation has been found to be very high and so, the cost of solar energy consumption unit is much more than conventional energy unit. And this is quite discouraging for general public to go for solar power plant. However, more interestingly, our estimation of the long term cost and area requirement of a standalone SPV power plant installation establishes our other objective that, contrary to general perception, it is a very much economical and cost effective system.

Before presenting the results and analysis of the case under study, i.e. of a typical residential building in Haryana in Section 4, we introduce different components of a standalone PV system and their functions in briefin the following section. In Section 3 the steps that are followed for designing the PV system and the method for determining the design parameters are described. Finally, in Section 5 we present the concluding remarks.

2. Standalone Photovoltaic System

Standalone photovoltaic system is a collection of interconnected electrical components, using which we can generate electricity from sun light and satisfy our daily energy requirement without worrying about any interval when the sunlight may not be available [3]. This type of system is useful only when there is requirement of load to run in night time or in other time when sunlight is unavailable for some period. The components of such a system are: 1) Solar PV array, 2) Charge Controller, 3) Inverter, 4) Battery, 5) Cables and 6) Protection devices. Depending on load requirement and radiation intensity at the location, the components of the system will have to be specified. Figure 1 gives a schematic diagram of interconnection of components of a typical stand-alone photovoltaic power system.

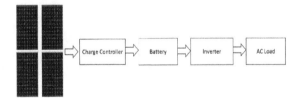

Figure 1. Standalone PV system components

In the following subsections we give a brief review of the functions of the components [5].

2.1. Solar Photovoltaic Panel

Photovoltaic cell or solar cell, which generates electricity from the sun light, is the main and primary component in a PV system. Current and voltage generated depend on the area of the cell. A 13.5"x13.5" size solar cell can generate voltage of about 0.55volt and a current density of 30–35mA/cm^2 [5]. A solar panel is made of a collection of these basic solar cells. To meet voltage and current requirements of a particular system, a number of panels are connected in series (to increase voltage) and in parallel (to increase current) combinations forming a solar PV array.

2.2. Storage Battery

Storage battery is the vital component a standalone PV system. Its function is to store energy during sunshine hours and supply current to load during non-sunshine hours. Lead Acid battery, VRLA battery, Lithium-ion battery, etc. are different types of batteries that can be incorporated in solar PV system.

2.3. Charge controller

To regulate and monitor current flow between PV array and battery, a device, called charge controller, is used. The main function of solar charge controller is to limit the flow at which electric current is added to or drawn from batteries. It prevents overcharging and protect battery from voltage fluctuation [6]. Two types of charge controllers are available: solar charge controller with PWM based technology and solar charge controller with MPPT based technology. In this paper MPPT design based charge controller is employed.

2.4. Inverter

Inverter (also known as power conditioning unit) is the heart of the system. Most of the applications in a residential building generally use AC current, whereas PV module and battery bank are power source of DC current. Inverter does the job of converting DC power to AC power in a PV system.

2.5. Balance of the System Components

Components such as protective devices, blocking & bypass diodes, lightning-protection system, fuses, bus bar and cable wiring constitute what is known as balance of system components [7]. These components are required to protect the system in an efficient way. Cable size should be chosen in such a way that voltage drop or cable loss is minimized.

2.6. Load

Power consumption units are load for a PV system to be planned. A proper load estimation is necessary for designing a standalone PV system. For the purpose of PV system design, electrical loads may be classified broadly as either resistive or inductive. Resistive loads do not necessitate any significant surge current when energized. Like light bulb, electric heater etc. are resistive loads. On the other hand, inductive load requires a large amount of surge current when first energized which is about three times the normal energy requirement. Fan, electric motor, air-conditioner etc. are inductive load. Depending on the load estimation of a building a proper design can be implemented.

3. Methodology for PV System Design

PV system design is a process of determining capacity (in terms of power, voltage and current) of each component of a stand-alone photovoltaic power system with the view to meeting the load requirement of the residence for which the design is made. The designing is done following the steps given below:
    Step 1: Site inspection and radiation analysis.
    Step 2: Calculation of building load requirement.
    Step 3: Choice of system voltage and components.
    Step 4: Determine capacity of Inverter.
    Step 5: Determine capacity of Battery.

Step 6: Charge controller specification.
Step 7: DC Cable Sizing.
Step 8: Solar PV array specification and design layout.
Step 9: PV Module orientation and land requirement.
Step 10: Cost Analysis.

3.1. Site Inspection and Radiation Analysis

The first step and the most important part of the design is site inspection and radiation analysis [8]. It will determine whether a stand-alone system is viable or not. According to the radiation data of the location we can find out the number of sunny days in a year. Amount of electrical energy that can be generated depends on the radiation intensity throughout the year. Maximum, minimum and average temperature is required to measure the cell temperature which will affect the module voltage and current output. Shadow analysis will help to find out the time duration for which solar radiation falls on solar

arrays. Azimuth angle and altitude angle is required to find out the sun path at that location [9].

3.2. Calculation of Building Load Requirement

The electrical load of a specific house will dictate how powerful a PV system has to be installed. The residence load profile is determined by listing all the residential applications with their power ratings and hours of operation at different seasons to obtain the total average energy demand in watt-hours. Inductive load and resistive load should be separately calculated to specify inverter rating. Table 1 gives an idea of how to estimate load. AP_n represents the name of electrical applications in a building, for example light, fan, TV etc. whereas N_n is its total quantity (n= 1, 2, 3, …). Here n is the representing serial number of applications. According to Table 1 total power rating (TP) is the summation of the rated power of individual load multiply by no of products, i.e.$\Sigma_n W T_n$.

Table 1. Load Estimation

Name of the Application (AP_n)	Rated Power of Application (W_n)	Quantity (N_n)	Wattage (W_{Tn})	Summer(A)		Autumn(B)		Winter(C)		Spring(D)	
				Hours used (AH_n)	Wh/day ($AWH_n =$)	Hours used (BH_n)	Wh/day ($BWH_n=$)	Hours used (CH_n)	Wh/day ($CWH_n=$)	Hours used (DH_n)	Wh/day ($DWH_n=$)
AP_1	W_1	N_1	W_1*N_1	AH_1	AH_1*W_{T1}	BH_1	BH_1*W_{T1}	CH_1	CH_1*W_{T1}	DH_1	DH_1*W_{T1}
AP_2	W_2	N_2	W_2*N_2	AH_2	AH_2*W_{T2}	BH_2	BH_2*W_{T2}	CH_2	CH_2*W_{T2}	DH_2	DH_2*W_{T2}
AP_3	W_3	N_3	W_3*N_3	AH_3	AH_3*W_{T3}	BH_3	BH_3*W_{T3}	CH_3	CH_3*W_{T3}	DH_3	DH_3*W_{T3}

Total energy required at the residence is $\Sigma_n AWH_n$ in summer time and similarly for winter, spring and autumn seasons. In the design we have to take the load profile which is maximum of these four seasons. Let us denote this by E_{daily}.

3.3. Choice of System DC Voltage and Components

Once the building load is determined, DC Voltage of the PV system has to be fixed. Generally, it should be taken as high as possible so that less current will be required to meet the high energy requirement [3]. Lower current through cables will reduce electrical energy loss, because cable has resistivity and high current will cause joule heating of cable. Otherwise, much thicker wires are required which will increase cost of the system. In a typical standalone system, in addition to PV panels, other subsidiary components required are battery, inverter, charge controller, cables and mounting structure.

3.4. Determining Capacity of Inverter

Solar PV system delivers DC voltage and power. So an inverter, which convert DC power to AC power, is needed as most of the applications used in a house require AC power. There are still some applications of DC power in some areas. But in this paper, to keep it simple, we have not considered them.An inverter is rated by its output power (P_{KVA}) and DC input voltage (V_{dc}). Power rating of the inverter should not be less than the total power consumed in different loads. On the other hand, it should have the same nominal voltage of battery bank that is charged by solar PV module. In a household consumption of power in appliances can be classified into two categories: resistive power (P_{res}), such as in light, heater,

iron, etc., and inductive power (P_{ind}), such as in fan, motor, etc.Typically, capacity of the inverter is taken to be the sum of all the loads running simultaneously and 3.5 times the total power of the inductive loads to take care of surge protection. Further, the obtained value is to be multiplied by 1.25 to get the requirement, if an option of 25% extra is kept for a reasonable future load expansion [3]. As this is not mandatory, so in our analysis we neglect this. We get the power(P_{inv1}) that should be deliveredby inverter as follows:

$$P_{inv1} = \left(TP + 3.5*P_{ind}\right) \quad (1)$$

Here, total power rating of all loads TP $= P_{res} + P_{ind}$. However, this is an ideal situation. This power calculation has to be corrected for power factor of inverter.

Power rating of inverter (P_{output}) is related to the real power that is delivered by inverter as output and it is given by the following expression of power factor (PF) [12].

$$PF = \frac{\text{Deliverable Real power}}{\text{Power rating of Inverter}} \quad (2)$$

Here 'Real power' is the power that is consumed for work on the load (P_{inv1} in this case) and it is as calculated from Equation (1).Value of PF is generally taken as 0.8 for most of the inverters. So,

$$P_{KVA} = \frac{P_{inv1}}{PF} \quad (3)$$

Inverter converts DC power to AC power. But this conversion is not 100% efficient. So, efficiency (η_{inv}) of inverter is an important parameter which has to be taken care of. Continuous AC power load, which is the total power (TP as obtained above) needed when all the appliances are running at steady state condition, has to come from a DC power source, such as battery. Therefore, the continuous power load to the inverter(TP1) is given by [11]

$$TP1 = \frac{TP}{\eta_{inv}} \qquad (4)$$

Now continuous (DC) input current (I_{dc}) to an inverter from PV modules can be determined, if the system DC voltage (V_{dc}) is specified, according to the following equation,

$$I_{dc} = \frac{TP1}{V_{dc}} \qquad (5)$$

This parameter is needed for battery selection and design.

In terms of energy, daily input energy to the inverter (E_{inv}) is daily maximum energy requirement (which is E_{daily} as stated in Sec 3.2) divided by the inverter efficiency, that is,

$$E_{inv} = \frac{E_{daily}}{\eta_{inv}} \qquad (6)$$

This much amount of energy must come from battery daily to fulfill the load requirement of inverter. System DC voltage also should be specified for the inverter that will allow further calculation to be done accordingly.

3.5. Determining of Capacity of Battery

The battery type generally suggested for use in solar PV power system application is deep cycle battery, specifically designed such that even when it is discharged to low energy level it can still be rapidly recharged over and over again for years. The battery should be large enough to store sufficient energy to operate all loads at night, cloudy or rainy days. Battery storage is conventionally measured in Ah (ampere hour) unit.

The charge storage capacity, which is essentially the energy storage capacity, of the battery bank (B_{Ah}) is determined by the daily energy requirement and number of days for backup power (N_{backup}) using the following equation [12],

$$B_{Ah} = \frac{E_{inv}*N_{backup}}{V_{dc}*DoD} \quad in\ AHr \qquad (7)$$

The percentage of total charge, that is, energy of battery that can be allowed for running the load is referred as depth of discharge (DoD) of the battery. C-rating is also an important part of choosing a battery. It tells us what will be the optimum charging and discharging rate of a battery. Typically C-10 rated batteries are available in the market. So optimum battery bank (BO_{Ah}) should bechosen at that rate according to the following formula,

$$BO_{Ah} = \frac{TP}{V_{dc}*\eta_{inv}}*C\_rating \qquad (8)$$

To meet requirements of the application load a number of batteries has to be connected in series for system voltage specification and in parallel for current specification. The number of batteries connected in series (B_S) is obtained by system DC voltage and voltage of individual battery using the following equation,

$$B_S = \frac{V_{dc}}{Voltage\ of\ a\ single\ battery} \qquad (9)$$

The number of batteries which will be connected in parallel (B_P) can be obtained by the following equation,

$$B_P = \frac{B_{Ah}}{Ah\ capacity\ of\ a\ single\ battery} \qquad (10)$$

The total number of batteries (N_B) can then be obtained by the following equation,

$$N_B = B_S*B_P \qquad (11)$$

If we take battery efficiency (η_{Bat})to be about 85% typically for lead acid battery [12], then energy required(E_{Bat}) from solar PV array to charge the battery bank is given by the following equation

$$E_{Bat} = \frac{V_{dc}*B_{Ah}}{\eta_{Bat}} \qquad (12)$$

3.6. Charge Controller Specification

The solar charge controller is generally sized in a way that will enable it perform its function of current control. A good charge controller must be able to withstand the array current as well as the total load current and must be designed to match the voltage of the PV array as well as that of the battery bank. MPPT charge controller is specified based on PV array voltage handling capacity. Now-a-days, MPPT charge controller usually comes with inverter. There is a recommend voltage range, within which we have to choose the PV array DC voltage. Let's take the PV array voltage to be CC_{volt} which should be greater than system DC voltage.

3.7. Solar PV Array Specification and Design Layout

The Solar PV array is the main component of a standalone PV system. When PV modules are connected in series in a small group it is called PV string and PV array is a collection of PV strings. According to the voltage and current rating PV array design should be done. From PV to battery there are long cable so we must consider the voltage and energy loss in it. Let us denote the cable efficiency by η_{Cable}. Typically in a standalone system 3% voltage loss is considered [11] giving $\eta_{Cable} = 97\%$. So, PV array voltage minimum should be V_{PV}, given by

$$V_{PV} = \frac{CC_{Volt}}{\eta_{Cable}} \qquad (13)$$

Similarly, energy required from the PV array (E_{PV}) can be calculated by the following equation

$$E_{PV} = \frac{E_{Bat}}{\eta_{Cable}} \qquad (14)$$

whereas current requirement from PV array per hour can be calculated from

$$I_{PV} = \frac{E_{PV}}{V_{PV}*Daily\ Sunshine\ hour} \qquad (15)$$

Here Daily sunshine hour is the average sun hour available per day at the installation site. It depends on average radiation at the site and temperature of the module.

A number of PV modules has to be connected in series (S_{PV}) to form a string and a number of strings then has to be connected in parallel (P_{PV}) to obtain the rated voltage and current of the array. Let maximum voltage of an individual PV module is V_m and maximum currentI_m. In our calculations we have taken Standard Test Condition (STC) and followed theIEC standard 60891(2009) to get the correct values of V_m and $I_m$according to ambient condition of the PV power plant installation site.

Now, if V_{PV} and I_{PV} are voltage and current requirements of PV array, respectively, then S_{PV} can be obtained from

$$S_{PV} = \frac{V_{PV}}{V_m} \qquad (16)$$

First, modules are connected in series. So, voltage of the PV string is equal to V_{PV}, but its current rating is only the current which is generated by individual PV module. Then, per PV string voltage and current rating is (V_{PV}, I_m).

Current requirement from PV array, I_{PV}may be high. But to protect the modules from current surge few strings are connected first in parallel and a fuse is put into the circuit. The number of such parallel strings (p) in a bunch depends on current fuse rating, let's say 20A, and depends on module manufacture.The value of 'p' is determined by dividing the fuse dc current of PV module by the rated current of one module as in equation (5).

$$p = \frac{20}{I_m} \qquad (17)$$

And the number of such bunch of strings required to complete the array is given by the following formula

$$P_{PV} = \frac{I_{PV}}{p * I_m} \qquad (18)$$

So total number of PV modules required are N_{PV}, given by

$$N_{PV} = P_{PV} * S_{PV} * p \qquad (19)$$

3.8. DC Cable Sizing

The design of a PV power system is incomplete until the correct size and type of cable is selected for wiring the components together. There are two types of DC cable: Inverter to Battery DC cable, SPV to inverter DC cable.

3.8.1. Inverter to Battery Cable Sizing

The maximum continuous input current (I_{BI}) that should be used as the basis for inverter cable wiring.

$$I_{BI} = \frac{TP}{\eta * V_{LB}} \qquad (20)$$

Lowest voltage of the battery (V_{LB}) is just above at the voltage at which it will get disconnected to prevent further discharging. This value is usually mentioned in the inverter specification sheet.

3.8.2. SPV to Inverter Cable Sizing

There are 3% voltage loss and energy loss in the cable is considered [11]. According to this assumption current rating of the cable (I_{DC}) is

$$I_{DC} = I_{PV} * 1.25 * 1.25 \qquad (21)$$

Usually safety practice is to oversize the wire by 25% above the continuous current that the wire might handle due to high radiation intensity. Voltage drop in the cable is $V_{DROP\_DC} = V_{PV} * 3\%$

$$V_{DROP\_DC} = \frac{2 * L_{DC_{cable}} * I_{DC} * \rho}{A_{DC\_Cable}} \qquad (22)$$

Here ρ is the resistivity of the cable material, L_{DC_CABLE} is the length of the cable and A_{DC_Cable} is the area of the cable. With the help of the area of the cable, cable diameter can be calculated. In market cables are available according to the AWG or SWG rating.

3.9. Energy of Solar PV System and Corresponding Inverter Rating: Summary

The essential features of the analysis we presented above can be summarized as follows:

Total energy needed from PV system per day can be calculated using the following equation

$$E_{PV} = \frac{E_{daily} * N_{backup}}{\eta_{inv} * \eta_{cc} * \eta_{Bat} * DoD * \eta_{Cable}} \qquad (23)$$

And the corresponding inverter rating comes as

$$\text{Inverter rating} = \frac{E_{PV}}{PF * \text{Daily Sunshine Hour}} \qquad (24)$$

4. Case Study of a Residential Building at Gurgaon in India

Gurgaon, a town located in the northern hemisphere part of the earth at latitude and longitude of 28.6°N and 77.2°E respectively, in the north-west part of India. This geographical location of Gurgaon implies that the solar array should be inclined at an optimal angle of about 26° facing southward for all year round to maximize solar energy receive if it is oriented fixed. The average radiation of this location is about 5.5 kWh/m²/day and average ambient temperature is about 30°C, whereas maximum and minimum ambient temperatures are 45.3°C and 7°C, respectively, if the location is devoid of overcasts from nearby trees and buildings. A residential building/house in this location is chosen for the analysis of a standalone PV system.

Table 2. Monthly average radiation data of the site

Month	Irradiance (kWh/m²/day)
January	4.26
February	5.47
March	6.49
April	6.55
May	6.35
June	6.03
July	5.44
August	5.25
September	6.05
October	6.17
November	5.01
December	4.32

4.1. Layout of the building

The layout of a building is shown in Figure 2

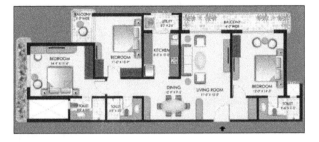

Figure 2. Floor plan of a typical bungalow in Gurgaon

It is a typical bungalow type building in Gurgaon having 3 bedrooms and 2 toilets plus hall, kitchen, etc. The solar power plant will be installed besides the building, about 20m away.

4.2. Load Profile of the Building

The load profile can be determined by summing up the power rating of all the appliances in the building. The energy requirement of the building can be determined based on the hours of usage per day. An estimation of the requirement done for the building is shown in Table 3. According to this calculation power consumption at different season has dissimilar requirements. As per the table it is observed that in summer time the energy demand is maximum compare to other seasons throughout the year. In our calculations, we have to consider the maximum energy demand which is at summer time.

The energy requirement per day in summer is about 26 kWh or 26 units. Total inductive load power rating is 4225W and resistive load power rating is 8865W. Extra load and energy requirement for future provision is included in this estimation. So the total load power rating is 13.01kW.

Table 3. Total energy requirement and power rating of the building

	Power Rating (watt)			Energy rating (kWh/day)			
	Inductive	Resistive	Total	Summer	Autumn	Winter	Spring
Per room	60	1357	1417	3.42	2.86	3.67	3.67
Per Bathroom	60	470	530	0.75	0.75	0.75	1.2
Kitchen	780	815	1520	3.75	4.15	3.79	3.79
Hall	1201	2978	4257	8.86	7.94	4.68	4.65
Balcony & Staircase	855	1072	1927	1.69	1.64	1.66	1.66
Total for (3 Bedroom+2 Bathroom)	3139	9876	13015	26.056	23.831	22.642	23.51

4.3. Inverter specification

Inverter should be specified according to the resistive load and inductive load requirements of the building which have be shown above in Table 3. According to this load estimation the summery of the inverter sizing is given in Table 4. In this system, specifications of a typical inverter, for example, ReneSola 30KVA/360VDC inverter, is used. As per the inverter data sheet [13] it has overload handling capacity of 150% for 10 seconds which is sufficient to run high inductive load of the building. Efficiency of the inverter is 90% at peak condition. System Voltage is 360VDC.

Recommended MPPT voltage range is 450-550 VDC [13]. According to inverter datasheet maximum voltage handling capacity of MPPT is 750 VDC and maximum current handling capacity is 120A [13]. We have taken PV array MPPT voltage of about 500VDC.

Table 4. Inverter summary

Parameters	Calculated Parameter Value	
Inductive Load(P_{ind})	3139W	As per Table 3
Resistive Load(P_{res})	9876W	As per Table 3
Total Continuous output power (TP)	13kW	As per Table 3
Efficiency (η_{inv})	90%	As per inverter data sheet [13]
Input power to the inverter (TP1)	14.5kW	Using Equation4
Input DC voltage (V_{dc})	360VDC	As per inverter data sheet [13]
Input DC current to inverter (I_{dc})	40ADC	Using Equation5
Total Inverter Power (P_{inv1})	24kW	Using Equation 1
Power factor (PF)	0.8	As per inverter data sheet [13]
KVA rating (P_{KVA})	30KVA	Using equation 3
Energy coming from inverter (E_{daily})	26kWh (in Summer)	From Table 3
Energy input to inverter (E_{inv})	28.8kWh	Using equation 6
Output AC voltage	240VDC	As per inverter data sheet [13]
Number of phase	Three phase	As per inverter data sheet [13]
Types	Solar PCU or Hybrid type	As per inverter data sheet [13]
MPPT voltage from PV (CC_{volt})	500VDC	As per inverter data sheet [13]

4.4. Battery Specification

In order to estimate the battery bank size, energy requirement should be known. For inverter total input energy needed is 28.8kWh or 28800Wh from battery. System Voltage is 360VDC. So Battery voltage should be near about 360VDC. Days of autonomy is also a vital part to determine the size of the battery bank. The summary of the battery sizing is given in the following table.

Table 5. Battery Summery

Usage/day	7 hours	Taken
Autonomy (N_{backup})	3	Taken
Battery type	Lead Acid type	As per Battery data sheet [14]
Depth of discharge (DoD)	80%	As per Battery data sheet [14]
Required capacity of battery bank (B_{Ah})	300Ah	Using equation 7
Battery bank operating voltage(V_{dc})	360VDC	As per Inverter data sheet [13]
Each battery voltage	2V	As per Battery data sheet [14]
Each battery capacity	400Ah	Using equation 8
Total number of strings (B_P)	1 (connected in parallel)	Using equation 10
No. of batteries in each strings (B_S)	180 (connected in series)	Using equation 9
Total no of battery required (N_B)	180	Using equation 11
C-rating	C-10	As per Battery data sheet [14]
Energy require to charge battery(E_{Bat})	127.33kWh	Using equation 12

4.5. PV Array Specification

PV array should be designed according to the energy requirement to charge the battery bank. There are different parameters required to design the PV array in proper way. In this designing we have chosen the SunPower 345W_P module. PV array layout will be designed according to maximum voltage (V_m), maximum current (I_m), temperature coefficient and fuse rating of the module. Cable loss is also require to calculate the proper voltage and energy requirement from the PV array. Let Wire loss is about 3% and MPPT voltage is 500V chosen. So, Array voltage minimum should be: 500/0.97 = 515Volt (V_{PV}) and energy from the SPV array (E_{PV}) should be: 127.33/0.97= 131.27 kWh, so current-hour required would be about 131270/515= 255Ah.

In Gurgaon, Haryana location daily average solar energy is above 5kWh/m²/day. Daily sunshine hour is 5Hrs. So per hour current requirement is 255/5 ≈ 51A (I_{PV}) from solar array. Solar PV Array specification should be

at least (515Volt, 51A). Let cell temperature average be 55°C. Taking into account the temperature coefficient of Module, we get V_m: 54.4V, I_m: 6.01A. Voltage will degrade when the cell temperate will increase.The detail calculation procedure is mentioned in the methodology section. The following table has given the summary of the PV array sizing.

Table 6. Solar module specification

Power rating per PV module	345W_P	
V_{OC}	68.2V	
V_m	57.3V	
I_{SC}	6.39A	
I_m	6.02A	
Module efficiency	21.5%	As per PV module data sheet [15]
Power tolerance	0-5%	
Technology	96 Mono-crystalline	
Maximum system voltage	1000V IEC	
Maximum series fuse	20A	
Power temperature coefficient	-0.3%/°C	
Generated power for first 5 years	95%	

Table 7. Solar PV array specification

Total PV array capacity	31kW$_P$	Total no of module * wattage rating per module
Array voltage output (V_{PV})	544V	Using equation 13
Array current output (I_{PV})	54A	Using equation 15
No. of strings (P_{PV})	3	Using equation 18
No. of modules in series (S_{PV})	10	Using equation 16
No. of modules in parallel connected in string (p)	3	Using equation 17
Total no of module	90	Using equation 19
DC wire length	20m x 2	Taken

4.6. System cable sizing

In a PV system choosing proper cable is a very important part in designing. If a DC cable draw more current than its current carrying capacity then it will damage the cable. That's why the cable current rating should be chosen more than the actual capacity. In this design aluminum cable is chosen.

4.6.1. Inverter to Battery Cable Sizing

As per equation (20) the maximum continuous input current that should be used as the basis for inverter cable wiring =14KVA/0.9/315V = 50A. So, current capacity of inverter wire is 50A * 1.25 = 63A

Wire should be chosen as per cable ampere rating.

4.6.2. SPV to Inverter Cable Sizing

Cable loss has taken 3% .Current rating as per equation (21) is 54A * 1.25 * 1.25 = 86A. So, in battery to get 360

VDC through MPPT voltage require is 500/0.97= 515Volt. DC voltage Drop is about (515-500)= 15Volt.

Then as per equation (22), the area of the cable is 1123×10^{-8} meter square. So diameter of the wire is 3.8mm.

4.7. PV Array Orientation and Land Requirement

As per the PV module specification sheet PV module dimension is 41.2" x 61.4". The row spacing between two modules which are facing towards sun and located just behind on another is measured by [14] the following equation.

$$\text{Row Spacing}(Y) = (X).\frac{\cos\cos(\text{Azimuth Angle})}{\tan\tan(\text{Altitude Angle})} \quad (25)$$

Where X is the height of the module from ground as shown in the Figure 3.

In this design module tilt angle is taken as the latitude of that location which is 28°. Shadow length will be taken according to December 22 because at that day shadow length of any object in northern hemisphere will be maximum compare to other days in a year [4].

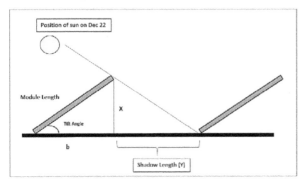

Figure 3. Calculating the minimum distance between rows

Azimuth and altitude anglesat Gurgaon are chosen according to 10AM of that day [19]. Module length is 41.2", as in [15]. So according to the above formula shadow length will be 28.8". Calculating the value of b as shown in Figure 3 one module installation length required is minimum(36.4"+28.8") =65.2".

So total length required for parallel connection is: 65.2*8 + 36.4= 588"≈ 14.2m and taking 2m extra spacing= 16.2m. Total length require for series connection is 61.4 * 10 = 614" ≈ 15.6m and 1.5m extra spacing = 17.2m. So total area requirement is 16.2m X 17.2m = 277 m² ≈ 2982 ft². Extra length is taken for maintenance work or for other purposes.

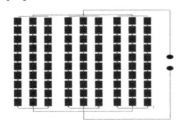

Figure 4. 17.2m x 16.2m Area of the PV array

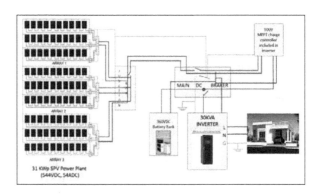

Figure 5. Wiring diagram of a standalone PV array

4.8. Summary of the PV System Components and Cost Estimation

Solar PV modules and the associated components are costly. So the generated electricity cost will be high. We have done a long term cost analysis [20] to see whether it would be cost effective in the long run, even if initial investment is high.

The components which are used in this system and the total cost are given in the following tabular form (as shown in Table 8). PV module cost per watt peak is taken Rs.50/-, and battery cost per watt peak is taken Rs.15/- per Ah. Now we want to see how much is the initial investment and what will be the payback period, if there is any. It will also establish financial feasibility of the PV system. The payback period is the time required to cover up or fulfill the initial investment including the installation cost. Life of the solar PV modules are typically 25 years and batteries have to be replaced after 5 years period because usually the life of the battery is not more than 5 years, usually. So taking this assumption 4 times the battery bank has to be replaced. We have assumethe price of the battery will remain same during this 25 years.So initial investment cost of the project throughout the 25 year is the summation of the first materials cost, installation cost, maintenance cost and the present value of the future investment for the battery cost

Table 8. Initial investment for the PV system

No.	Name of The Component	Quantity	Price per quantity(Rs.)	Total Price(Rs.)
1	Solar Panel (SunPower) Model No.X21-345W$_p$	90	17,250/-	15,52,500/-
2	Exide Battery LMXT (400AH, 2V)	180	6,000/-	10,80,000/-
3	ReneSolar Inverter (30KVA,360VDC)	1	7,88,400/-	7,88,400/-
	Total cost of above materials			34,20,900/-
5	Cost of cable, design, Labor, Metering, junction box and Control Device etc. are lamp together as 20% of equipment cost and add on			6,84,180/-
	Total Initial investment			41,05,080/-

Price of the land is not included in this calculation. In this calculation, inflation rate and discount rate is taken 5% and 7% respectively.

Following is the step by step procedure to find the payback period and financial feasibility of the system.

Step1 : Determine no. of units generated by PV system during day.

The average number of daily sun shine hour in the location is about 5.5 hours. This 31kW$_p$ solar PV system has efficiency near about 75%. According to module specification sheet PV module efficiency is 95% and it will last for first 5 years and after that it will degrade at

the rate of 0.4% per year. So per day unit generation for first year will be (5.5 * 31 * 75% * 95%) = 121 units. Similarly 25 years calculation can be done which is shown in Figure 6.

Step 2: Find out rate of electricity from conventional grid connection:

Rate of electricity in India varies in different locations and so also energy consumption rate per month. In Haryana where the system is supposed to be installed has conventional electricity rate of Rs.6.75/unit [16]. It has a tendency to increase at the rate 6% per annum. Also by diesel generator per unit generation cost is Rs.20/. Diesel rate also changing day by day. In this calculation we assume that the diesel rate is increasing 2% per annum.

Step 3: Determine present value of future investment for battery bank.

First we have to calculate the present value of the future investments using the following equation [17].

$$\text{Present value} = \text{Future value}^* \left(\frac{1 + \text{inflation rate}}{1 + \text{discount rate}} \right)^n \quad (26)$$

where, present value is the present valuation of the future investment and future value is cost of product after n years. Inflation rate is the rate at which value of money decrease with time and discount rate is the rate at which the value of money increases with time due to interest it can earn. Here n is number of years.

According to such calculation the total present value of the 25 years battery investment cost is about Rs.34lakhs. If we invest half of that money as a fixed deposit in a bank at interest rate of 8.75%, then at each time of battery purchase we can get back that money. Also after 25 year we can get back Rs.2.4lakhs of this investment. So we can say that the total initial investment including the battery cost is Rs.58lakhs.

Step 4: Determine the savings per year.

After installing the PV system there will be no electricity cost. So that will be the savings per year deducting the maintenance cost per year. In this calculation 10% of the yearly savings is used for maintenance purpose of PV system. For example in first year electricity consumption will be (121*6.75*365) = Rs.2,99,504/- and out of that 10% will be used for maintenance purpose, so overall savings will be Rs.2,69,554. In that way after 25 year savings will be about 1.4 crore and the present value of that savings is 1.07 crore. So, the saving after 25 years is (commutative savings + extra bank savings for battery – maintenance cost), which is, according to the calculation, about Rs.85lakhs and net present value of that amount is Rs.50lakhs. This is a positive value (shown in Figure 6). So this project is feasible and acceptable because according to the financial strategy if the NPV of any system is positive then the project is feasible and acceptable. The payback period of the system is 14 years. The detailed calculation is shown in Figure 7.

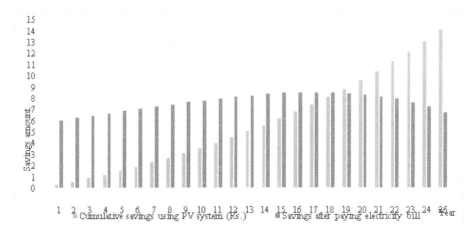

Figure 6. Savings after Investment in PV system vs fixed deposit in bank (Comparing with conventional grid connection)

Here, we assume that the rate of electricity will grow at the rate of 6.0% per annum till the life of the system which is 25 years.

Inflation rate 5.00% 0.9813=(1.05/1.07)
Discount rate= 7%

Year	Daily average sun hours (Hours)	Wattage of solar Array (KW)	System is 75% efficient (KWh)	Module efficiency (%)	Units generated Number (KWh)	Rate of electricy (Rs./ unit)	Yearly savings (Rs.)	Maintenance cost @10% of the yearly savings (Rs.)	Net yearly savings (Rs.)	Cumulative savings (Rs.)	Let, Inflation rate is 5.0% and discount rate is 7%	The present value the savings	Real saving in nth year	Bank Interest rate 8.75% on fixed deposit	Savings after paying electricity bill
1	5.5	31	128	95	121	6.75	299,504	29,950	269,554	269,554	0.98	269,554	269553.99	6,329,883	6,030,379
2	5.5	31	128	95	121	7.16	317,475	31,747	285,727	555,281	0.96	280,384	549938.11	6,558,037	6,240,562
3	5.5	31	128	95	121	7.58	336,523	33,652	302,871	858,152	0.94	291,649	841587.51	6,786,611	6,450,088
4	5.5	31	128	95	121	8.04	356,715	35,671	321,043	1,179,195	0.93	303,367	1144954.80	7,014,471	6,657,756
5	5.5	31	128	95	121	8.52	378,117	37,812	340,306	1,519,501	0.91	315,556	1460510.78	7,240,310	6,862,193
6	5.5	31	128	94.6	121	9.03	399,117	39,912	359,205	1,878,706	0.89	326,852	1787363.13	7,462,634	7,063,518
7	5.5	31	128	94.2	120	9.58	421,275	42,128	379,148	2,257,854	0.88	338,547	2125910.19	7,681,575	7,260,300
8	5.5	31	128	93.8	120	10.15	444,655	44,466	400,190	2,658,043	0.86	350,654	2476564.06	7,895,577	7,450,921
9	5.5	31	128	93.4	119	10.76	469,325	46,932	422,392	3,080,436	0.84	363,187	2839751.10	8,102,877	7,633,552
10	5.5	31	128	93	119	11.40	495,354	49,535	445,818	3,526,254	0.83	376,161	3215912.37	8,301,488	7,806,134
11	5.5	31	128	92.6	118	12.09	522,816	52,282	470,535	3,996,789	0.81	389,592	3605504.15	8,489,171	7,966,354
12	5.5	31	128	92.2	118	12.81	551,792	55,179	496,612	4,493,401	0.80	403,494	4008998.43	8,663,410	8,111,619
13	5.5	31	128	91.8	117	13.58	582,362	58,236	524,125	5,017,527	0.78	417,885	4426883.45	8,821,385	8,239,024
14	5.5	31	128	91.4	117	14.40	614,613	61,461	553,152	5,570,679	0.77	432,781	4859664.24	8,959,939	8,345,325
15	5.5	31	128	91	116	15.26	648,639	64,864	583,775	**6,154,454**	0.75	448,199	5307863.20	9,075,541	8,426,902
16	5.5	31	128	90.6	116	16.18	684,535	68,454	616,082	6,770,536	0.74	464,157	5772020.64	9,164,256	8,479,720
17	5.5	31	128	90.2	115	17.15	722,404	72,240	650,163	7,420,699	0.73	480,675	6252695.40	9,221,696	8,499,292
18	5.5	31	128	89.8	115	18.18	762,352	76,235	686,117	8,106,816	0.71	497,770	6750465.47	9,242,980	8,480,628
19	5.5	31	128	89.4	114	19.27	804,494	80,449	724,044	8,830,861	0.70	515,463	7265928.62	9,222,683	8,418,189
20	5.5	31	128	89	114	20.42	848,948	84,895	764,053	9,594,914	0.69	533,774	7799703.07	9,154,781	8,305,833
21	5.5	31	128	88.6	113	21.65	895,840	89,584	806,256	10,401,170	0.67	552,725	8352428.13	9,032,593	8,136,753
22	5.5	31	128	88.2	113	22.95	945,304	94,530	850,773	11,251,944	0.66	572,337	8924764.95	8,848,718	7,903,415
23	5.5	31	128	87.8	112	24.32	997,478	99,748	897,730	12,149,674	0.65	592,632	9517397.20	8,594,963	7,597,486
24	5.5	31	128	87.4	112	25.78	1,052,509	105,251	947,258	13,096,932	0.64	613,635	10131031.83	8,262,266	7,209,756
25	5.5	31	128	87	111	27.33	1,110,554	111,055	999,499	**14,096,431**	0.62	635,368	10766399.84	7,840,610	**6,730,056**
											Total=	**10,766,400**			

Initial Investment for system= **4105080** Plus Present cost of future invertment for Battery= **5,820,582**

Net saving after 25 years = total saving in 25 years - initial investment====>> **8,520,253**

<u>After 25 years</u> <u>NPV of B.savings</u>

Save in Bank= 6,730,056 **4,198,205**

<u>NPV of savings using PV =</u> **5,098,277**

Present Battery cost= 1,080,000

	Present Value	Bank interest after		
Replacement cost of battery for first time after 5yr =	982,727	2,609,382		
Replacement cost of battery for second time after 10yr=	894,215	2,326,281		
Replacement cost of battery for third time after 15yr=	813,675	1,895,668		
Replacement cost of battery for fourth time after 20yr PV=	740,389	1,240,681	SAVINGS	NPV
Total present cost for battery=	**3431004.44**		244,405	152,460

Figure 7. Cost analysis of standalone PV array

Similar way, we compare cost of diesel generated electricity with that of PV system to get the payback period and feasibility study. If we follow the same procedure, according to diesel generator electricity cost, the payback period of the system will be 8 years and

savings will be much more than grid connected system. It is also shown in Figure 8 that if we invest this amount of money in bank at an interest rate of 8.75%, then after paying the electricity bill the saving after 25 years will be less than the savings we can get installing the PV system.

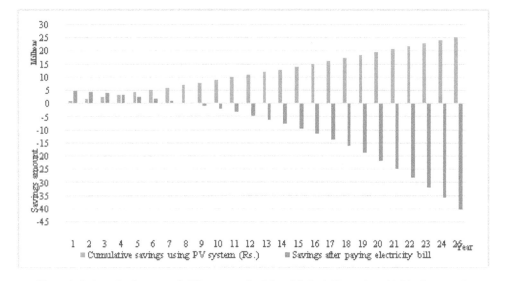

Figure 8. Savings after Investment in PV system vs fixed deposit in bank (Comparing with Diesel generator)

5. Conclusion

Solar Photovoltaic standalone system is a clean source of energy. Such systems are generally envisaged for use in rural remote areas where grid system is not available. Urban areas hardly utilize this for energy needs as conventional energy available from grid is much cheaper and initial investment of PV system is quite high. In the urban area of Haryana, for example in Gurgaon, the solar radiation varies from 400W/m^2 to 1100W/m^2, depending on different climatic conditions. This is enough to provide the energy requirement of a building in this area if efficiently tapped. In this paper we present a complete design of a solar PV system step by step and its life cycle cost analysis. The results of this study indicate that at the optimal configuration for electrifying a typical building about 13 kW of power is needed. The initial installation cost of the standalone PV system is high, about Rs.42 lakh. However, it is beneficial and suitable for long term investment as the payback period is less than 15 years while the system life expectancy period is about 25 years. If the initial prices of the PV systems are decreased, which is expected with the advent of technological uplift and the increase in production volume, then the payback period will further be reduced. So standalone PV energy source is a viable energy solution even for urban areas. It is a concept contrary to the general perception. With the help of this system we can fulfil our daily energy requirement at any scale. In urban areasgrid connection system is readily available, but there also people can invest money initially for fulfilment of their daily energy need with standalone PV system and enjoy it free for long years without power interruption. It is also profitable compare to saving this investment amount in a bank and pay for the ever increasing cost of energy needs and inflation. In this study, cost estimation of the whole system including

cabling, design, labor, control devices and maintenance has also been provided. The same design procedure can be applied to other locations. As a final remark, respective governments should get involved in providing financial support for procurement and installation of PV system, make it a popular choice and propagate this energy solution.

Acknowledgement

We express our gratefulness to the members of the solar photovoltaic department of National Institute of Solar Energy, Gurgaon for their moral support. Our thanks and appreciations also go to those people who willingly helped us out with their abilities whenever we needed. Also I want to thank the American Library in Delhi for their help.

References

[1] Shahriar Shafiee, Erkan Topal,*"When will fossil fuel reserves be diminished?"*ScienceDirect, Energy Policy, Volume 37, Issue 1, January 2009, Pages 181-189.

[2] Ambuj D. Sagar, *"Alleviating energy poverty for the world's poor"* ScienceDirect, Energy Policy, Volume 33, Issue 11, July 2005, Pages 1367-1372.

[3] Guda, H. A and Aliyu U.O, *"Design of a standalone photovoltaic system for a residence in Bauchi"*, International Journal of Engineering and Technology, Volume 5 No. 1, January 2015

[4] Oloketuyi S Idowu, Oyewola M Olarenwaju and Odesola I Ifedayo, *"Determination of optimum tilt angles for solar collectors in low-latitude tropical region"*, Springer,International Journal of Energy and Environmental Engineering, 4:29, 2013.

[5] Chetan Singh Solanki, *"Solar Photovoltaic: Fundamentals, Technologies and application"*, PHI learning Pvt., 2011.

[6] Manju, B.S., Ramaprabha, R; Mathur, B.L., *"Modelling and control of standalone solar photovoltaic charging system"*, IEEE, Page(s) 78-8, March 2011.

[7] Miro, Zeman, "Photovoltaic *system*",TU Delft Open Course Ware, 2014. Available: http://ocw.tudelft.nl/fileadmin/ocw/courses/ SolarCells/res00029/CH9_Photovoltaic_systems.pdfnetLibrary e-book.

[8]   T.Markvart, A. Fragaki, J.N. Ross, *"PV system sizing using observed time series of solar radiation"* ELSVIER Solar energy 80 (2006) 46-50.

[9]   Lurwan, S.M., Mariun, N. Hizam, H., Radzi, M.A.M. ,Zakaria, A., *"Predicting power output of photovoltaic systems with solar radiation model"* IEEE, Page(s) 304-308, 2014.

[10] Mekhilef, S., Rahim, N.A., *"A new three-phase inverter power-factor correction (PFC) scheme using field programmable gate array"* ,IEEE, Page(s) 1004-1007 vol.2, Dec 2002.

[11] GSES India sustainable energy Pvt. Ltd., *"Standalone Photovoltaic system design & installation"*, Second edition September 2014.

[12] Khaled Bataineh, Doraid Dalalah *"Optimal Configuration for Design of Stand-Alone PV System"*, Scientific Research, Smart Grid and Renewable Energy, 2012, 3, 139-147

[13] Renesola RFE series 30KVA inverter manual: http://www.renesola.com/renesola-380v-30kva-solar-charge-inverter-1231/ (May, 2015).

[14] Exide LMXT series tubular Lead Acid Solar battery catalogs:http://www.exide4u.com/solar(May 2015).

[15] Sunpower X-series Solar Panels product sheet: http://us.sunpower.com/solar-panels-technology/x-series-solar-panels/(May 2015).

[16] Commercial Electricity Tariff Slabs and Rates for all states in India in 2015: https://www.bijlibachao.com/news/commercial-electricity-tariff-slabs-and-rates-for-all-states-in-india-in-2015.html andhttp://www.uhbvn.com/Tarriff.aspx.

[17] Mohanlal Kolhe, Sunita Kolhe, J.C. Joshi, *"Economic viability of standalone solar photovoltaic system in comparison with diesel-powered system for India"* ELSEVIER Energy Economies 24 155-165 (2002).

[18] F. John Hay, *"Economics of Solar Photovoltaic Systems"*, University of Nebraska-Lincoln Extension, Institute of Agriculture and Natural Resources, G1282, Issued January 2013.

[19] Abhik M. Pal, Subhra Das, *" Analytical Model for Determining the Sun's Position at all Time Zones"*, Int. J. Energy Engineering Vol.5, No.3 (2015).

[20] Abhik M. Pal, *"Designing of a Standalone Photovoltaic System for a Residential Building at Gurgaon, Haryana"*, M. Tech. Project Report, Amity University, India (2015).

23

Design Optimization and Performance Evaluation of a Single Axis Solar Tracker

Mohammed Ben OUMAROU*, Abdulrahim Abdulbaqi TOYIN, Fasiu Ajani OLUWOLE

Department of Mechanical Engineering, University of Maiduguri, PMB: 1069 Borno State, NIGERIA
*Corresponding author: mmbenomar@yahoo.com

Abstract The paper presents the optimization in design, construction and performance test of a microcontroller-based, single axis solar panel tracking system, using locally available recoverable materials. The tracking system consists of two light sensors and an automated microcontroller to drive the motor and three batteries. Three parameters were considered: solar intensity, voltage and time of alignment/exposure of solar panel to solar radiation. Current and power were obtained and compared with those of a fixed axis solar panel of same specifications. The solar tracker provided a constant alignment, better orientation of the solar panel relative to the sun; and ensured production of more energy by capturing the maximum of sun rays hitting the surface of the panel from sunrise to sunset. The present study has shown that the solar tracking system could both be optimized in terms of design with a performance increment of 47.5% and cost. The solar tracking system is affordable and found to cost $ 154.00. It is also a sustainable energy solution which would assist in reducing both solid and liquid wastes as well as noise and air pollution.

Keywords: solar energy, tracking, optimisation, material re-use, low cost, sustainability

1. Introduction

Developing countries across the world are presently dealing with various problems ranging from poverty, hunger, population increase as well as lack of electricity for their basic primary needs. Some imported technologies do not simply fit; they are either too costly to purchase or are very difficult to repair once faulty. In the selection process of an appropriate source, form of renewable energy or technology, factors of major importance to be considered include: availability of parts and raw material, location, ease of installation, ease of maintenance, reliability, capacity, cost and environmental impact. Solar energy is readily available in most semiarid parts of tropical Africa with Nigeria receiving a yearly average of 5.61 kWh/m^2 [1] among others. However, solar power depends directly on light intensity, duration of sunshine time, geographical position and prevailing climate. To optimize the amount of energy received, a solar panel must be perpendicular to the light source; and since the sun moves both throughout the day as well as throughout the year, a solar panel needs to be able to follow the sun's movement to produce the maximum possible power.

On one hand, Asmarashid et al. [2] designed a low power single axis solar tracking system regardless of motor speed while Okpeki and Otuagoma [3] designed and constructed a bi-directional solar tracking system. Hemant et al. [4] on the other hand, presented the design

and experimental study of a two axis (azimuth and Polar) automatic control solar tracking system to track solar PV panel according to the direction of beam propagation of solar radiation. The designed tracking system consists of sensor and Microcontroller with built-in ADC operated control circuits to drive motor. The results indicate that the energy surplus becomes about (45-56%) with atmospheric influences. In case of seasonal changes of the sun's position there is no need to change in the hardware and software of the system. Solar tracking systems design has received considerable attention throughout the world in recent years [5-11].

This paper presents the design optimization and performance evaluation of a sustainable single axis solar tracker for use in semi arid regions of the world with Maiduguri (Nigeria); located on latitude 11.85° North and longitude 13.08° East and an annual mean daily global solar insolation of 6.176W/m^2 – day; as the study area.

2. Materials and Methods

Two prototypes solar panels were designed and constructed for the purpose of this study; one fixed and the other one, able to move and track the sun movement from sunrise time to the sunset time of the study area (12 hrs 25 minutes = 44,700 seconds). The main components of solar tracking system are as follows: two solar panels (monocrystaline photovoltaic module. Model type= SE-20M maximum power 20watts), a used electric glass door raising mechanism from an old car, a stepper motor designed, two sensors (light dependent resistor), three batteries (sealed lead acid battery 12volts, initial current: 2.1Amperes) and an electronic circuit (controlled by a microcontroller PIC). Other equipment include two digital thermocouples and a CASIO DATABANK stopwatch.

Sizes and other important physical characteristics of the solar tracking system and fixed axis system are determined using the formulas and correlations from the literature.

2.1. Determination of the Speed of the Stepper Motor [12]

Taking θ =260˚S (where θ is the angle tilted or covered by the solar tracker).

$$\theta = \frac{2\pi N}{60} = \omega t \qquad (1)$$

ω = θ/t where t = 12hours, 25minutes = 44700 seconds
θ in radians = 260˚/180˚ × π = 4.538rads
where N is the speed of the stepper motor.

2.2. Determination of the Number of Revolution of the Electric Motor Per Teeth

Number of revolution of electric motor in 12hrs 25minutes.

2.3. Determination of the Thickness of the Teeth

Taking the diameter of the pinion to be 92 mm (pitch circle diameter PCD)

$$Module = \frac{pitch\ circle\ diameter\ PCD}{Number\ of\ teeth\ (T)}. \qquad (2)$$

2.4. Determination of the Power Required by the Stepper Motor to Drive the Mechanism [13]

$$Power = Current(I) \times Voltage(V) \qquad (3)$$

2.5. Efficiency of the Solar Tracker

$$= \frac{Output power}{Input power} \times 100\% \qquad (4)$$

The power required by the electric motor (consumption)

$$P = I \times V \qquad (5)$$

3. Results and Discussion

Two solar collection systems (Plate 1) were constructed and tested. The solar tracking system (Figure 1 and Figure 2) was powered by a motor and controlled by a microcontroller PIC, whose circuit diagram is shown in Figure 3. Tests were conducted with the solar tracker and

results were compared to the fixed type from 19th January 2014 to 24th July 2014, within the same test periods. Table 1 shows a summary of calculated characteristics of the designed solar panels. Table 2 shows the summary of the monthly average results from the performance evaluation tests conducted during the period of study, while Figure 4, Figure 5, Figure 6 and Figure 7 depict the graphical behaviour of both the fixed solar collector and the one equipped with the tracking system, during the months of January through April of the test period.

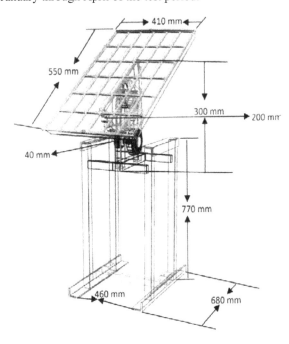

Figure 1. Isometric View of the Designed Solar Tracker

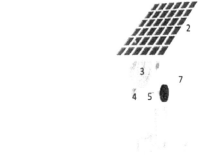

No of	List of parts
1	Sensor
2	Panel
3	Spur gear
4	Pinion gear
5	Shaft
6	Stepper motor
7	Circuit board
8	Frame
9	batteries

Figure 2. Exploded View of the Solar Tracker

Table 1. Summary of Calculated and Generated Characteristics of the Designed Solar Panels

Calculated & Test Parameters	Fixed Axis Type	Solar Tracker Equipped type
N (rpm)	Nil	9.693×10^{-4}
Revolution of electric motor per teeth (rev/teeth)	Nil	0.04012
Power (P) required to drive mechanism (Watts)	Nil	11.5
Efficiency of collection (%)	41.25	88.66

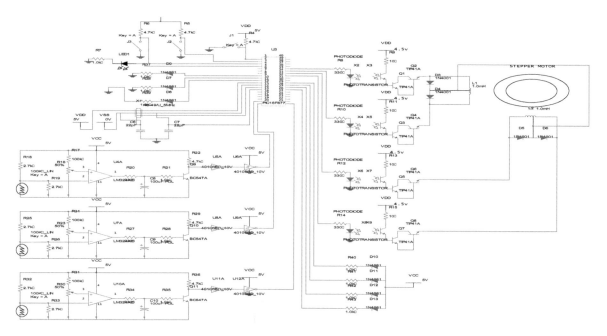

Figure 3. Solar Tracker Circuit Diagram

Table 2. Summary of the Monthly Averages of the Performance Evaluation Test Results

Time	Month One		Month Two		Month Three		Month Four		Month Five		Month Six	
	Tracker	Fixed	Tracker	Fixed	Tracker	Fixed	Tracker	Fixed	Tracker	Fixed	Tracker	Fixed
8 am	0.83	0.58	0.8	0.6	0.76	0.5	0.85	0.61	0.73	0.5	0.68	0.5
9am	0.88	0.61	0.87	0.71	0.8	0.63	0.88	0.66	0.87	0.64	0.82	0.6
10 am	0.93	0.76	0.88	0.74	0.83	0.65	0.98	0.72	0.9	0.72	0.85	0.62
11 am	0.96	0.75	0.96	0.77	0.98	0.82	1.16	0.9	0.92	0.73	0.92	0.74
12 am	1.15	0.88	0.98	0.79	1.2	0.8	1.28	1	0.94	0.74	0.96	0.76
1 pm	1.26	1.08	0.99	0.83	1.26	0.96	1.48	1	0.97	0.76	1.02	0.81
2 pm	1.26	0.96	0.85	0.74	1.57	1	1.66	1.28	1.02	0.78	1.26	0.94
3 pm	1.18	0.76	0.82	0.7	1.59	0.99	1.48	1.24	0.90	0.68	1.01	0.82
4 pm	0.91	0.69	0.8	0.69	1.36	0.96	1.29	0.92	0.83	0.61	0.95	0.70

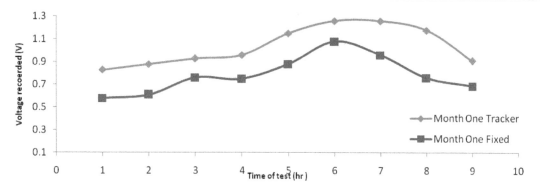

Figure 4. Average Voltage produced by the Solar collection systems in January/ February

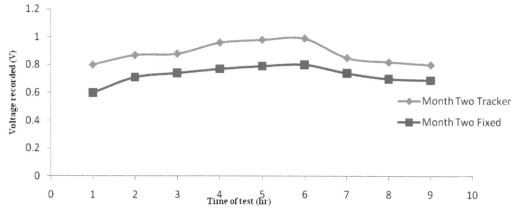

Figure 5. Average Voltage produced by the Solar collection systems in February / March

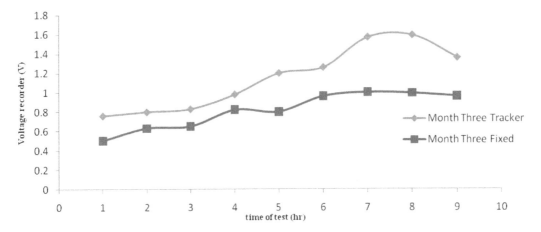

Figure 6. Average Voltage produced by the Solar collection systems in March/ April

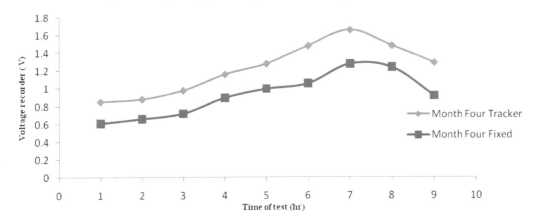

Figure 7. Average Voltage produced by the Solar collection systems in April/ May

The patterns of all graphs (i.e.: fixed axis solar collection system and solar tracker system) are closely similar, in Figure 4, Figure 5, Figure 6 and Figure 7, which shows the direct impact of the tracking system, bearing in mind the fact that the two solar collection systems share the same basic mechanical and physical characteristics. There is an average increase in voltage of 1.85 volts from the fixed solar panel to the tracked system during the period of testing as shown in Figure 4, Figure 5, Figure 6 and Figure 7. This difference in voltage is not inclusive of the battery being charged and powering of the tracking system. It is a direct comparison with the voltage being generated and produced by the fixed solar panel. Tests results (Table 2) favourably compare with those previously obtained [10], with some slight differences which are attributed to climatic conditions, experimental errors as well as other parameters, such as more sun and less clouds, moisture, haze, dust, and smog, with climate being the most important factor. This climate makes the soils dryer and the winds are lesser during that period. The period also constitutes a transitional period to the dry and hot seasons of the year in the study area. Overall, a 93% agreement was recorded between the two sets of records. However, noticeable drops in performance were mostly observed in the months of March and April (Table 2). These low performances recorded in the months of May, June and July are mainly due to the raining season which is accompanied with heavy and day long clouds as well as relatively low temperature when compared to the daily ambient temperature of 30 to 40°. The results are also in good agreement with those of Small Power Systems [14] in California where annual gains between 30 percent and 40 percent are typical. They went further to ascertain that the gain on any given day may vary from almost zero to nearly 100 percent in a generally good area.

Cost had always been an important aspect of a project. It determines the affordability and sometimes the viability which is a quality dependent parameter. Since costs are subject to timely fluctuations, the analysis may only be useful for the present. In the future, increases in cost may occur.

Table 3. Bill of Engineering Measurement and Evaluation

S/No.	MATERIAL	Quantity	Cost per Unit(N)	Total Cost (N)
1	Solar panel	2	10,000	20,000
2	Material*		29,320	29,320
3	Direct labour cost			9,728
5	Total cost of production			59,048

The cost of this work (Table 3) depends on the prices of material with which the parts of the solar tracker were made of and the prevailing market price. As the main objective of this work is to design of a low cost device, the cost of the panels needs to be removed from the overall cost, to obtain the cost of the tracking system:

Cost of tracking system

= Overall cost − Cost of solar panels   (6)

$= 59,048 - 20,000$

$= N39,048$

However, making use of the used electric glass window raising mechanism from the imported used cars; some developed countries of Africa (i.e.: Nigeria, Niger, Chad, Cameroon, Benin, Senegal, Mali, etc...) are being flooded with on weekly basis from Europe and America; the material cost (*) from Table (3) is reduced by =N=7,000.00. This used electric glass window raising mechanism replaces items 3, 4 and 5 of Figure (2). At the present Dollar ($) to Naira (=N=) exchange rate of 208.00 naira to a US dollar, the newly designed and improved constructed solar tracking system costs $ 154.00; $ 80 lower than the previous design proposed by Oumarou and Abdulrahim [10] even though the exchange rate of the local currency to the American dollar keeps fluctuating. This design would help in solid waste management by making use of some parts from the cars and other items brought from developing countries.

With the present situation of lack of electricity in some parts of Nigeria, it was found that 10 million naira ($48,076.00) is weekly required to provide electrical energy to the University of Maiduguri in Northern Nigeria, with over 3 million naira ($14,423.00) for lighting only. The proposed solar tracking system requires the same amount of money for the provision of nearly 100 units of the solar trackers lighting facilities. To this cost, there is need to include the influence of some implicit and explicit parameters (i.e.: transport,....) The proposed solar tracking system would, if adopted, assist in reducing solids wastes in terms of used filters, engines parts replaced, effluents as spent oil, air and noise pollution which would have been caused while using Diesel fuel power generators; bearing in mind that universities are institutions of learning where noise and other solid wastes having an immediate effect.

4. Conclusion

The present study has shown that the solar tracking system could both be optimized in terms of design, cost by making use of previously used mechanisms; recovered from cars and other similar systems. The power gained by the solar tracker system over the fixed horizontal solar collection system was 47.5%. The solar tracking system is

affordable and found to cost $ 154.00. Favourable climatic conditions also play an important role in increasing the performance of solar trackers. The designed solar tracker would assist in reducing both solid and liquid wastes as well as noise and air pollution.

References

[1]   Apricus Solar Company "Insolation Levels in Africa"; [online] Available: http://www.apricus.com/html/insolation_levels_africa.htm, [Accessed Jan. 5, 2012].

[2]   Asmarashid Ponniran, Ammar Hashim and Ariffuddin Joret "A design of low power single Axis solar tracking system regardless of motor speed"; *International Journal of Integrated Engineering,* Vol.3, 2011(No. 2), pp:5-9.

[3]   Okpeki U. K. and Otuagoma S. O. "Design and Construction of a Bi-Directional Solar Tracking System"; *International Journal of Engineering and Science,* Vol. 2, 2013 (Issue 5), pp: 32-381.

[4]   Hemant Kumar Nayak, Manoj Kumar, Nagendra Prasad and Rashmi Rekha Behera " Fabrication and Experimental Study on Two-Axis Solar Tracking", *International Journal of Applied Research in Mechanical Engineering, Vol.1, 2011(Issue-1), pp.: 123-126.*

[5]   Hamilton S. J. *"Sun-tracking solar cell array system",* Department of Computer Science and Electrical Engineering, University of Queensland. 1999, *Bachelors Thesis.*

[6]   Han Wan Siew "Solar Tracker", *SIM University,* 2008.

[7]   David Appleyard "Solar Trackers: Facing the Sun", *Renewable Energy World Magazine,* UK: Ralph Boon, June 1, 2009.

[8]   Amin N., Yung W. C. and Sopian K. "Low Cost Single Axis Automated Sunlight Tracker Design for Higher PV Power Yield" *ISESCO Science and Technology Vision,* 2008 (Vol.4), November.

[9]   Adrian Catarius and Mario Christener *"Azimuth-Altitude Dual Axis Solar Tracker";* Published Master Qualifying Project: submitted to the faculty of Worcester Polytechnic Institute, December 16, 2010.

[10] Oumarou M. Ben and Abdulrahim A. T. "Design and Testing of a Low Cost Single Axis Solar Tracker"; *Proceedings of the 30th International Conference on Solid Waste Technology and Management;* March 15-18, Philadelphia, 2015, PA U.S.A. pp: 926-936.

[11] Elham Ataei, Rouhollah Afshari, Mohammad Ali Pourmina, Mohammad Reza Karimian "Design and Construction of a Fuzzy Logic Dual Axis Solar Tracker Based on DSP"; *2nd International Conference on Control, Instrumentation and Automation (ICCIA).*2011, pp: 185-189.

[12] Shigley J.E. and Mischle C.R. *Mechanical Engineering Design,* McGraw Hill Companies Inc. 6th Edition, New York. 2001, USA.

[13] Abbott AF. *"Ordinary Level Physics";* 3rd edition, Heinemann Educational Books, London, 1977.

[14] Small Power Systems "Should you install a solar tracker?", [Online], Available: www.helmholz.us/smallpowersystems/Intro.pdf, [Accessed April 9, 2015.

24

Study of the Reynolds Number Effect on the Aerodynamic Structure around an Obstacle with Inclined Roof

Slah Driss, Zied Driss[*], Imen Kallel Kammoun

Laboratory of Electro-Mechanic Systems (LASEM), National School of Engineers of Sfax (ENIS), Univrsity of Sfax, Sfax, TUNISIA
*Corresponding author: zied.driss@enis.rnu.tn

Abstract In this work, we are interested on the study of the Reynolds number effect on the aerodynamic structure around an obstacle with inclined roof. Different Reynolds numbers equals to Re=2666, Re=10666, Re=24000 and Re=32000 are particularly considered. The software "SolidWorks Flow Simulation" has been used to present the local characteristics. The numerical model considered is based on the resolution of the Navier-Stokes equations in conjunction with the standard k-ε turbulence model. These equations were solved by a finite volume discretization method. The numerical model is validated with experimental results conducted on an open wind tunnel equipped by an adequate model.

Keywords: CFD, modeling, airflow, obstacle, inclined roof, wind tunnel

1. Introduction

The charactristics of the flows around building-shaped obstacles are different from those in laminar boundary immersed in turbulent boundary layers. It is important for the appropriate estimation of wind loading to investigate these characterestics of flows observed under oncoming turbulence [1,2]. Computational simulations have an advantage to clarify the process and mechanism of these localized phenomena. In recent years, several methods to generate turbulent inflow data were proposed and some of them were used as inflow data [3,4,5]. In this context, Nasrollahei et al. [6] used a builder software to study the thermal and cryogenic performance. The results of the research show that the temperature of Shevaduns in summer is less than the average maximum temperature and the average minimum temperature of outdoor, and the thermal condition of the Shevadun space during the measurement is lower than the thermal comfort limit which is specified in the standard. Ould Said et al. [7] dedicated to the numerical simulation of thermal convection in a two dimensional vertical conical cylinder partially annular space. The governing equations of mass, momentum and energy are solved using the CFD FLUENT code. Tominaga and Stathopoulos [8] reviewed current modeling techniques in computational fluid dynamics (CFD) simulation of near-field pollutant dispersion in urban environments and discussed the findings to give insight into future applications. De Paepe et al. [9] simulated five different wind incidence angles

using a turntable, in order to quantify their effect on indoor air velocities. The responses in local air velocities could largely be attributed to the relative position of the end walls of the scale models orientated towards the wind. This crucial position allows the measured air velocity trends to be explained. The estimated airflow rates gradually decreased for larger wind incidence angles. Lim et al. [10] presented a numerical simulation of flow around a surface mounted cube placed in a turbulent boundary layer which, although representing a typical wind environment, to match a series of wind tunnel observations. The simulations were carried out at a Reynolds number, based on the velocity at the cube height, of 20,000. The results presented include detailed comparison between measurements and large eddy simulation (LES) computations of both the inflow boundary layer and the flow field around the cube including mean and fluctuating surface pressures. Melo et al. [11] developed two Gaussian atmospheric dispersion models, AERMOD and CALPUFF. Both incorporating the PRIME algorithm for plume rise and building downwash, are intercompared and validated using wind tunnel data on odour dispersion around a complex pig farm facility comprising of two attached buildings. The results show that concentrations predicted by AERMOD are in general higher than those predicted by CALPUFF, especially regarding the maximum mean concentrations observed in the near field. Comparison of the model results with wind tunnel data showed that both models adequately predict mean concentrations further downwind from the facility. However, closer to the buildings, the models may over-predict or under-predict concentrations by a factor of two, and in certain cases even larger, depending on the conditions. Meslem et al. [12] observed changes in the prediction of local and global mean-flow quantities as a function of the considered turbulence model and by the lack of consensus in the literature on their performance to predict jet flows with significant three-dimensionality. The study reveals that none of the turbulence models is able to predict well all jet characteristics in the same time. Reynolds stress turbulence model leads to a better agreement between the numerical results and the experimental data for the local jet flow expansion, whereas global flow expansion and ambient air induction are better predicted by the shear stress transport k-ω turbulence model. All linear (Low Reynolds and Renormalization Group) and nonlinear (quadratic and cubic) k-ε turbulence models overestimate local and global expansions and ambient air induction. The k-ω turbulence model underestimates on one hand the global expansion and the ambient air induction and on the other hand the transverse jet deformation is not well predicted. The turbulence kinetic energy increases unrealistically in the jet near field for all k-ε turbulence models and Reynold's Stress Models (RSM). In this region shear-stress transport (SST) k-ω model was in close agreement with measurements. Ntinas et al. [13] predicted the airflow around buildings is challenging due to the dynamic characteristics of wind. A time-dependent simulation model has been applied for the prediction of the turbulent airflow around obstacles with arched and pitched roof geometry, under wind tunnel conditions. To verify the reliability of the model an experiment was conducted inside a wind tunnel and the air velocity and turbulent kinetic energy profiles were measured around two small-scale obstacles with an arched-type and a pitched-type roof. Luo et al. [14] studied models of cuboid obstacles to characterize the three-dimensional responses of airflow behind obstacles with different shape ratios to variations in the incident flow in a wind-tunnel simulation. Wind velocity was measured using particle image velocimetry (PIV). The flow patterns behind cuboid obstacles were complicated by changes in the incidence angle of the approaching flow and in the obstacle's shape ratio. Gousseau et al. [15] used Large-Eddy Simulation (LES) to investigate the turbulent mass transport mechanism in the case of gas dispersion around an isolated cubical building. Close agreement is found between wind-tunnel measurements and the computed average and standard deviation of concentration in the wake of the building. A detailed statistical analysis of these variables is performed to gain insight into the dispersion process. In particular, the fact that turbulent mass flux in the stream wise direction is directed from the low to high levels of mean concentration (counter-gradient mechanism) is explained. The large vortical structures developing around the building are shown to play an essential role in turbulent mass transport Smolarkiewicz et al. [16] performed large-eddy simulations (LES) of the flow past a scale model of a complex building. Calculations are accomplished using two different methods to represent the edifice. The results demonstrated that, contrary to popular opinion, continuous mappings such as the Gal-Chen and Somerville transformation are not inherently limited to gentle slopes. Calculations for a strongly stratified case are also presented to point out the

substantial differences from the neutral boundary layer flows. Ahmad et al. [17] provided a comprehensive literature on wind tunnel simulation studies in urban street canyons/intersections including the effects of building configurations, canyon geometries, traffic induced turbulence and variable approaching wind directions on flow fields and exhaust dispersion. Jiang et al. [18] studied three ventilation cases, single-sided ventilation with an opening in windward wall, single-sided ventilation with an opening in leeward wall, and cross ventilation. In the wind tunnel, a laser Doppler anemometry was used to provide accurate and detailed velocity data. In LES calculations, two subgrid-scale (SS) models, a Smagorinsky SS model and a filtered dynamic SS model, were used. The numerical results from LES are in good agreement with the experimental data, in particular with the predicted airflow patterns and velocities around and within, and the surface pressures over, the models.

According to these anteriors studies, it's clear that the study of the aerodynamic around the obstacle is very interesting. Indeed, the literature review confirms that there is a paucity on the inclined roof obstacle study. For thus, we are interested on the study of the Reynolds number effect.

2. Geometrical System

The computational domain is shown in Figure 1. It is defined by the interior volume of the wind tunnel blocked by two planes. The first one is in the tranquillization chamber entry and the second one is in the exit of the diffuser. The test vein is equipped by the considered inclined roof obstacle.

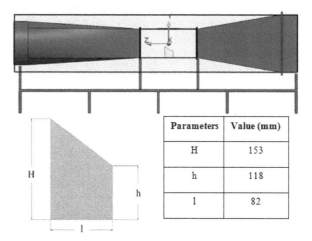

Parameters	Value (mm)
H	153
h	118
l	82

Figure 1. Geometrical arrangement

3. Numerical Model

The software "SolidWorks Flow Simulation" has been used to present the local characteristics. The numerical model considered is based on the resolution of the Navier-Stokes equations in conjunction with the standard k-ε turbulence model. These equations were solved by a finite volume discretization method [19,20,21].

3.1. Boundary Conditions

The boundary condition is required any where fluid enters the system and can be set as a pressure, mass flow, volume flow or velocity. Since we study in this application the effect of Reynolds number, we will give different values to the inlet velocity. For the outlet pressure we take a value of 101325 Pa which means that at this opening the fluid exits the model to an area of an atmospheric pressure. Knowing that the obstacle is suspended in our domain, both the roof top obstacle and the wall of our domain are considered as a wall boundary condition.

| Reynolds number | Velocity inlet | Pressure outlet |
Re	V (m.s⁻¹)	p (Pa)
2 666	0.1	101325
10 666	0.4	101325
24 000	0.9	101325
32 000	1.2	101325

Figure 2. Boundary conditions

3.2. Mesh Resolution

Figure 3 shows the initial mesh of the model. It is named initial since it is the mesh that the calculation starts from and it could be further refined during the calculation if the solution-adaptive meshing is enabled. The initial mesh is constructed from the basic mesh by refining the basic mesh cells in accordance with the specified mesh settings. The basic mesh is formed by dividing the computational domain into slices by parallel planes which are orthogonal to the global coordinate system's axes. Flow simulation options permit the computational mesh adjustment. In the near wall of roof top obstacle, the "initial" mesh corresponds to a cell of 5 cm. However, the "refined" mesh corresponds to a cell of 0.5 cm. In these

cases, the number of hexahedral cells is respectively equal to 5621 and 37038.

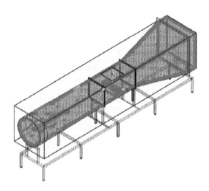

Figure 3. Meshing

4. Numerical Results

4.1. Magnitude Velocity

Figure 4 presents the distribution of the magnitude velocity in the longitudinal planes defined by X=0 mm. According to these results, it has been noted that the velocity is weak in the inlet of the collector. It is indeed governed by the boundary condition values of the inlet velocity. In this region, the velocity field is found to be uniform and increases progressively downstream of the collector. At the test vein, an important increase has been noted due to the reduction of the tunnel section that causes the throttling of the flow. While the upper side of the obstacle is characterized by the high velocity, a brutal drop is located behind the obstacle and this is due to the deceleration of the velocity field while passing through the obstacle. In the test vein, the velocity keeps increasing till the out of the test section. Then, a decrease has been noted through the diffuser where the minimum velocity values are recorded in the lateral walls of the diffuser. Indeed, it's clear that the Reynolds number has a direct effect on the increase of the maximum value of the velocity. For example, with Re=2666 the maximum value of the velocity is equal to V=0.88 m.s⁻¹. However, with Re= 32000 the maximum value of the velocity is equal to V=10.49 m.s⁻¹.

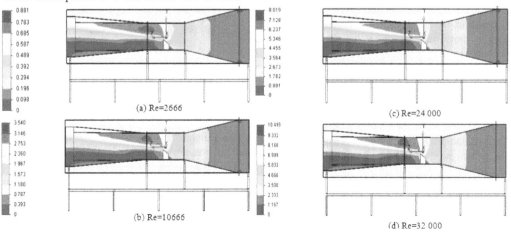

Figure 4. Velocity in the longitudinal plane X=0 m

4.2. Velocity Streamlines

Figure 5 presents the distribution of the velocity streamlines in the longitudinal plane defined by X=0 mm. According to these results, it has been noted that the velocity is weak in the inlet of the collector. It is indeed governed by the boundary condition value of the inlet velocity. In this region the velocity is found to be uniform and increases progressively downstream of the collector. At the test vein, an important increase has been noted due to the reduction of the tunnel section that causes the throttling of the flow. While the upper side of the obstacle is characterized by the high velocity, a brutal drop is located behind the obstacle. This is due to the deceleration of the velocity field while passing by the obstacle. In the test vein the velocity keeps increasing till the out of the test section. Then, a decrease on the maximum values has been noted through the diffuser where the minimum velocity values are recorded in the lateral walls of the diffuser. The flow circulation appears in the dead zones where the velocity presents a weak value. In our case, the flow circulations are located behind the obstacle and in the diffuser outlet. Besides, the maximum value of the velocity streamlines increases with the increase of Reynolds number value.

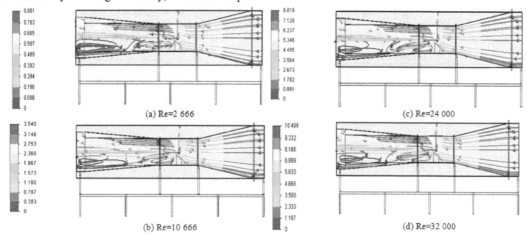

Figure 5. Velocity streamlines in the longitudinal plane x=0 mm

4.3. Static Pressure

Figure 6 presents the distribution of the static pressure in the longitudinal planes defined by X=0 mm. According to these results, it can easily be noted that the total pressure is on its maximum in the inlet of the collector. Besides, it has been observed a depression above the obstacle. The pressure continues decreasing the way out of the test vein. A brutal drop of the pressure has been noted just behind the obstacle. The distribution of the static pressure in the plane y=0 mm shows that a depression zone is located in the second half of the wind tunnel through the diffuser. Indeed, it's clear that the Reynolds number has a direct effect on the static pressure distribution. In fact, the maximum value of the static pressure increases with the increase of the Reynolds number value. For example, in the longitudinal plane X=0, the maximum value of the static pressure is equal to p=101325 Pa for Re=2666 and becomes equal to p=101372 for Re= 32000.

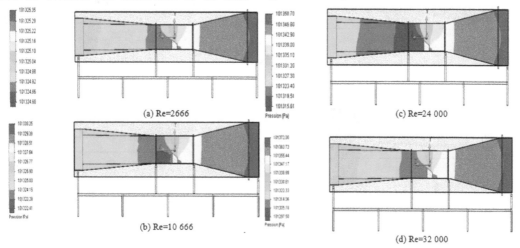

Figure 6. Static pressure in the longitudinal plane X=0 mm

4.4. Dynamic Pressure

Figure 7 presents the distribution of the dynamic pressure respectively in the longitudinal planes defined by X=0 mm According to these results, the dynamic pressure is found to be weak in the collector inlet and increases gradually through the collector as long as the tunnel section gets smaller. When it gets to the test section, the dynamic pressure keeps increasing in the upstream of the obstacle. A compression zone is recorded in the region located behind the obstacle and is developed through the diffuser. The distribution of the dynamic pressure in the transverse plane shows a minimum zone located in the downside of the wind tunnel and a maximum zone located in its upper side. Indeed, it's clear that the Reynolds number has a direct effect on the distribution of the

dynamic pressure. In fact, the maximum value of the dynamic pressure increases with the increase of Reynolds number value. For example, at longitudinal plane X=0 mm, the maximum value of the dynamic pressure is equal to p_d=0.43 Pa for the Reynolds number equal to Re= 2666. However, it becomes equal to p_d=60 Pa for the Reynolds number equal to Re=32000.

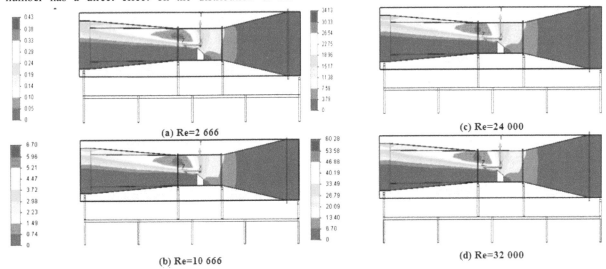

Figure 7. Dynamic pressure in the longitudinal plane X=0 mm

4.5. Turbulent Kinetic Energy

Figure 8 presents the distribution of the turbulent kinetic energy in the longitudinal plane defined by X=0 mm. From these results, it has been noted that the turbulent kinetic energy is found to be very weak in the first half of the wind tunnel in the obstacle upstream. A wake characteristic of the maximum value of the turbulent kinetic energy appears upstream of the obstacle. This wake starts in the obstacle corner until the outlet of the

diffuser. Indeed, it's clear that the Reynolds number has a direct effect on the turbulent kinetic energy distribution. In fact, the maximum value of the turbulent kinetic energy increases with the increase of the Reynolds number value. For example, in the longitudinal plane X=0 mm, the maximal value of the turbulent kinetic energy is equal to k=0.03 $m^2.s^{-2}$ for a Reynolds number equal to Re=2666. However, it becomes equal to k=4.44 $m^2.s^{-2}$ for a Reynolds number equal to Re= 32000.

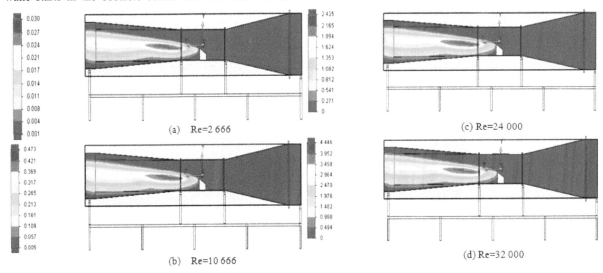

Figure 8. Turbulent kinetic energy in the longitudinal plane X=0 mm

4.6. Turbulent Dissipation Rate

Figure 9 presents the distribution of the turbulent dissipation rate on the longitudinal plans defined by X=0 mm. From these results, it has been noted that the turbulent dissipation rate is found to be very weak in the

first half of the wind tunnel in the obstacle upstream. A wake characteristic of the maximum value of the turbulent dissipation rate appears upstream of the obstacle. This wake starts in the obstacle corner until the outlet of the diffuser. Indeed, it's clear that the Reynolds number has a direct effect on the turbulent dissipation rate distribution.

In fact, the maximum value of the turbulent dissipation rate increases with the increase of the Reynolds number value. For example, in the longitudinal plane X=0 mm, the maximal value of the turbulent dissipation rate is equal to ε=0.04 W/Kg for a Reynolds number equal to Re=2666. However, it becomes equal to ε=72.2 W/Kg for a Reynolds number equal to Re= 32000.

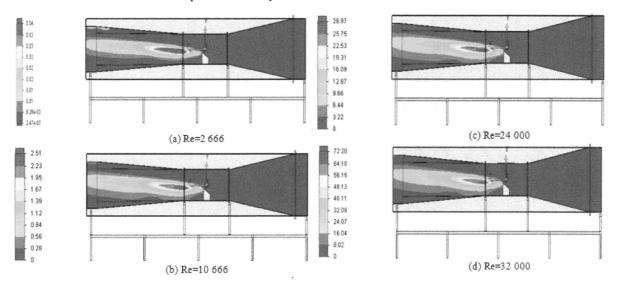

Figure 9. Turbulent dissipation rate in the longitudinal plane x=0 mm

4.7. Turbulent Viscosity

Figure 10 presents the distribution of the turbulent viscosity in the longitudinal planes defined by X=0 mm. According to thes results; it's clear that the viscosity is at its minimum in the collector region but starts increasing after crossing the obstacle. Its maximum is located in the outlet of the diffuser. Indeed, it's clear that the Reynolds number has a direct effect on the turbulent viscosity. In fact, the maximum value of the turbulent viscosity increases with the increase of the Reynolds number value. For example, in the longitudinal plane X=0 mm, the maximal value of the turbulent viscosity is equal to μ_t=0.005 Pa.s for a Reynolds number equal to Re=2666. However, it becomes equal to μ_t=0.06 Pa.s for a Reynolds number equal to Re= 32000.

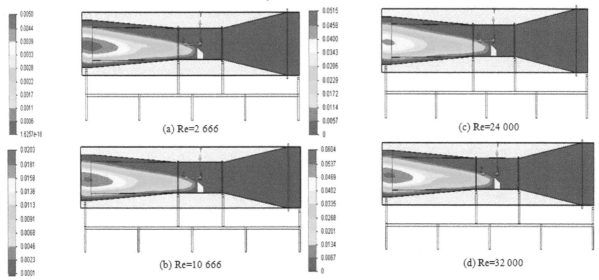

Figure 10. Turbulent viscosity in the longitudinal plane X=0 mm

4.8. Vorticity

Figure 11 presents the distribution of the vorticity respectively in the longitudinal plane defined by x=0 mm. According to these results, the vorticity is at its minimum in the collector region and starts increasing after crossing the obstacle. The greatest vorticity values are reached after hitting the obstacle blade. Indeed, it's clear that the Reynolds number has a direct effect on the vorticity distribution. In fact, the maximum value of the vorticity increases with the increase of the Reynolds number value. For example, in the longitudinal plane X=0 mm, the maximal value of the vorticity is equal to 6.2 s^{-1} for a Reynolds number equal to Re=2666. However, it becomes equal to V=96.5 s^{-1} for a Reynolds number equal to Re= 32000.

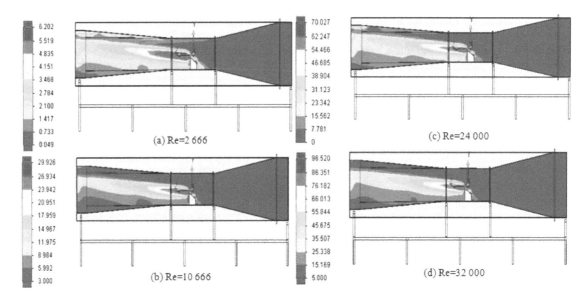

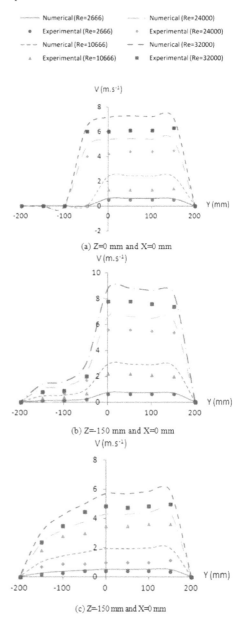

Figure 11. Vorticity in the longitudinal plane X=0 mm

5. Comparison with Experimental Results

In this section, we are interested on the comparison of the numerical results with the experimental results conducted in the LASEM laboratories using a wind tunnel (Figure 12). The velocity profiles are chosen for points situated in the test section. The considered planes are defined by Z=0 mm, Z=150 mm and Z=-150 mm. The results for each plane are shown respectively in Figure 13. For each transverse plane, values are taken along the directions defined by X=0 mm. Near the obstacle, it's clear that the velocity value is very weak. Outside, the velocity has a maximum value. Indeed, it's clear that the Reynolds number has a direct effect on the maximum value. For example, for the direction defined by Z=0 mm and X=0 mm, the maximum value is equal to V=0.6 m.s^{-1} for the Reynolds number equal to Re= 2666. However, this value increases to V=7 m.s^{-1} for the Reynolds number equal to Re=32000. The comparison between the numerical and experimental velocity values leads us to the conclusion that despite some unconformities, the values are comparable. The numerical model seems to be able to predict the aerodynamic characteristics of the air flow around the inclined roof obstacle.

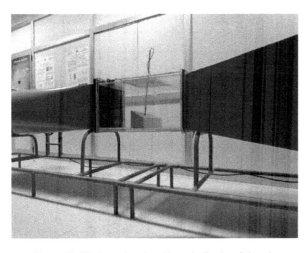

Figure 12. Wind tunnel equipped by an inclined roof obstacle

Figure 13. Velocity profiles

6. Conclusion

In this paper, we have performed a numerical simulation to study the Reynolds number effect on the aerodynamic characteristics around an obstacle with inclined roof. Numerical results, such as velocity fields, pressure and turbulent characteristics are presented in the wind tunnel, considered as a control volume. According to the obtained results, it has been noted that the Reynolds number has a direct effect on the local characteristics. In fact, the maximum values of the velocity and the turbulent characteristics increase with the increase of the Reynolds number. Indeed, the circulation zones of the velocity streamlines and the wakes characteristics of the maximum values of the turbulent characteristics are located after hitting the obstacle. Use of this knowledge will assist the design of packaged installations of the wind rotors in the buildings. In the future, we propose to study the geometrical effect of the obstacle.

Nomenclature

H: big high, m

h: small high, m

k: turbulent kinetic energy, $J.kg^{-1}$

l: length, m

p: pressure, Pa

Re: Reynolds number, dimensionless

t: time, s

V: magnitude velocity, $m.s^{-1}$

X: Cartesian coordinate, m

Y: Cartesian coordinate, m

Z: Cartesian coordinate, m

ε: dissipation rate of the turbulent kinetic energy, $W.kg^{-1}$

μ_t: turbulent viscosity, Pa.s

References

[1] Nozawa, K., Tamura, T., *Large eddy simulation of the flow around a low-rise building immersed in a a rough-wall turbulent boundary layer*, J. Wind Eng. Ind. Aerodyn., 90. 1151-1162. 2002.

[2] Kawai, H., *Local peak pressure and conical vortex on building*, Proceedings of the 10th International Conference on Wind Engineering. 1807-1812. 1999.

[3] Kondo, K., Murakami, S., Mochida, A., *Generation of velocity fluctuation for inflow boundary condition of LES*, J. Wind Eng. Ind., 67&68. 51-64. 1997.

[4] Kataoka, H., Mizuno, M., *Numerical flow computation around 3D square cylinder using inflow turbulence*, J. Archit. Plan environ. Eng., AIJ 523. 71-77. 1999.

[5] Noda, H., Nakayama, A., *Reproducibility of flow past two-dimensional rectangular cylinders in a homogeneous turbulent flow by LES*, J. Wind Eng., 89. 321-324. 2001.

[6] Nasrollahei, N., Mahdavinejad, M., Hadiyanpour, M., *Studying the Thermal and Cryogenic Performance of Shevadun in Native (Local) Buildings of Dezful Based on Modeling and Environmental Measuring*, American Journal of Energy Research, 1 (3). 45-53. 2013.

[7] Ould said, B., Retiel, N., Bouguerra, E.H., *Numerical Simulation of Natural Convection in a Vertical Conical Cylinder Partially Annular Space*, American Journal of Energy Research, 2 (2). 24-29. 2014.

[8] Tominaga, Y., Stathopoulos, T., *CFD simulation of near-field pollutant dispersion in the urban environment: A review of current modeling techniques*, Atmospheric Environment, 79. 716-730. 2013.

[9] De Paepe, M., Pieters, J. G., Cornelis, W. M., Gabriels, D., Merci, B., Demeyer, P., *Airflow measurements in and around scale-model cattle barns in a wind tunnel: Effect of wind incidence angle*, Biosystems Engineering 115, 211-219. 2013.

[10] lim, H.C., Thomas, T.G., Castro, I.P., *Flow around a cube in a turbulent boundary layer: LES and experiment*, Journal of Wind Engineering and Industrial Aerodynamics, 97. 96-109. 2009.

[11] De Melo, A. M. V., Santos, J. M., Mavrroidis, I., Reis Junior, N. C., *Modelling of odour dispersion around a pig farm building complex using AERMOD and CALPUFF. Comparison with wind tunnel results*, Building and Environment, 56. 8-20. 2012.

[12] Meslema, A., Bode, F. Croitorub, Cristiana C., *Comparison of turbulence models in simulating jet flow from a cross-shaped orifice*, European Journal of Mechanics B/Fluids, 44. 100-120. 2014.

[13] Ntinas, G.K., Zhangb, G., Fragos, V.P., Bochtis, D.D., Nikita-Martzopoulou, Ch., *Airflow patterns around obstacles with arched and pitched roofs: Wind tunnel measurements and direct simulation*, European Journal of Mechanics B/Fluids, 43. 216-229. 2014.

[14] Luo, W., Dong, Z., Qian, G., Lu, J., *Wind tunnel simulation of the three-dimensional airflow patterns behind cuboid obstacles at different angles of wind incidence and their significance for theformation of sand shadows*, Geomorphology, 139-140. 258-270. 2012.

[15] Gousseau, P., Blocken, B., Van Heijst, G.J.F., Large-Eddy Simulation of pollutant dispersion around a cubical building, *Analysis of the turbulent mass transport mechanisim by unsteady concentration and velocity statistics*, Environmental pollution, 167. 47-57. 2012.

[16] Piotr, K.S., Sharman, R., Weil, J., Steven G. Perry, Heist, D., Bowker, G., *Building resolving large-eddy simulations and comparison with wind tunnel experiments*, Journal of Computational Physics, 227. 633-653. 2007.

[17] Ahmed, K., Khare, M., Chaudhry, K.K. *Wind tunnel simulation studies on dispersion at urban street canyons and intersections- a review*, J. Wind Eng. Ind. Aerodyn., 93. 697-717. 2005.

[18] Jiang, Y., Alexander, D., Jenkins, Huw. Rob Arthur, Qingyan Chen, Natural ventilation in buildings: *measurement in a wind tunnel and numerical simulation with large-eddy simulation*, Journal of Wind Engineering and Industrial Aerodynamics, 91. 331-353. 2003.

[19] Driss, Z., Bouzgarrou, G., Chtourou, W., Kchaou, H., Abid, M.S., *Computational studies of the pitched blade turbines design effect on the stirred tank flow characteristics*, European Journal of Mechanics B/Fluids, 29. 236-245. 2010.

[20] Ammar, M., Chtourou, W., Driss, Z., Abid, M.S., *Numerical investigation of turbulent flow generated in baffled stirred vessels equipped with three different turbines in one and two-stage system*, Energy, 36. 5081-5093. 2011.

[21] Driss, Z., Abid, M.S., *Use of the Navier-Stokes Equations to Study of the Flow Generated by Turbines Impellers*. Navier-Stokes Equations: Properties, Description and Applications, 3. 51-138. 2012.

25

Policies Enhancing Renewable Energy Development and Implications for Nigeria

Nnaemeka Vincent Emodi[1,*], **Nebedum Ekene Ebele**[2]

[1]College of Business, Law and Governance, James Cook University, P. O. Box 6811, Cairns Qld 4870, Australia
[2]Department of Climate Change, Hallym University, Hallym University 1 Hallymdaehak-gil, Chuncheon, Gangwon-do, South Korea
*Corresponding author: emeka50@snu.ac.kr

Abstract The main objective of this study was to review the various policies and strategies promoting renewable energy development around the world. The success and failures of each country and regions were examined through a case study so as to learn some valuable lessons and derive useful implications for the development of renewable energy in Nigeria. The study initially reviewed the current renewable energy policies and identified the barriers to the development of renewable energy technology in Nigeria. The lessons from the case study were classified into support mechanisms which include; capital, fiscal, tax incentives, legislative, political, technological and environmental support. The lessons from case study were used to develop implications in addressing the development of renewable energy technologies through effective policies and strategies in Nigeria. Furthermore, some future perspectives of renewable energy development in Nigeria were discussed. This study intends to support the Nigerian government and policymakers in decisions making and policy formulation on the short-, medium- and long-term.

Keywords: renewable energy policies, nigeria, projects, case study, tax incentives, barriers

1. Introduction

Access to clean modern energy services is an enormous challenge facing the African continent because energy is fundamental for socioeconomic development and poverty eradication. Today, 60% to 70% of the Nigerian population does not have access to electricity [1]. There is no doubt that the present power crisis afflicting Nigeria will persist unless the government diversifies the energy sources in domestic, commercial, and industrial sectors and adopts new available technologies to reduce energy wastages and to save cost [2].

In diversifying energy sources that is sustainable, renewable energy (RE) becomes the best option. A lot of attention have recently been focused on RE and energy efficiency (EE) measures as the only way to address the problem of clean energy provision [3]. The development of RE projects is considered as a huge opportunity from a strategic and financial point of view, as well as from a technological and environmental aspect [4]. In the development of RE projects, the government is a major player in ensuring the success of RE development and deployment [5].

This is achieved through the establishment of strategic policies, plans and the adoption of proper mechanisms. These have the capacity to alter the price of RE technologies through subsidies, taxes, funds power production and grid access for RE electricity generators.

In order to develop an effective strategy to promote RE development in Nigeria, it is essential to examine the situation of other countries in terms of RE technology development so as to learn from their success and failures. This will help present some important implications for the Nigeria government to learn from the international experiences.

This study aims to answer the following research questions;
•   What are the success and failures of renewable energy development from international experiences?
•   What lessons can be derived from the experience so as to provide implications for Nigeria?

In order to address the research questions, this study first examine the current RE policies in Nigeria and the barriers to its development. A literature review is then used to identify the success and failures of renewable energy policies and strategies employed in various countries and regions of the world. This study presents the following research objectives which are;
•   To identify the reasons for the success and failures of various RE policies and strategies from international experiences.
•   To derive lessons and implications for Nigeria in the development of RE policies and strategies.

The rest of this study is organized as follows. Section 2 presents an overview on the Nigerian energy policies and strategies. The barriers to Re development in Nigeria are identified in Section 3. Section 4 presents the literature review summary of lessons from international experiences on RE development. Section 5 provides some policy implications for Nigeria based on the lessons from international experiences. Section 6 discuss some future perspectives of renewable energy development in Nigeria. Section 7 concludes the study.

2. The Nigerian Energy Policies and Strategies

The Nigerian energy policies and strategies are summarized to provide an insight on the status quo of renewable energy policies, and they are listed from the earliest to the latest.

2.1. National Electric Power Policy (NEPP), 2001

The National Electric Power Policy (NEPP) was the first of its kind in the wake of reform in the Nigerian power sector. Its development was due to the recommendations of the Electrical power implementation Committee (EPIC), which was the body in charge of reforms and transformation of the power sector in 1999. The NEPP was created in March 2001, and presented three bold steps in achieving the goal of reforming the power sector. The first step was to privatize NEPA which was state owned and introduce Integrated Power Producers (IPPs) of electricity. The next step was to increase competition between participants in the market, gradually remove subsidies and sale excess power to the DISCOs. In the last step, it was expected that the market and competition would have been more intense and allow for full cost pricing of supply, and liberalization of the electricity market would have been complete [6].

2.2. National Energy Policy (NEP), 2003, 2006, 2013

Before the Federal Government of Nigeria approved the energy policy in the year 2003, there was no comprehensive energy policy. The established energy policy was called the National Energy Policy (NEP) which was developed by the Energy Commission of Nigeria (ECN). The National Energy Policy (NEP) sets out government policy on the production, supply and consumption of energy reflecting the perspective of its overall needs and options. The main goal of the policy is to create energy security through a robust energy supply mix by diversifying the energy supply and energy carriers based on the principle of "an energy economy in which modern renewable energy increases its share of energy consumed and provides affordable access to energy throughout Nigeria, thus contributing to sustainable development and environmental conservation" [7,8,9].

2.3. National Economic Empowerment and Development Strategy (NEEDS), 2004

The National Economic Empowerment and Development Strategy (NEEDS) was developed by the National Planning Commission (NPA) in 2004 and was intended to develop and alleviate poverty in the country. This involves the action of human resources on the natural resources to produce goods necessary to satisfy the economic needs of the community. On infrastructure, NEEDS promotes the privatization of government infrastructure and was one of the key instrument in achieving a revamped service delivery. The Nigerian government will however, fund projects that have very low attractiveness and high investment cost to investors such as those in rural areas. Furthermore, the increased share of renewables in the national energy mix was further encouraged in the NEEDS. This involves the suggestion for the creation of renewable energy agency and technologies which will be funded under the National Power Sector Reform Act. This was the milestone towards the adoption of renewables in the power sector and its utilization for rural electrification [10,11].

2.4. National Power Sector Reform Act (EPSRA), 2005

The National Power Sector Reform Act established in 2005 ensured the liberalization of the Nigerian power sector. The Act was due to the NEPP developed in 2001 and made provision for new legal and regulatory framework for the power sector. The Act gave way to unbundling and privatization of the power sector, which intends to introduce competition in the electricity market, enhance rural electrification, while protecting consumer rights and developing performance standards in the power sector [12].

2.5. Renewable Electricity Policy Guidelines (REPG), 2006

Developed by the federal Ministry of Power and Steel in December, 2006, the Renewable Electricity Policy Guidelines (REPG) mandated the Nigerian government on the expansion of electricity generation from renewables to at least 5% of the total electricity generated and a minimum of 5 TWh of electricity generation in the country (REPG, 2006). This policy document presents the Nigerian government's plans, policies, strategies and objectives for the promotion of renewables in the power sector [13].

2.6. Renewable Electricity Action Programme (REAP), 2006

Developed in relation to the REPG by the Federal Ministry of Power and Steel in 2006, the Renewable Electricity Action Programme (REAP) set out a roadmap for the implementation of the REPG. The document presents an overview of the Nigerian electricity sector and relates it to renewable energy development. The documents also reviews government targets and provides strategies for renewable energy development such as; leveling the playing field for renewable electricity producers, multi-sector partnerships, demonstration projects, supply chain initiatives, etc. The study also made provision for financing renewable programs and explored the roles of government ministries and agencies, then concludes with a risk assessment, monitoring and evaluation [14].

2.7. Nigerian Biofuel Policy and Incentives (NBPI), 2007

The aim of this policy was to develop and promote the domestic fuel ethanol industry through the utilization of agricultural products. This was in line with the government's directive on an Automotive Biomass Programme for Nigeria in August 2005. The NNPC was mandated to create an environment for the take-off of the ethanol industry. The policy further aimed at the gradual reduction of the nation's dependence on imported gasoline, reduction in environmental pollution, while at the same time creating a commercially viable industry that can precipitate sustainable domestic jobs. The benefits of this policy was to create additional tax revenue, provision of jobs to reduce poverty, boost economic development and empower those in the rural areas, improve agricultural activities, energy and environmental benefits through the reduction of fossil fuel related GHGs in the transport sector [15].

2.8. Renewable Energy Master Plan (REMP) 2005 and 2012

The Renewable Energy Master Plan (REMP) was developed by the Energy Commission of Nigeria (ECN), in collaboration with the United Nations Development Programme (UNDP) in 2005 and was later reviewed in 2012. The REMP expresses Nigeria's vision and sets out a road map for increasing the role of renewable energy in achieving sustainable development. The REMP is anchored on the mounting convergence of values, principles and targets as embedded in the National Economic Empowerment and Development Strategy (NEEDS), National Energy Policy, National Policy on Integrated Rural Development, the Millennium Development Goals (MDGs) and international conventions to reduce poverty and reverse global environmental change [16,17].

The REMP stress the need for the integration of renewables in buildings, electricity grids and for off-grid electrical systems. Further, the importance of solar power in the country's energy mix was also highlighted in the policy document. According to the REMP, Nigeria intends to increase the supply of renewable electricity from 13% of total electricity generation in 2015 to 23% in 2025 and 36% by 2030. Renewable electricity would then account for 10% of Nigeria's total energy consumption by 2025. However, the REMP have not been approved by the National Assembly to be passed into law.

2.9. National Renewable Energy and Energy Efficiency Policy (NREEEP), 2014

The National Renewable Energy and Energy Efficiency Policy (NREEEP) outlines the global thrust of the policies and measures for the promotion of renewable energy and energy efficiency. The FMP developed the NREEEP in 2014 and is awaiting the approval of the Federal Executive Council [18].

2.10. Multi-Year Tariff Order (MYTO), 2008 and 2012

In 2008, a 15-year roadmap towards cost reflective tariffs called the Multi-Year Tariff Order (MYTO 1) was developed by the Nigerian Electricity Regulatory Commission (NERC). The first two phases, 2008-200 and 2012-2017 were designed to keep consumer prices relatively low, through still affecting the price increases in a gradual manner. The final regime is intended to provide the necessary incentives for power producers and investors to operate and maintain electricity infrastructure.

The NERC has released the Multi-Year Tariff Order 2 (MYTO 2), which has similar features to MYTO 1 but includes some improvements, and will be effective from 1st June 2012 to 31st May 2017. The retail tariff in MYTO 2 will be reviewed bi-annually and changes may be made for all electricity generated at wholesale contract prices, adjusted for the Nigerian inflation rate, US$ exchange rate, daily generation capacity, and accompanying actual CapEx and OpEx requirements that will vary from those used in the tariff calculation [19].

2.11. Draft Rural Electrification Strategy and Implementation Plan (RESIP), 2014

The Power Sector Reform team initially prepared the Rural Electrification Strategy and Implementation Plan in 2006 (RESIP, 2006). However, a committee involved in the power sector reviewed and redrafted the RESIP in 2014. It was expected to establish a clear institutional step-up for the sector and set a roadmap which will result in the development of an enabling framework for rural electrification in Nigeria. The primary objective of the RESIP is to expand access to electricity as rapidly as can be afforded in a cost-effective manner. This includes the use of on-grid and off-grid means of electricity supply. The draft is ready and awaiting approval from the government [20].

Figure 1. An Overview of the Nigerian Energy Policies, Laws, Programmes

A general overview of the various Nigerian energy policies, legislations, regulations, standards, programmes and incentives, their achievements, gaps, alignment, overlaps and opportunities are presented in Figure 1. From

Figure 1, it can be observed that various gaps exist in the Nigerian energy policies, while some policies such as the ERSP Act of 2005 is long overdue for revision. There is also no need for the development of an energy efficiency law, instead the development should be on voluntary renewable energy and energy efficiency standards. Besides this, the follow-ups of all the Nigerian energy policies, strategies and targets are lacking, as well as commitment from the state and local government in renewable energy development. Some overlaps also exist in some activities of some ministries and government agencies, while duplication of the same activities is observed in some Federal ministries in the country. Alignment required in the Nigerian energy policies include the harmonization of National renewable energy, energy efficiency and rural electrification policy objectives, coordination of government agencies and effectiveness in the fulfillment of the mandate to coordinate policy issues.

3. Barriers to the Development of Renewable Energy in Nigeria

In order to achieve success in the development of renewable energy in Nigeria, some obstacles have to be identified. It is however, not easy to identify a single factor that can have a long positive impact on the development of renewables. We should then look at the supportive measures that can lead to its full exploit or can hinder its development. The factors in the Nigerian case are summarized in this section to list out the essential components affecting renewable energy development. The factors are classified into; capital investment, fiscal incentives, legislation and regulation, politics, policy and strategy, technology and innovation, and environmental support programs.

3.1. Financial Investment

Even though renewable energy sources have low operational and maintenance costs, most renewable energy technologies have high up- front capital cost compared to their conventional energy alternatives[1]. Apart from the higher capital costs most renewable energy technologies face the barrier of being perceived as untested technologies. Given these twin barriers to renewable energy technologies, investors face higher risks and uncertainties when making investment decisions. Therefore in a capital constrained economy like Nigeria, where there are many competing demands for available scarce capital resources, the promoters of Renewable energy technology face the problems of high transaction costs and restricted access to capital. On the other hand the end users of renewable energy technology, especially the poor, face problems of access to credits. Lack of access to micro financing, high interest rates, poor business development skills by renewable energy system vendors and unsupportive climate for investments are some of the primary barriers to market growth [21].

3.2. Power Purchase Agreement

Currently there is no Power Purchase Agreements plan for renewable energy generation to the national grid. A system of rational expectations between renewable electricity producers and the grid operators are imperative for the growth in grid-based renewables [22]. The Power Purchase Agreements set the terms by which power is marketed and/or exchanged. It determines the delivery location, power characteristics, price, quality, schedule, and terms of agreement and punishments for breach of contract [23]. Legally binding long-term Power Purchase Agreements are a must since they provide comfort for the developers of renewable as well as lenders, and would also encourage the expansion of renewable electricity development through investments[2].

3.3. Legislation and Regulation

Achieving adequate energy supply where renewables play a role necessitates the creation of appropriate policy framework of legal, fiscal and regulatory instruments that would attract domestic and international investments [24]. Clear rules, legislation, roles and responsibilities of various stakeholders along every stage of the energy flow from supply to end-use are key elements of the overall policy framework needed to promote renewable energy technologies [21]. Such policy, legal and institutional frameworks are at their beginning stage in Nigeria and are being developed under the reform program[3].

3.4. Politics, Policy and Strategies

Strong and long-term political support at the federal, state and local government level is a consistent component in the successful development of renewable energy [25]. The political support includes the proper implementation of policies and strategies, price support mechanisms, and provision of funds for R&D activities, as well as the deployment of renewable energy technologies [26-27]. The Nigerian energy policies and strategies discussed in the last section presented the government's plan for development of renewable energy [28]. However, the follow-up and active implementation of these policy is lacking from the Nigerian government.

3.5. Technology and Innovation

The development of renewable energy requires support in all stages of research, demonstration and deployment so as to achieve a competitive local industry [29]. Although most renewable energy technologies such as solar PV, wind power, small hydropower, etc. have become popular in developing countries, it's still a new technology in most developing countries such as Nigeria [30]. There is need for the Nigerian government to invest in R&D activities in renewable energy so as to enhance technological innovation. Sometimes, public sector funding may not be enough to increase R&D activities and as such, private sector investment is vital for the realization of technological innovation in renewable energy.

---

[1] http://www.ucsusa.org/clean_energy/smart-energy-solutions/increase-renewables/barriers-to-renewable-energy.html#.VjIbnSsbjMw

[2] http://www-wds.worldbank.org/external/default/WDSContentServer/WDSP/IB/2006/04/27/000012009_20060427113507/Rendered/PDF/36986.pdf

[3] http://www.un.org/esa/earthsummit/nigeriac.htm.

3.6. Environmental Support Program

The increase in global climate change has had its impact in many countries of the world, and this is due to the increase in energy consumption, especially fossil fuels as about 84% of the global CO2 emissions and 64% of the world's GHG emissions are attributed to energy consumption [31]. This has drawn the attention of the international communities to see to its reduction. Countries with high carbon emission from coal, gas and oil power plants have been under intense pressure to limit emission rate [32]. This countries have been tasked to provide funds to which comes in the form of environmental support programs. This support programs has a big impact into the drivers for technological support and thus, the development of renewables. This environmental support programs have not been fully exploited as a potential in Nigeria for the increase in renewable energy development.

3.7. Public Awareness

Awareness of the opportunities offered by renewable energies and their technologies is low among public and private sectors. This lack of information and awareness creates a market gap that results in higher risk perception for potential renewable energy projects [33]. The general perception is that renewable energy technologies are not yet mature technologies, hence are only suited for niche markets and as such will require heavy subsidy to make it work [34]. There is therefore a need for dissemination of information on renewable energy resource availability, benefits and opportunity to the general public in order to raise public awareness and generate activities in the area. Such process is paramount to building public confidence and acceptance of renewable energy technology. Providing information to selected stakeholder groups like the investors can help mobilize financial resources needed to promote renewable energy technology projects [21]. The draft Renewable Energy Master Plan proposes the set-up of a National Renewable Energy Development Agency (NREDA), which can assist in increasing public awareness and providing information and assistance to interested stakeholders. This is to be done together with non-governmental organizations (NGOs) [17].

4. Lessons from International Experiences

In some countries, the method of renewable energy policies comes as a "package" and not a "stand-alone" kind of policies. The packaged policies works in an interactive mode, where the success or failure of a single policy will depend on the effectiveness of other complementary policies. These section deals with the lessons learnt from international experiences on the development of renewable energy in relation to the Nigerian case as presented in Section 3.

4.1. Capital Support

One of the most important experiences in tackling capital investment provision is observed in Spain. The Spanish experience through its wind energy proved that the need for financial support is not permanent issue, since

the cost gradually reduced as the project becomes successfully done. With the implementation of legislative support through FiTs or tendering arrangement, there is less need for investors to receive financial aid in installation. Sometimes when prices are guaranteed, the investors will have enough confidence to invest in renewable energy projects without financial support. In a situation where FiTs and tendering arrangement is absent, subsides presents itself as the main mechanism as can be observed in the Swedish wind energy schemes [35].

The provision of capital for investment goes hand-in-hand with technological support, since the establishment of the RE technology needs initial push to the market. This is the situation in Germany's Solar PV market where in the beginning of PV installations, a lot of financial support is given by means of favorable loans [36]. It is therefore important to properly identify the type of mechanism (usually legislative) works best for a particular RE technologies in order to ensure an effective, efficient and successful completion of the project. In stand-alone RE technologies used for rural electrifications off-grid areas, lessons learnt from some Asia-Pacific countries showed that financial incentives are very effective which comes in the form of subsidies achieve a high success rate [37]. A summary of some measures employed by some countries/regions to address capital investment support in RE is shown in Figure 2.

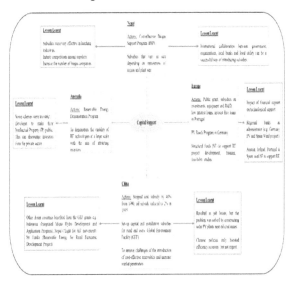

Figure 2. Capital Support

4.2. Fiscal Support

The provision of fiscal incentives can address financial barriers to the development of renewables. Fiscal support can come in various forms which may be from tax exemptions on imported RE equipment, to tax holidays on generation incomes [38]. Some tax incentive policies includes; import duty/excise duty concession, VAT concession, tax credit, production tax concession, and tax holiday on generation income. Furthermore, environmental taxes have proved to be an effective fiscal support in some countries. An example is observed in Denmark which was one of the first European Union countries to implement an environmental tax. Since 1992, Denmark energy consumers have been charged a CO2 tax and some of the revenue generated goes to generators of electricity from renewable energy [39]. Another country is Sweden which used the same approach and this added in the expansion of biomass. This made energy from fossil fuel such as coal very expensive [40]. Other forms of fiscal incentives are presented in Figure 3.

4.3. Tax Incentives

The reduction of complete exemption on tax can stimulate investment private sector investors into investing in renewables. An example is cited in Sweden and Germany where investment in wind schemes was an offset against individual tax, while tax relieve on renewable energy investment was received by investors in Ireland, Netherlands and Spain [41]. This was the same situation in Greece where the installation of solar thermal water-heating systems was stimulated by tax exemptions for residential buildings [42]. In the Netherlands, industries received accelerated depreciation of investment in equipment if they invest in renewables and energy efficiency projects [43]. This is also shown in Figure 3 under fiscal support.

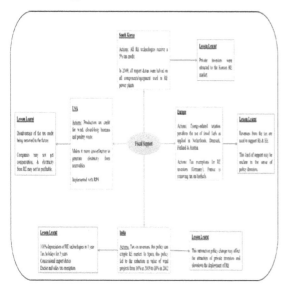

Figure 3. Fiscal Support

4.4. Legislative Support

In ensuring the accessibility of renewable energy market, some legislative support which comes in form of different strategies have been used by some countries. The main objective of these strategies is to let the market decide the tariff structure and the tariff to be paid to the renewable electricity generators based on a bidding system or in some cases, certificate markets [44]. Some legislative strategies are feed-in-tariff (FiT) and tendering arrangements, net-metering programs, and green certificates. FiT system have been proved to be a more cost effective than a quota system in some countries, while the reverse is the case as was experienced in the South Korean case [45]. Generally, FiT resulted in the reduction in electricity prices. Most countries that were successful in the FiT system include countries in the EU such as Denmark, Germany and Spain [46]. These countries recorded high percentage share of electricity from renewables through the FiT system. On the other hand, the quota system employed by the United Kingdom and Italy

have not been too successful. In the United States (USA), the net-metering programs were employed and the rules varies from a program to another. However, in the USA case, household participants are credited according to the offset in energy consumption, which is at an essential subsidized tariff rate [47]. Usually in a net metering program, a maximum offset of zero dollar monthly electricity bill is allowed, but in some cases, the credit of excess feed-in is allowed to roll over the next month [48]. The net metering programs also limit the amount of electricity from renewables. Countries such as Belgium, Italy, Poland, Romania, Sweden and the UK have applied for green certificates [49]. In the case of Belgium, a minimum price (s) was set and this varies across regions, while a price was imposed in the case of Poland which is the average price of the year before the current one. The UK and Sweden did not guarantee prices unlike their counterpart countries. Other legislative mechanisms are shown in Figure 4.

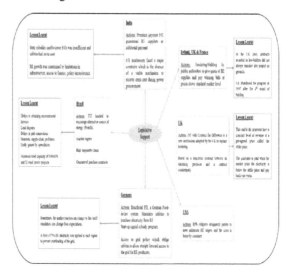

Figure 4. Legislative Support

4.5. Political Support

In order to achieve an expanded RE market, the creation of a favorable legal and regulatory framework is essential. Thus, the development of RE technologies requires a strong political support through an easy bureaucratic procedure (some cases are shown in Figure 5).

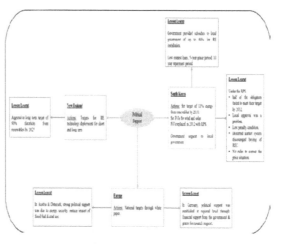

Figure 5. Political Support

Clear and simple regulations about the required license for the construction and operation of RE power plants is very important. Another means of political support is through the setting of national target as was carried out in the case of New Zealand, Austria and Denmark with strong support from the government. In Germany, the government provided financial support through research grants at the regional level [50-51]. Similar situation was observed in South Korea where the government provided subsidies to local government of up to 60% of RE installation. The South Korean government also offered low interest loans, 5 year grace period and a 10 year repayment period for all RE technologies [52].

4.6. Technological and Environmental Support

In order to support the development of RE, the technological and environmental support needs to be explored. For technological aspect, R&D support comes in mind. Most R&D programs are typically in the form of grants or loans and sometimes provided with no expectations of financial gains. However, it is very important that the particular R&D program receiving the funds for the R&D develops patentable technologies. In Australia, the government competitive grant scheme called the Renewable Energy demonstration Programs (REDP) was used to support the development of commercial RE projects and mini-grid project [53]. The German Federal Ministry of Economic Cooperation and Development (BMZ) commissioned the Renewable Energy Support Programme for ASEAN (RESP-ASEAN) to enable the sharing of expertise and policies in order to improve framework conditions for RE in ASEAN countries. This produced corresponding guidelines for bioenergy projects in Indonesia bioenergy project and Philippines solar project [54].

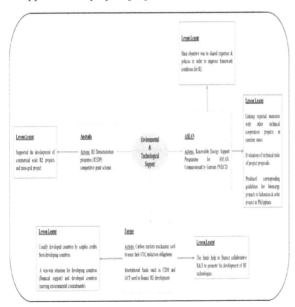

Figure 6. Environmental and Technological Support

On environmental support, crediting mechanisms such as the Clean Development Mechanism (CDM) and Global Climate Fund (GCF) can provide financial and technological incentives to aid in the reduction of

greenhouse gas emission levels in energy intensive sectors [55]. Another support is through environmental audits which is generally used to access the current environmental performance of an industry and to identify measures to improve energy efficiency [56]. However, the provision of funds is usually the problem of the organization carrying out the audit. This has been observed in Canada, Denmark, Netherland and Sweden, but their own form of assistance is usually through other voluntary schemes [57]. Other countries that applied this scheme are Costa Rica, Israel, Portugal, Taiwan, Thailand and Tunisia, while low success rate was observed in Egypt [58]. Other experiences are presented in Figure 6.

5. Implications for Nigeria

The review of the various country case and the lessons presented some important implications for Nigeria and this can be used to address the barriers to RE development and they are as follows.

5.1. Subsidies and Grants

In order to achieve a successful subsidies and grant mechanism for RE development in Nigeria, a strong commitment is required from the government in not only the provision of funds, but also making them efficient enough through a cost recovery mechanism so as to ensure sustainability. The absence of a cost recovery system make the receiving entity vulnerable to financial critical situation and they may fail to maintain their operational efficiency.

5.2. Loans

On the issue of loans, the role of the Nigerian government is vital for a successful financial support on RE technologies. The first step for the government in recovering loans like revolving funds may help encourage banks to finance RE projects and hence overcome the barrier of RE investment. The government can further support micro-finance banks by financing loans, providing soft loans and long loans for RE projects.

5.3. Feed-in-Tariff

Although FiT proved successful in some countries and unsuccessful in others, some measures that could be employed by policy makers in Nigeria include; provision of higher price to generators to stimulate increased supplies of RE electricity, ensure a shorter payback period on investments, avoid market monopoly by removing market entry barriers of small investors. Others are; the provision of a flexible FiT system based on technology type, market structure, and location. The government should ensure a secured return over years for investors so as to reduce the risk in RE projects. Finally and most importantly, the government should ensure that the FiT system is sustainable on the long run through an incremental cost recovery system.

5.4. Renewable Portfolio Standards

Considered as the least-cost option for RE development in many countries, RPS has a reputation for bringing down the cost of RE technologies and it creates a competitive market. The RPS system has shown in many countries to be a sustainable policy to RE generators since the government will compensate them for their extra costs through subsidies. This option if considered by the Nigerian government has the potential to create a RE market in Nigeria that will be very competitive since Nigeria have the market and large RE resource potential. However, the Nigerian government should ensure the flexibility of targets and adjust it on the short-term bases. The government should also take note of regional imbalance in pricing as some locations may have higher RE resource potential than others, while some may have the problems of electricity transmission and distribution.

5.5. Competitive Bidding

This mechanism is considered to be the most favorable policy for end users of electricity and the government in most countries, since it reduces the price of RE technologies through market-based pricing. The downside of competitive bidding is that it may face the risk of price bids that are unsustainable on the long-run. Also extreme low price of energy/electricity may prevent investors and this should be taken into account by the Nigerian government if bidding system should be implemented.

5.6. Fiscal Support

It should be noted that tax incentives on investments can led to reduction in operational efficiency. This sometimes occur when the owners of the RE project receive the benefit of investment tax credit, after which the owner lose interest in maintenance and operation of the RE power plants. A more feasible approach will be production tax credit which motivates the RE owners to ensure an effective operation of the power plant. Another effective fiscal support is the structural tax policy with strict enforcement mechanism followed by a good administrative system. In the Nigerian case, this will ensure that the incentives are efficiently utilized and the problems of tax payment and bad practices are avoided.

6. Future Perspectives of Renewable Energy Development in Nigeria

There is the need for the Nigerian government to develop the renewable energy industry within the country boarders. It is important to harmonize the Nigerian energy sector rather than fragmentize it. Mutual cooperation between government agencies and ministries is the first step in a successful renewable energy development. The government and policymakers need to explore renewable energy options such as small hydropower plants across the country, especially in the rural areas. If this is done in the rural areas with flowing small rivers, the communities could be transformed into semi-urban centers. Thus, enhancing various economic activities that will improve the socio-economic wellbeing of those in the rural areas. On solar energy awareness, the rate has increased in recent times and the acceptance is expected to increase in the future [59]. Some successful research such as the implementation of solar PV water pump and

electrification, solar cooking stoves, crop drying facilities, etc. have been undertaken out in the Sokoto Energy Research Center and the national Center for Energy Research and Development under the supervision of the Energy Commission of Nigeria. On wind development, the government have completed the installation of the 20 MW wind farm in Katsina state. However, the progress of wind development in Nigeria is still relatively low as compared to other developing countries in Africa. Since the northern part of Nigeria have high potential for wind energy and the communities have low access to energy, the government most do more regarding the exploitation of wind energy to meet the needs to the populace.

Biomass have been an important renewable energy source in Nigeria, but its sustainability is still a big issue in Nigeria. The government should continuously use municipal waste, oil palm product, sugar can and rice husk for the sustainability of biomass energy production. Also the government incorporation with the private sector should establish some biogas plants to help the country's energy sector through energy production [60]. On policy aspect for renewable energy development, effective policy options that can attract foreign direct investment into the country should be a top priority for the Nigerian government. This will not only aid in the production of energy from renewable energy source, but also create jobs and enhance the transfer of technology and knowledge to the local partners. This should be done with some fiscal incentives to encourage indigenous and foreign investors in the renewable energy industry [61]. Furthermore, the development of a ten-year energy plan that will be review in a short time frame. Other policy options to enhance renewable energy development in the future include the introduction of a National innovation System with a special focus on renewable energy development, and the development of a renewable portfolio standards to increase electricity generation from renewable energy sources. The Net Metering Policy and Biofuels Obligation Policy that have proved successful in other countries should be adopted in the Nigerian context to see if effectiveness, efficiency and successful integration. All these present a successful prospect for renewable energy development in Nigeria. It is important to state that Nigeria's long decades of dependence on conventional energy resources may create a "lock-in" for the transition to renewable energy economy. Also, this transition will be expensive in the short term due to the huge investments in renewable energy investment, and the transfer of technology and knowledge. However, the long-term benefit of this move to an economy focused on renewable energy development will outweigh the short-term disadvantages.

7. Conclusion

This study reviewed the current RE and energy efficiency policies in Nigeria and also interrelate them in order to identify the gaps, overlaps and barriers that exist. Then a case study on various country cases which were both successful and unsuccessful were reviewed with the aim of deriving some lessons from the international experiences. This provided useful implication for Nigeria on how to address subsidies and grants, loans, FiT, RPS, competitive bidding and fiscal support for renewable energy development. RE development have been successful in most countries through government interventions in form of fiscal incentives and political will of the government. However in other countries, the same policy and strategies have proved unsuccessful or a failure. The lessons from the success and failure cases are vital in the attainment of suitable renewable energy policy development for Nigeria. Hence, the Nigerian government and policy makers alike should look into these lessons in order to improve RE development in Nigeria. The information provided in this study is intended to support the Nigerian policymakers in making proper decisions that are based on international best practice, while taking caution from past mistakes of other countries.

References

[1]   Oyedepo, S. O. Energy and sustainable development in Nigeria: the way forward. Energy, Sustainability and Society, 2(1), 1-17. 2012.

[2]   Oyedepo, S. O. Towards achieving energy for sustainable development in Nigeria. Renewable and Sustainable Energy Reviews, 34, 255-272. 2014.

[3]   Ahuja, D., & Tatsutani, M. Sustainable energy for developing countries. SAPI EN. S. Surveys and Perspectives Integrating Environment and Society, (2.1). 2009.

[4]   Springer, R. A Framework for Project Development in the Renewable Energy Sector. Contract, 303, 275-3000. 2013.

[5]   Ölz, S. Renewable energy policy considerations for deploying renewables. 2011.

[6]   Maduekwe, N. C. Unbundling and Privatization of the Nigerian Electricity Sector: Reality or Myth? University of Dundee, Scotland, United Kingdom. (2011). Retrieved August 6, 2015, from www.dundee.ac.uk/cepmlp/gateway/files.php?

[7]   National Energy Policy (NEP) (2003). Energy Commission of Nigeria (ECN). Abuja: Federal Republic of Nigeria. Retrieved August 6, 2015, from www.energy.gov.ng. Accessed on 6th August, 2015.

[8]   National Energy Policy (NEP) (2006). Energy Commission of Nigeria (ECN). Abuja: Federal Republic of Nigeria. Retrieved August 6, 2015, from www.energy.gov.ng.

[9]   National Energy Policy; Draft Revised Edition (NEP) (2013). Energy Commission of Nigeria (ECN). Abuja: Federal Republic of Nigeria. Retrieved August 6, 2015, from www.energy.gov.ng

[10]   National Planning Commission, National Economic Empowerment, & Development Strategy (NEEDS) (2004). National Economic Empowerment and Development Strategy NEEDS (Vol. 2). International Monetary Fund.

[11]   Marcellus, I. O. Development planning in nigeria: reflections on the national economic empowerment and development strategy (NEEDS) 2003-2007. J. Soc Sci, 20(3), 197-210. 2009.

[12]   The Federal Government of Nigeria (FGN) (2005).Electric Power Sector Reform Act (EPSRA). Retrieved July 6, 2015, from www.nercng.org/index.php/nerc-documents/func-startdown/35/.

[13]   Iwayemi, A., Diji, C., Awotide, B., Adenikinju, A. and Obute, P. Towards Sustainable Universal Electricity Access in Nigeria. CPEEL. (2014). Retrieved August 6, 2015, from Available online at: www.cpeel.ui.edu.ng/sites/default/files/monograph-2.pdf.

[14]   Renewable Electricity Action Program (REAP). Federal Ministry of Power and Steel (2006). Federal Republic of Nigeria. Retrieved August 6, 2015, from www.iceednigeria.org/backup/workspace/uploads/dec.-2006-2.pdf.

[15]   Nigerian National Petroleum Corporation (NNPC) Draft Nigerian Bio-Fuel Policy and Incentives. Nigerian National Petroleum Corporation, Abuja. 2007.

[16]   Renewable Energy master Plan (REMP) (2005). Energy Commission of Nigeria (ECN) and United Nations Development Programme (UNDP). Retrieved August 6, 2015, from Available online at: www.spidersolutionsnigeria.com/wp-content/uploads/2014/08/Renewable-Energy-Master-Plan-2005.pdf.

[17]   Renewable Energy master Plan (REMP) (2012). Energy Commission of Nigeria (ECN) and United Nations Development Programme (UNDP). Retrieved August 6, 2015, from www.energy.gov.ng.

[18]   National Renewable Energy and Energy Efficiency Policy (NREEEP) (2014). Energy Commission of Nigeria (ECN) and Federal Ministry of Science and Technology (FMST). Retrieved August 6, 2015, from www.energy.gov.ng.

[19]   Emodi, N. V., & Yusuf, S. D. Improving electricity access in Nigeria: obstacles and the way forward. International Journal of Energy Economics and Policy, 5(1), 335-351. 2015.

[20]   Rural Electrification Strategy and Implementation Plan (RESIP) (2014). Power Sector Reform Team. Retrieved August 6, 2015, from www.power.gov.ng/National%20Council%20on%20Power/Rural%20Electrification%20Committe%20Recommendation.pdf.

[21]   Efurumibe, E. L. Barriers to the development of renewable energy in Nigeria. Scholarly Journal of Biotechnology, 2, 11-13. 2013.

[22]   Ajobo, J. A., & Abioye, A. O. An Expository on Energy with Emphasis on the Viability of Renewable Energy Sources and Resources in Nigeria: Prospects, Challenges, and Recommendations.

[23]   Renewable Electricity Policy Guidelines (REPG). Federal Ministry of Power and Steel (2006). Federal Republic of Nigeria. Retrieved August 6, 2015, from www.iceednigeria.org/backup/workspace/uploads/dec.-2006.pdf.

[24]   Chineke, C., Nwachukwu, R., Nwafor, O., Ugboma, E. & Ndukwu, O. Much Ado About Little: Renewable Energy and Policy. Journal of International Scientific Publications. Vol9. 2015. http://www.scientific-publications.net/get/1000015/1432901545984409.pdf.

[25]   Birmingham ECOTEC Research and Consulting Limited. Renewable energies: success stories. Office for official publications of the european communities. 2001.

[26]   Stokes, L. C. The politics of renewable energy policies: The case of feed-in tariffs in Ontario, Canada. Energy Policy, 56, 490-500. 2013.

[27]   IRENA Policy Brief Evaluating Policies in Support of the Deployment of Renewable Energy. International Renewable Energy Agency. 2012. www.irena.org/DocumentDownloads/Publications/Evaluating_policies_in_support_of_the_deployment_of_renewable_power.pdf.

[28]   Sambo, A. S. Renewable energy development in Nigeria. In Energy commission of Nigeria paper presented at the World's future council and strategy workshop on renewable energy, Accra, Ghana. 2010, June.

[29]   Sagar, A., & Gallagher, K. S. Energy technology demonstration and deployment. Ending the Energy Stalemate: A Bipartisan Strategy to Meet America's Energy Challenges, 117. 2004.

[30]   Nigeria, N. Continental J. Social Sciences 3: 31-37, 2010 ISSN: 2141-4265© Wilolud Journals, 2010 http://www. wiloludjournal. com THE ROLE OF RENEWABLE ENERGY RESOURCES IN POVERTY ALLEVIATION AND SUSTAINABLE DEVELOPMENT IN NIGERIA. 2010.

[31]   International Energy Agency Staff. CO2 emissions from fuel combustion. OECD. 2012.

[32]   Reducing Greenhouse Gas Emissions: The Potential of Coal. International Energy Agency, 2005.

[33]   Painuly, J. P. Barriers to renewable energy penetration; a framework for analysis. Renewable energy, 24(1), 73-89. 2001.

[34]   Oghogho, I. SOLAR ENERGY POTENTIAL AND ITS DEVELOPMENT FORTAINABLE ENERGY GENERATION IN NIGERIA: A ROAD MAP TO ACHIEVING THIS FEAT. International Journal of Engineering and Management Sciences, 5(2), 61-67. 2014.

[35]   Meyer, N. I. Learning from wind energy policy in the EU: lessons from Denmark, Sweden and Spain. European Environment, 17(5), 347-362. 2007.

[36]   Frondel, M., Ritter, N., Schmidt, C. M., & Vance, C. Economic impacts from the promotion of renewable energy technologies: The German experience. Energy Policy, 38(8), 4048-4056. 2010.

[37]   United Nations (UN) Policies to Promote Renewable Energy Technologies (RETs) in the Asia-Pacific Region. Desk Study Report. Asian and Pacific Center for Transfer of Technology. 2012.

[38]   KPMG Taxes and Incentives for Renewable Energy. Energy & Natural Resources. 2011.

https://www.kpmg.com/Global/en/IssuesAndInsights/ArticlesPubl
ications/Documents/Taxes-Incentives-Renewable-Energy-
2011.pdf.

[39] Fouquet, D., & Johansson, T. Energy and environmental tax
models from Europe and their link to other instruments for
sustainability: policy evaluation and dynamics of regional
integration. In Presentation at the Eighth Senior Policy Advisory
Committee Meeting, Beijing, China, November (Vol. 18). 2005,
November.

[40] Joelsson, J. On Swedish bioenergy strategies to reduce CO2
emissions and oil use. 2011.

[41] European Environmental Agency Renewable Energies: Success
Stories. Chapter 6: Analysis of Member State/technology
examples of successful penetration. 2015.
www.eea.europa.eu/publications/...issue...27/Issues_No_27_06.pd
f.

[42] Menanteau, P. Policy measures to support solar water heating:
information, incentives and regulations. World Energy Council.
2007.

[43] Ruijs, A., & Vollebergh, H. R. Lessons from 15 Years of
Experience with the Dutch Tax Allowance for Energy Investments
for Firms. 2013.

[44] Winkel, T., Rathmann, M., Ragwitz, M., Steinhilber, S., Resch, G.,
Panzer, C., ... & Konstantinaviciute, I. Renewable energy policy
country profiles. Prepared within the Intelligent Energy Europe
project RE-Shaping (Contract no.: EIE/08/517/SI2. 529243), www.
reshaping-res-policy. eu. 2011.

[45] Kim, D. Y. Introduction of RPS and phase-out of FIT in
renewable energy policy. INT'L FIN. L. REV.(Aug. 1, 2012),
http://www. iflr. com/Article/3072471/Introduction-of-RPS-and-
phase-out-of-FIT-in-renewableenergy-policy. html.

[46] Ragwitz, M., Winkler, J., Klessmann, C., Gephart, M., & Resch, G.
Recent developments of feed-in systems in the EU-A research
paper for the International Feed-In Cooperation. Nature
Conservation and Nuclear Safety (BMU), Bonn: Ministry for the
Environment. 2012.

[47] Hetter, J., Gelman, R., & Bird, L. Status of Net Metering:
Assessing the Potential to Reach Program Caps. National
Renewable Energy Laboratory. 2014.

[48] American Public Power Association Public Power Utilities: Net
Metering Programs. 2014.
https://www.publicpower.org/files/PDFs/Public_Power_Net_Mete
ring_Programs.pdf.

[49] Leflaive, X. Water Outlook to 2050: The OECD calls for early and
strategic action. In Discussion paper 1219. Global Water Forum
Canberra. 2012.

[50] United Nation Development Programme. International Financing
Mechanisms for Renewable Energy in the Pacific Island Countries.
2010.
http://www.climateparl.net/cpcontent/pdfs/TJensen%20UNDP%2
0International%20Financing%20Mechanisms.pdf.

[51] KPMG. Taxes and Incentives for Renewable Energy. 2013.
https://www.kpmg.com/Global/en/IssuesAndInsights/ArticlesPubl
ications/taxes-and-incentives-for-renewable-
energy/Documents/taxes-and-incentives-for-renewable-energy-
2013.pdf.

[52] Choung, Y. Quick Look: Renewable Energy Development in
South Korea. Renewable Energy World. 2010.
http://www.renewableenergyworld.com/articles/2010/12/quick-
look-renewable-energy-development-in-south-korea.html.

[53] Hogg, K., & O'Regan, R. Renewable energy support mechanisms:
an overview. PricewaterhouseCoopers LLP, Globe Law and
Business. 2010.

[54] Deutsche Gesellschaft für Internationale Zusammenarbeit (GIZ)
GmbH. Renewable energy in South-East Asia (ASEAN-RESP).
2015. https://www.giz.de/en/worldwide/16395.html.

[55] Gallo, F. Exploring the complementarities between the Green
Climate Fund and the CDM: Developing the GCF's Project
Certification and Credit Issuance Process. CDM Policy Dialogue.
2012.
http://www.cdmpolicydialogue.org/research/1030_complementarit
ies.pdf.

[56] Gordić, D., Babić, M., Jelić, D., Konćalović, D., & Vukašinović,
V. Integrating Energy and Environmental Management in Wood
Furniture Industry. The Scientific World Journal, 2014.

[57] Organisation for Economic Co-operation and Development
(OECD). Promoting Sustainable Consumption: Good practices in
OECD countries. (2008).
http://www.oecd.org/greengrowth/40317373.pdf.

[58] Hogg, D., Elliot, T. & Stonier, C. International Review Of fiscal
instruments for low carbon development. 2011.
http://trpenvis.nic.in/test/doc_files/International_Review_Fiscal_I
nstruments.pdf.

[59] Shehu, A. I. (2012). Solar energy development in northern Nigeria.
J Eng Energy Res: 2 (1).

[60] Opeh, R, & Okezie, U. (2011). The significance of biogas plants
in Nigeria's Energy strategy. J Phys Sci Innov: 3.

[61] Ahmadu, M. L. (2012). Renewable Energy: Challenges and
Prospects of adopting a New Policy and Legal Paradigm in
Nigeria.
https://ases.conference-
services.net/resources/252/2859/pdf/SOLAR2012_0040_full%20p
aper.pdf.

26

Scenario Based Technology Road Mapping to Transfer Renewable Energy Technologies to Sri Lanka

Amila Withanaarachchi*, Julian Nanayakkara, Chamli Pushpakumara

Department of Industrial management, University of Kelaniya, Sri Lanka
*Corresponding author: amilaw@kln.ac.lk

Abstract As per the International Energy Agency (IEA) 2009 report, rapidly growing energy demand in developing countries is projected to double by 2030. After ending the three decades of civil war, the Sri Lankan economy has also shown a robust growth; hence the country has shown a continuous growth in energy demand. In 1995 Sri Lanka met 95% of the total electricity demand using major hydro power plants. But due to the escalating demand for electricity and government policies favouring the coal powered power plants, a completely different power mix exists today. By the end of year 2012, more than 70% of the total electrify requirements of the country was met with fossil based energy sources. Today, as responses to the threats of climate change manifest, following many other nations, Sri Lanka also considered renewable energy in their energy mix. However, the lack of technological capabilities has hindered the development of renewable energy technologies in the Country. The solution to such constrains lies with effective technology transfer and cooperation of renewable energy technologies. Technology Road-mapping and scenarios are two widely used future techniques to support strategic and long-range planning. This paper provides a combined approach of technology road mapping and scenario planning in order to foster the renewable energy technologies of Sri Lanka via effective technology transfer mechanisms. The combined approach consists with six steps, and foresight analysis tools such as literature review, expert's interviews, STEEPV analysis, and Delphi technique were used along the process. Four scenarios named 'Land of Republic', 'Green Paradise,' 'Drowning Island' and 'Black Island' were developed. Out of the four scenarios, 'Green Paradise' was considered as the most favourable scenario for Sri Lanka and technology roadmap was developed targeting this scenario. The proposed technology roadmap consists with six steps and the roadmap suggests effective technology transfer mechanism to foster the renewable energy technologies of the country.

Keywords: renewable energy, road-mapping, scenario planning, sri lanka, technology transfer

1. Introduction

Energy is undoubtedly the basic need for continuity of economic development and human welfare. In modern societies, electrical energy proves to be one of the crucial forms of energy used by human beings in manufacturing products and providing service. As the human population increases, the amount of electricity usage grows as well. It is projected that the world electricity usage will reach 32,922TW-hours by 2035[1].

In 1995 Sri Lanka produced 95% of its grid electrical energy needs from conventional hydro power plants. However, expansion of household electricity consumption and the boost in the industrial sector of the country have forced the country to depend on alternative energy resources such as fossil fuels. The total amount of electricity generated during 2012 was 11,878.8 GWh out of which 70.9% was from thermal power plants (both oil and coal), while 23.0% was from major hydro and the balance 6.2% was from Non-Conventional Renewable Energy (NCRE) which comprised of small hydro, wind power, biomass and solar [2].

1.1. Rising Demand for Local Electricity Generation

In parallel to many emerging Asian countries, Sri Lanka has been struggling to meet the rising demand for power. After ending the 30years of civil war, the country's economy has been showing robust growth, in turn accelerating the demand for power. Electricity demand growth rate in the past has most of the time revealed a direct correlation with the growth rate of the country's economy.

The Central Bank of Sri Lanka expects an average GDP growth rate of 8% in real terms in the four years from 2012 to 2015. The demand for power is expected to accelerate on the back of the expanding economy, and the current statistics show that the country's electricity demand has been growing at an average rate of 5.9% per annum [3]. The overall favourable economic prospects, increased investments in the industrial and manufacturing sectors, coupled with the government's long-term vision of providing electricity to all households, is expected to increase the demand further in the medium and long term.

1.2. Need of Clean Energy

Considering the fact that Sri Lanka's large reserves of hydro power have already been utilized, the CEB had diversified to thermal power, resulting in a gradual shift in the industry power mix [4]. Based on World Bank reports, Table 1 exhibit the per capita carbon dioxide emissions of different regions compared with Sri Lanka.

Table 1. Average co$_2$ emission metric tons per capita in year 2010

Country/ Region	2010 CO2 Emissions (MT/capita)
World	4.9
Low middle income countries	1.6
South Asia	1.4
Sri Lanka	0.6

The per capita carbon dioxide emission in Sri Lanka at present is only 0.6 metric tons per year, which is far below the global average of 4.9. This indicates that Sri Lanka has adequate carbon space for establishing fossil fuel power plants [5]. As evident by many international publications, the unprecedented levels of economic growth emerging in the developing nations will make them responsible for future growth in energy demand [6]. In catering to the rapid demand growth, current electricity generation expansion plan of Sri Lanka is mainly concentrated on imported coal. Coal has been identified as the least cost option taking into consideration mainly the cost of production [7]. Based on the published data, energy sector is the main contributor to the GHGs emission [8]. Thus clean energy is an essential requirement in combating the climate change.

1.3. Why Technology Transfer & Collaboration?

In today's context the technology behind renewable energy has been evolved so rapidly and the advancements gained by many developed and rapidly developing nations like China and India provide good examples for developing countries like Sri Lanka. The two countries have seen Technology transfer as a cornerstone in reaching a global solution to climate change.

Concerning the limited technological capabilities, firms in developing countries find that internal development of technology needed for new products, new processes and operational improvement is somewhat difficult. Hence, technological progress in developing countries relies heavily on imported technology, which is based on the transfer, absorption and adaptation of existing knowledge [9]. The success of technology transfer implementation and utilization, however, relies very much on the internal capacity to absorb and accumulate foreign knowledge by technology acquirers.

1.4. Significance of the Study

As stated earlier countries like China and India teach us a valuable lessons as they have managed to bypass most of the mistake made by European countries and have managed to reach the current level of developments in just

a couple of years' time. Literature reveals that a clear vision on the future along with robust technological roadmap aided these nations to reach the current level of development. Technology road mapping and scenarios are two widely used future techniques to support strategic and long-range planning. This paper provides a combined approach of technology road mapping and scenario planning in order to foster the renewable energy technologies of Sri Lanka via effective technology transfer mechanisms.

2. Literature Review

In this section of the paper we present a comprehensive literature review conducted in both scenario planning and technology road mapping during the course of this study. The two methods are widely used approaches in technology foresight. This section of the paper also discusses the limitations of scenario planning and technology road mapping and also shows how the combined approach has managed to overcome the weaknesses of the two individual approaches.

2.1. Scenario Planning

In literature we could find much evidence where scenario planning was used for future development routes. For instance, the International Panel of Climate change (IPCC) has created scenarios focusing on the future development of energy and environment on the future developments of green house gas emissions [10]. In addition, Brown et al. (2001) have studied scenarios focusing on clean energy future. Also Shell has utilized scenarios to identify opportunities and challenges in the global business environment [11]. These examples states that most successful Corporations and Nations have used scenarios as a tool to lead the energy sector, as when people can visualize a future, they can begin to create it [12].

2.1.1. Scenarios – A brief History

Like many other early forecasting techniques, the scenario method is a post-war planning concept. Following the work of Herman Kahn and others at RAND and the Hudson Institute in the 1960s, scenarios reached a new dimension with the work of Pierre Wack in Royal Dutch/Shell [13]. From there on throughout the history scenarios has been used by many industries and nations to shape their future for the betterment of all the stakeholders. Consequently many scholars have attempted to define and developed much scholarly literature on this method.

2.1.2. Definitions of Scenarios

According to the classic reference of Kahn and Wiener, scenarios are hypothetical sequences of events, built with the intent of attracting attention to causal processes and points of decision. Some authors in turn define scenarios as archetypical descriptions of alternative images of the future, created from mental maps or models which reflect different perspectives on past, present and future developments [14]. Combining all these views, Finland study on future prospects of alternative agro-based bio energy suggest that the scenario will be developed towards an internally consistent story about the path from the present to the future [14]. In parallel to the above arguments International Energy Agency defined scenarios as a tool for helping us to take a long view in a world of great uncertainty [15]. Similarly scenarios are stories that anticipate the future. They are narratives created by researchers, or by participants in a workshop [13].

2.1.3. Characteristics of Scenarios

According to literature, well written scenarios have the following characteristics:

01. A scenario is not a forecast of the future [14].

02. At least two scenarios are needed to reflect uncertainty and each of the scenarios must be plausible [13].

03. It is also notable that a scenario planning approach does not make a stand on, e.g. the most probable scenario [14].

04. This means that they must grow logically (even cautiously) from the past and the present and they must be internally consistent [13].

05. Typically there will be a mixture of quantifiable and non-quantifiable components. But by integrating quantitative models to the scenario building, we will not be able to predict the future nor to quantify the probabilities of a scenario. By quantifying these scenarios will be able to make the results more tangible [16].

06. Scenarios must also be relevant to the issues under scrutiny. In other words, they should provide useful and comprehensive idea generators and test conditions, against which the plans and strategies can be considered [14].

07. They may be presented in discursive, narrative ways (illustrated with vignettes, snippets of fiction and imitation newspaper stories, etc.) or tabulated in the form of tables, graphics, and similar systematic frameworks [17].

08. In order to be challenging, scenarios must take under consideration potential surprises which may cause discontinuities in future [14].

09. Link historical and present events with hypothetical events in the future [13].

2.1.4. Different Types of Scenarios

In literature, scenario has been divided into two main approaches. As per UNIDO Technology Foresight Manual, scenarios can be both exploratory and normative. Exploratory scenarios start from the present and ask questions such as "What next?" and "What if?" for the development of explorative scenarios. Normative, or inward scenarios, involve back casting, typically starting with the most desirable future. The main questions are "Where to?" and "How to?" [17]. While exploratory scenarios are designed to explore several plausible future configurations of the world, IEA suggest a forecasting type scenario which we are pretty much familiar to us. This type of scenario is often referred to as a "business-as-usual (BAU) scenario", which assumes the continuation of historical trends into the future and that the structure of the system remains unchanged or responds in predetermined forms [15].

2.1.5. Drawbacks of Scenarios

Traditionally scenarios have been developed to support formulation of a vision and mission statement driven by the most desired vision. However in literature scenarios have been criticized for several reasons. These criticisms comprises with:

01. Scenarios are too distant to support strategy development [18].

02. Though scenarios are multidimensional, there is no real 'search' for possible futures in the plural sense [19].

03. History of forecast in Norway state that some of the predictions made were so wrong that a scenario builders who, by accident, were able to suggest the right price of oil prices at the time being when the forecasts were made, immediately would have been categorized as an illusionary wizard telling fairytales or creating wild cards [19].

These drawbacks suggest that we need an approach that is not focusing at what seems probable today but which may open up for surprises, ideas and perspectives that are a bit more "far out" and "political incorrect." Scenarios should be developed in such a way that breaks the demand for probability.

2.1.6. Advantages of Scenarios

Irrespective of such limitations in scenario process, literature backed the advantages of scenario as a very strong tool in any futuristic study. The following list comprises with some of such evidences.

01. Through the development of scenarios and other related approaches, foresight techniques add value to long term planning by explicitly transferring the complexity and contingency of future into commonly understood decision points, challenges and potentials [16].

02. Among the futurists, it is common sense that the future is an open space. In case of predictions which simply extrapolate past developments into the future fails to recognize disruptions. Scenarios have proven to be very effective instrument to organize the option space and include disruption [16].

03. National attempts in developing scenarios (Norway 2030) have provided an early basis for long-term thinking in relation to a broad range of challenges for the country. Thus scenarios have contributed to readjustments in important areas where changes usually take longer time to accomplish [19].

04. The use of scenarios in research related studies have shown that it is especially useful when the uncertainty of the research question is high and there are multiple resolutions to the issue [20].

05. Scenarios were also considered as a tool that can be used to bridge the gap between thinking and actions when addressing a specific question [12].

In today's competitive world successful firms act in advance rather than react, by going ahead rather than by running behind. Identifying the advantages of this approach scenario planning has been used in various organizations for decades. General Electric and Shell are well known examples of large firms that have been among the pioneering adopters' of scenario approaches as strategic management tools. Thus in strategic planning scenarios are arguably one of the most effective tools especially when the uncertainties related to business and future plays a significant role in the industry [20].

2.2. Technology Road-mapping

In the introduction of this paper we looked at adverse impact of non-renewable energies such as oil and coal to a sustainable energy future. In this backdrop renewable energies are gaining importance in the discussion in the new paradigm of sustainable development. To guide these developments it can make use of technological foresight tools in order to pave the path for a sustainable development. Among examples of such tools it could cite the Technology Roadmap, which is a widely used technique in the industry for the development of long term planning strategies, making it possible to align market, product and technology over time. Literature provides ample examples on the use of technological road-mapping in countries like United States, countries of the European Union (EU), and also emerging economies in East-Asia. Further various types of organizations perform or commission studies road map, which features businesses, universities, research centres, industry associations, government departments, and ministries. Among the organizations, one can emphasize the participation of the United State Department of Energy (USDOE) and the International Energy Agency (IEA) [21].

2.2.1. Technology Road-mapping: A Brief History

Technological forecasting has a long history as a management technique. However it is Motorola who brought this technology prominence in management planning. In early 1980s Motorola used the technique 'technology road-mapping' as a way of anticipating technology needs alongside further production developments. Since then road-mapping has been used in a variety of contexts, particularly in the industry at corporate level (e.g. BP, Philips Electronics) and sector level (e.g. US Aluminum), as well as in Government Laboratories (e.g. Sandia Labs) [13].

2.2.2. Definitions of Road-mapping

In literature, scholars have taken multiple approaches to define the term road-mapping. A very general definition has been given by Schaller (2004) and referred roadmap as layout of paths or routes exists or may exist in a particular geographic area to help travellers in planning the trip in order to reach a particular destination [21]. A more futuristic approach has been taken by Robert Galvin, former Motorola chairman, and he defined the roadmap as "an extended look at the future of a chosen field of inquiry composed from the collective knowledge and imagination of the brightest drivers of the change" [13]. Combining this general and futuristic approach, Lizaso and Reger defined road-mapping as a process of assuming a given future(s) and providing paths to get it, by means of a certain amount of foresight and a certain amount of consensus. Then, they further defined the aim of road-mapping as to evaluate what is technically possible, desirable and expected, and to understand what needs to happen for moving ahead [12].

Apart to this more generic approach, number of scholars have also tried to defined road-mapping in a more managerial approach. For example Lizaso and Reger (2004) portray roadmaps as the evolution of markets, products and technologies to be explored, together with the linkages between the various perspectives. They are time based plans that help organisations to determine where they are, where they want to go and how to get there. Similarly dos Santos et al (2013) define road-mapping as a flexible method whose main objective is to

assist in strategic planning for market development, product and technology in an integrated manner over time.

2.2.3. Types of Roadmaps

Due to a high number of application areas, types of roadmaps have varied. Some examples include science/research roadmaps, cross-industry roadmaps, industry roadmaps, technology roadmaps, product roadmaps, product-technology roadmaps and project/issue roadmaps [13]. In terms of intended purpose Phaal, et al (2004) has identified eight types of roadmaps. The list comprises with product planning, service/capability planning, strategic planning, long-range planning, knowledge asset planning, program planning, process planning, and integration planning. Out of the listed types of roadmaps by far the most common type of technology roadmap, is the product planning, which relates to the insertion of technology into manufactured products [22].

2.2.4. Drawbacks of Roadmap

The above list of applications indicates the highest potentiality of road-mapping as a foresight tool in the diverse range of fields. However just like any foresight method, road-mapping also comprises with shortcomings which needs to be overcome. Saritas and Aylen (2010) list the following weaknesses of road-mapping approach. Road-mapping is normative, rather than exploratory; they encourage linear and isolated thinking, thus there is little scope for creativity, communication and collaboration; Dissemination is difficult, thus only experts can understand the output, especially if it is couched in technical terms [13].

2.2.5. The Advantages of Roadmap

The advantage of road-mapping has been extensively discussed in literature by number of scholars. Saritas and Aylen suggest that the road-mapping is an effective tool for; (a) portraying structural relationships among science, technology and applications, thus (b) improving coordination of activities and resources in increasingly complex and uncertain environments, (c) identifying, evaluating, and selecting strategic alternatives that can be used to achieve desired science and technology objectives, (d) communicating visions to attract resources, (e) stimulating investigations, (f) monitoring progress [13].

More specific advantages of technology road-mapping has been discussed by the Phaal et al (2003) and they specify that the technology road-mapping has the grate potential with the following; (i) Supporting the development and implementation of integrated strategic business, product and technology plans, providing companies have the information, process and tools to produce them. (ii) Provide a means for enhancing an organization's 'radar', in terms of extending planning horizons, together with identifying and assessing possible threats and opportunities in the business environment [22].

3. Methodology

The methodology adopted in study encompasses two distinct foresight techniques. The scenario building process was combined with the technology road mapping and the proposed methodology comprises with six steps.

Following table (Table 2) illustrate the research architecture used in our study.

3.1. Step 01

The preliminary phase covers preparatory work for technology road mapping process. The preliminary step was obtained by conducting a detailed literature survey in the renewable energy sector. Literature pertaining to the renewable energy sector of Sri Lanka the world developments was thoroughly studied in order to understand the current developments and constrains in the sector.

The influencing factors pertaining to the renewable energy sector of Sri Lanka was further studied through experts' interview. The policy makers, academics, Independent Power Producers (IPPs), and government officials were interviewed to narrow the factors that were identified in literature review. As stated in Table 2, the broad picture of Sri Lanka's renewable energy sector was the outcome of stage 01.

Table 2. Research architecture which combine scenario planning with technology road mapping

Architecture	Method	Outcome
Step 01: Road mapping preparation	01. Literature review 02. Expert interview	Current status of the renewable energy sector
Step 02: Tine assessment	01. Literature review	Year 2030
Step 03: System analysis	01. STEEPV analysis 02. Two round Delphi survey	Most important factors which are crucial for the development of the renewable energy sector of Sri Lanka.
Step 04: Scenario projection	01. Framework scenarios based on the prioritized key factors in Step 03.	The four scenarios of Sri Lanka's renewable energy future
Step 05: Scenario Building	01. Outcome of Step 04 02. Literature reviews 03. Expert Interviews	Elaborated scenarios
Step 06: Road-mapping	01. Focused group interviews 02. Literature on Sri Lanka's policy papers	Technology roadmap.

3.2. Step 02

Assessing the time for the road-mapping process is one of the crucial factors, as most of the work thereafter depends on the defined time horizon. Thus the selected time horizon should be far enough to develop scenarios and at the same time should be a feasible to attain the required goals. However some literature on technology road mapping do not emphasize on a particular time horizon. But Kappel, researcher who has contributed heavily in the technology road mapping literature have stated that the roadmap should contain the explicit time scope [23]. Literature indicates that about 32% of the roadmaps identified make predictions for the year 2020, using a long-term period range from 15 to 20 years. However, further studies shows that some road maps have targeted year 2050, which highlights the importance given to global sustainable energy in the energy analysis of countries, institutions, companies and universities.

When looking into the Sri Lankan perspective, the government of Sri Lanka has already obtained targets for year 2020 to generate 20% of the total electricity generation using renewable energy power options [24]. The 'Long term generation expansion plan', developed by the Ceylon Electricity Board (CEB) of Sri Lanka has already predicted the country's renewable energy scenario till year 2032. However the stated scenario was purely a technical analysis and has failed to look in to the qualitative characteristics and factors [25]. Concerning all these facts, this study has decided to develop scenarios for year 2030.

3.3. Step 03

The purpose of the 3rd step in the scenario building and technology road mapping process is to identify the most important factors which are crucial for the development of the renewable energy sector of Sri Lanka. In order to attain said objectives technology foresight techniques such as STEEPV and Delphi surveys were used at this stage. STEEPV is an acronym for Social, Technological, Economical, Environmental, Political and Values [26]. Thus a workshop was conducted with twelve experts representing these six thematic areas of the country and eight factors were identified, which are crucial for the development of the renewable energy sector of Sri Lanka. Thereafter two round Delphi survey was conducted to prioritize the outcomes of STEEPV and finally two most important factors were identified.

3.4. Step 04

The four scenarios are constructed based on two identified uncertainties and prioritized factors from the previous stage [18]. This will be the pillars of scenarios, as this will creates the foundation to develop and elaborate the scenarios of Sri Lanka's renewable energy future.

3.5. Step 05

The framework developed in the previous step is rather skeletal and, in order to be called scenarios, they are complemented with the qualitative arguments. The qualitative content analysis of the arguments gives this option by revealing the rationale and context around the quantitative statements. The contexts can be interpreted and written down as a more narrative 'story' of the future. The narrative interpretation is here written as four scenarios of the future of renewable energy in Sri Lanka up to 2030 [14].

3.6. Step 06

As stated by Lizaso and Reger (2004), road mapping is about assuming a given future(s) and providing paths to get it, by means of a certain amount of foresight and a certain amount of consensus [12]. Thus, once we identified the equally plausible futures of Sri Lanka's renewable energy sector by year 2030, next step is to evaluate what is technically possible, desirable and expected, and to understand what needs to happen for moving ahead. In order to attain this goal number of literature pertaining to the technical evaluation in Sri Lanka as well as the policy papers on the local energy sector was evaluated. Finally the local experts who are already heavily engaged in the renewable energy sector of the county were interviewed to sharpen the technology roadmap.

4. Results & Discussion

The objective of conducting the scenario planning process was to develop scenarios of Sri Lanka's renewable energy future by year 2030. As the outcome of the environmental scanning which we conducted under the six thematic areas of the country (STEEPV Analysis) following eight factors were highlighted.

1. Public concern on environmental impact
2. Energy intensive lifestyle of the public
3. Government supportive policies
4. Private and public sector participation
5. High cost of fossil based thermal energy
6. International pressure on emission of GHGs
7. Technical limitations in the grid
8. Global technological advancements

Two round Delphi survey was conducted with a panel of 10 experts and the results were analyzed using a combined Delphi, AHP model called Delphic Hierarchy process or DHP. Eight key factors were taken into the Delphi survey and outcomes were analyzed based on the level of importance and the expected probability of occurrence of each of the factor. The results were presented in Table 3.

Table 3. Factors ranked based on their contribution to the development of renewable energy sector of Sri Lanka by 2030

Factors	Weighting of factors based on the level of importance	Probability of occurrence
Government moving towards rational policies to promote renewable energy	0.2828	Probable
Real cost of the fossil fuels going up	0.2724	Probable
Global technological advances are conducive to renewable energy advancements	0.2334	Highly Probable
Growing energy intensive lifestyle of the public	0.0678	Highly Probable

As illustrated by the above table 'Government moving towards rational policies to promote renewable energy' and 'rising real cost of the fossil fuels' are the two most important factors that were derived from the experts survey. However the experts have considered these two factors to be 'probable' in respect of occurrence of the two events by 2030. Thus while they are highly important compared to other factors they also associated with a level of uncertainty as well. Literature state that the scenarios should be frame based on key factors which also associate with a certain level of uncertainty as well [13]. Thus these two factors have created the ideal foundation for us to frame scenarios of Sri Lanka's renewable energy future.

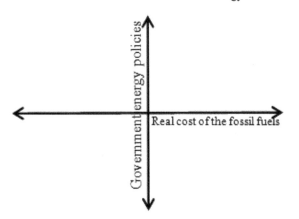

Figure 1. Scenario sketch

On the other hand, the last two factors 'global technological advances are conducive to renewable energy advancements and 'growing energy intensive lifestyle of the public' are less important compared to the government policies and rising real cost of the fossil fuels. But experts have considered these two events to be highly probable. This implies that irrespective of which ever scenario that's going to occur in the future. The stated two factors will occur. Thus concerning its probability we will consider 'Global technological advances' and 'Growing energy intensive lifestyle of the public' as given assumptions in writing scenarios. The frame working of scenarios under step 04 was accomplished by taking the stated important and uncertain factors in to the scenario grid. This is illustrated in Figure 1.

Then we identified and defined the extremes of the key uncertainties to visualize a better picture of the stated scenarios. The four extremes of each of the two factors can be presented as below.

- ***Government moving towards rational policies to promote renewable energy:*** Government will be formulating objective policies towards the right balance between renewable and non-renewable energy.
- ***Government take policy decisions to promote fossil based energy sources ignoring renewable energy options:*** Government will be formulating policies favouring the fossil based energy options ignoring the renewable based power generation options.
- ***Real cost of the fossil fuels going up:*** Real cost (without including government subsidies and taxes) of fossil fuels going up and making investments in renewable becoming more attractive in economic terms.
- ***Real cost of the fossil fuels barely increase:*** Real cost (without including government subsidies and taxes) of fossil fuels barely increases. Thus investments in renewable will not be that attractive in economic terms.

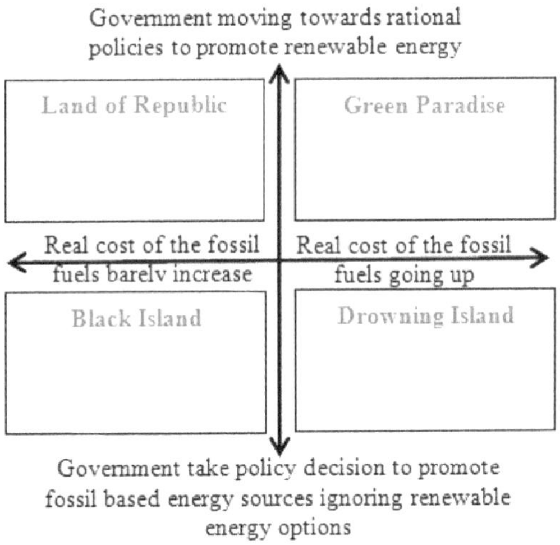

Government moving towards rational
policies to promote renewable energy

Land of Republic	Green Paradise

Real cost of the fossil ← Real cost of the fossil →
fuels barely increase   fuels going up

Black Island	Drowning Island

Government take policy decision to promote
fossil based energy sources ignoring renewable
energy options

Figure 2. Name the Scenarios

Based on the four extremes defined above, we sketched the four scenarios of Sri Lanka's renewable energy future. The results are shown below in Figure 2.

Moving forward to the next stage of the study (Step 05), scenarios were elaborated based on three key variables. The path to the scenario, the outcome of the scenario and finally the early warning signals were presented when describing the scenarios. Table 4; summarize the outcome of scenario planning process.

Considering the four scenarios the obvious choice was the second scenario (Green Paradise) in which the scenario was developed based on the rising fossil fuel prices and favourable government policies towards renewable energy. The study claims that the nation should target the "Green Paradise" as our ultimate destination by year 2030. Thus the technology roadmap was developed to reach said destination.

Table 4. Outcome of scenario planning process

(W01) Land of Republic	
Irrespective of slowly rising (Real) cost of the fossil fuels, Government moving towards rational policies to promote renewable energy.	
Path	The energy intensive lifestyle of the public will continue to grow. Real cost of fossil fuels barely increases. Thus investments in renewable will not be that attractive in economic terms. Irrespective of such concerns Sri Lankan government move towards rational policies to promote renewable energy.
Outcome	With some effort Sri Lanka will meet 20% of electricity supply from NCRE sources by 2020. By 2030 country will retain the 20% share. Global technological advancements will conducive to the development of renewable energy sector of the country making such renewable energy applications affordable, usable, safe and available.
Intended Energy Production	Approximated electricity demand by 2030 = 5,800MW Total generation from NCRE by 2030 (20%) = 1,160MW Energy production from different sources by 2030 {table}
Early warning signals	Balance of payment will reduce as there is lesser foreign exchange requirement to import fossil fuels. However payback period will also be extended for RE power projects as savings on fossil fuel is less.

Intended Energy Production table (W01):

NCRE Source	Total Genera(MW)	Share from NCRE
Small Hydro	812	70%
Wind	267	23%
Biomass	58	5%
Solar	23	2%
Other	0	0%
Total	**1,160**	**100%**

(W02) Green Paradise	
When (Real) cost of the fossil fuels going up, Government moving towards rational policies to promote renewable energy.	
Path	The energy intensive lifestyle of the public will continue to grow. Government will be formulating objective policies towards the right balance between renewable and non-renewable energy. Rising cost of fossil fuels will encourage the government and private sector to invest in renewable energy power projects.
Outcome	Thus Sri Lanka will easily meet country's target of generating 20% of electricity supply from NCRE sources by 2020. By 2030, 25% to 30% of the country's electricity mix will be catered via NCRE sources. Global technological advancements will conducive to the development of renewable energy sector of the country making such RE applications affordable, usable, safe and available.
Intended Energy Production	Approximated electricity demand by 2030 = 5,800MW Total generation from NCRE by 2030 (30%) = 1,740MW Energy production from different sources by 2030 {table}
Early warning signals	Balance of payment will reduce as there is lesser foreign exchange requirement to import fossil fuels. The saving of the fuel cost of the fossil fuel plants by the replacement by renewable power plants will make it possible to recover the RE (capital) cost within a few years.

Intended Energy Production table (W02):

NCRE Source	Total Genera (MW)	Share from NCRE
Small Hydro	696	40%
Wind	609	35%
Biomass	252	14.5%
Solar	174	10%
Other	9	0.5%
Total	**1,740**	**100%**

(W03) Drowning Island	
Government take policy decision to promote fossil based energy sources ignoring renewable energy options. Low cost of fossil fuels further encourages government decision.	
Path	The energy intensive lifestyle of the public will continue to grow. Real cost of fossil fuels going up and making investments in renewable becoming more attractive. Irrespective of such benefits, Sri Lankan government will formulate policies favouring the fossil based energy options ignoring the renewable based power generation options.

| Outcome | Global technological advancements will conducive to the development of renewable energy sector of the country making such renewable energy applications affordable, usable, safe and available. |
| | But due to the lack of policy support only 12% to 13% of electricity will generate from NCRE sources by 2020. Due to the rising demand, the share of NCRE will further reduce to 10% to 11% by 2030. |

Intended Energy Production

Approximated electricity demand by 2030　= 5,800MW
Total generation from NCRE by 2030 (10%) =　580MW

Energy production from different sources by 2030

NCRE Source	Total Genera (MW)	Share from NCRE
Small Hydro	383	66%
Wind	162	28%
Biomass	32	5.5%
Solar	3	0.5%
Other	0	0.0%
Total	**580**	**100%**

| Early warning signals | Balance of payment will be highest of all the scenarios, as the government encourages investments in fossil based power plants while the cost of fossil sources is advancing. But payback period will be less for renewable energy power projects as savings on fossil fuels are high. |

(W04) Black Island
Irrespective of rising (Real) cost of fossil fuels, Government takes policy decision to promote fossil based energy sources ignoring renewable energy options.

| Path | The energy intensive lifestyle of the public will continue to grow. Taking advantage of barely increasing real cost of fossil fuels government will formulate policies favouring the fossil based energy options ignoring the renewable based power generation options. Global technological advancements will conducive to the development of renewable energy sector of the country making such RE applications affordable, usable, safe and available. |
| Outcome | But due to the lack of policy support county will generate less than 10% share from NCRE sources. By 2030 the share of NCRE will further reduce to 6% to 7% of total generation. |

Intended Energy Production

Approximated electricity demand by 2030　= 5,800MW
Total generation from NCRE by 2030 (6%)　=　348MW

Energy production from different sources by 2030

NCRE Source	Total Genera (MW)	Share from NCRE
Small Hydro	226	65%
Wind	117	33.5%
Biomass	3	1.0%
Solar	2	0.5%
Other	0	0.0%
Total	**348**	**100%**

| Early warning signals | Balance of payment will be widening as the government spending on fossil sources is very much high over the price advantage in fossil fuels. However payback period will also be high for the renewable energy power plants as the as savings on fossil fuel is less. |

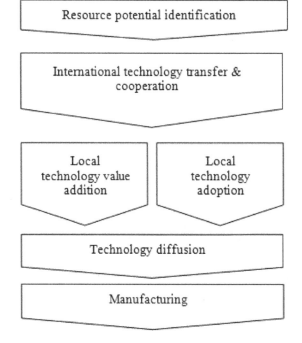

Figure 3. Renewable energy technology roadmap to Sri Lanka

The presentation of roadmap is very important in technology road mapping process. Roadmaps presented in multiple layers, bars, tables, graphs, pictures, flow charts, single layers, and text are some of such approaches [22]. Literature revels that the multiple layer roadmap with short medium and long term time horizons, clear definition of critical factors and suggested actions to be implemented are the most common form of technology road mapping available. However in our study we used flow charts to graphically represent the flow of activities that needed to manifest "Green Paradise" by year 2030. Thus the outcome of this study is presented in Figure 3 and technology roadmap for Sri Lanka to reach desired destination was developed in six sequential steps.

- **Resource potential identification**

The future of the renewable energy sector depends on the availability of the resources and the exploitation capacity. Presently the investors initiate the renewable energy power projects by identifying the resources potential by themselves. In our technology roadmap we propose to change this system, in which the authorities should initiate the renewable energy power project by identifying and allocating the resources to the IPPs based on resource optimization criteria. Thus the resources should be allocated under the condition that the investors have to optimally use the resource to generate the power.

Interviews conducted with the wind power experts of the country claim that the lack of information in the respective power sector is one of the major challenges faced by the renewable energy power producers of the country. Experts claim that the country has a comprehensive data base on the hydro power sector as it's been used comprehensively in the country. But still the authorities are trying to map the other renewable energy potential such as wind, solar and biomass to develop an accurate data base. Without comprehensive and consistent database, country will not be able to use the resources optimally to generate power.

Thus concerning all these facts together "Resource potential identification" was considered as the first step in technology road mapping process.

- **International Technology Transfer & Cooperation**

There are two ways in which a country can acquire the technology. Either develops the technology in-house of transfer already developed technology in other part of the world. Concerning current technological capabilities in the renewable energy sector of the country, experts claim that the next stage of the technology roadmap should be the effective transfer of technologies.

The Intergovernmental Panel on Climate Change (IPCC) provides a frequently quoted definition of climate change technology transfer as 'a broad set of processes covering the flows of know-how, experience and equipment for mitigating and adapting to climate change amongst different stakeholders such as governments, private sector entities, financial institutions, non-governmental organizations, and research or education institutions' [10].

Literature claims that the technologies can be transferred through several channels, mainly trade, foreign direct investment (FDI) in the form of either wholly owned subsidiaries or joint ventures, licensing agreements, official development assistance (ODA), hiring of foreign human resources, and person-to-person pathways such as training programmes, conferences or scientific exchanges.

A careful study of existing successful renewable energy technologies in the country such as hydro power sector provides us evidence of technology transfer in its initial stages. Local hydro power sector initiated with international trade in which turbines were imported from foreign sources. But today technology transfer mechanisms in the hydro power sector have extended to technology licensing and even to in-house R&D. However in commercial scale solar power sector, technology transfer mechanisms are only limited to international trade. Booming technologies such as wind power sector still depends on international trade. Technology licensing agreements could also be seen in the wind power sector.

However experts claim, that Sri Lanka still lacking with the expertise in the core engineering of renewable energy sector. Experts further claim the technology transfer needs to be carried out gradually in order to transfer the technologies sustainable. Example drawn from China and India claim that these countries did not transfer the renewable energy technologies overnight. Years of technology transfer process have strengthened the production capacities of these emerging countries to compete with many developed nations.

- **Local Technology value addition**

The technology transfer does not guarantee that the country will develop the capabilities to manufacture technologies in-house just after acquiring the technology. But small steps like local technology value addition and technology adoption will enable the country to reach the said destination. Under technology value addition, authorities should take appropriate measures to initially transfer and develop operation and maintenance capabilities. Such measures will help the country to continue and retrain the existing transferred technologies from foreign sources. In addition most of such imported technologies comprises with many small units (such as outer casing, wire mesh, etc...) which can be manufactured in-house. Such small steps in local technological value additions will enable us to gradually step towards more advance technologies to develop renewable energy technologies in-house.

- **Local Technology Adoption**

Most of the time the technology acquired to the country does not suit as it is not to local climatic and environmental condition. For example in the biomass sector gasification was used in many part of the world to generate power. But in the Sri Lankan context the technology needs to be adapted to suite local conditions such as humidity, particle size and so on. Thus these technologies are subjected to technology adoption to suite local conditions. Thus in the projected technology roadmap we propose to conduct technology adoptions in parallel to the technology value addition activities.

- **Technology diffusion**

The effective diffusion of these transferred technologies within the country is a very important step in technology roadmap. Experts believe that the government bodies such as the Sustainable Energy Authority in Sri Lanka (SLSEA) could play a prominent role in this respect. For example SLSEA could perform as the intermediate in between the end users and the technology providers. Such measures could be combined with the existing measures taken by the SLSEA. For example SLSEA already have a technology transfer division which already conducts regular workshops and seminars to educate general public. The information that disseminates to the general public also varies with the respective technologies as well. For example while SLSEA educate the rural households on biomass applications, the urban households were also educated on net-metering solutions. Hybrid systems such as combined solar and biomass applications were promoted as direct heat energy applications for the fuel switching program. In parallel to such measure the authorities should also engage in knowledge transfers to local components manufactures to sustain the transferred knowledge that were transferred from foreign sources..

- **Manufacturing**

This is the final destination the country should reach by year 2030. In other words country should be able to develop its own technologies by the stated time period. In doing so, the country should capitalize on the already established renewable energy solutions to reach the said target. Hydro power sector is one of such successful case in Sri Lanka.

The experience gained in the major hydro power sector has strengthened the small hydro power sector in many ways. For example 'VS Hydro' one of the leading firms in the small hydro power sector of Sri Lanka presently provides services as a one stop shop for this sector.

Starting from civil construction, followed by manufacturing of turbines, installation of plant and equipments, surveys and even operation and maintains are done by this local firm. The services of this firm has now even been extend to foreign countries, in which VS Hydro now transfer their technological assistance to number of small hydro power projects in Africa.

Similar examples could also found in direct energy application in the rural sector of the country. For instance SLSEA along with many NGOs have facilitated technology innovation opportunities such as cooking stoves. The cooking stoves are mainly used in the rural households as a very efficient biomass application in domestic cooking and so as an extended version that were used industry biomass heating purposes as well.

The experts believe that government institutes such as 'National Engineering Research & Development Center of Sri Lanka" or NERD center, can also play a prominent role by providing their technological capabilities. The fact that the NERD center has already engaged with a number of small scale renewable energy power projects throughout the country is an opportunity that we should excel in the future. For example biomass power plants, which operates from the strong and sustainable fuel food supply from the users of the technology, micro hydro power systems, and small scaled (100W) two blade wind turbine developed in a village in Nikawaratiya to charge batteries are some examples.

5. Conclusion

Scenario planning and technology road mapping are strong technology foresight approaches that have been widely used in the world renewable energy sector. However each of these approaches has their own limitation when applying in real life business scenarios. By combining scenario planning with technology road mapping, we will be able to overcome unique weaknesses of these two methods. This study has attempted to combine both the techniques and have overcome the critical limitations of each of these tools. For example the four scenarios developed in this study were used to describe the future circumstance of Sri Lanka's renewable energy sector. Thus the scenario alone will not give pathways into the desired scenario. By combining road map we have connect the future with the present and informed that action to reach the said destination. On the other hand scenario planning has also contributed in overcoming the limitations that we face in road mapping process.

The roadmap alone is more target oriented, and it suggest linear and isolated thinking. Thus roadmap alone is more difficult to communicate to a non-participant of the process as the results are too technical. The combination of scenario into the technology road mapping process opens up critical thinking and its highly participative and interactive nature makes the process more reachable even to a non-participant of this process.

Finally, it shall be highlighted that neither road mapping nor scenarios provides a silver bullet to the struggling renewable energy issues of Sri Lanka [18]. Scholars such as Saritas and Aylen, have argued that the road mapping process does not end with the roadmap itself. The true value of road mapping lies in an on-going process [13]. The authors fall in line with this advice, and we believe the linking scenarios to road mapping processes should be followed up as an on-going process,

understanding that it is a learning process that usually reveals more questions than answers.

References

[1] Bagheri Moghaddam, N., Mousavi, M., Moallemi, A. and Nasiri, M, "Formulating directional industry strategies for renewable energies in developing countries: The case study of Iran's wind turbine industry", *Renewable Energy*, 39 (1). 299-306. 2012.

[2] SLSEA, *Sri Lanka Energy Balance 2012: An Analysis of Energy Sector Performance*, Sustainable Energy Authority of Sri Lanka., Colombo, 2013.

[3] Ceylon Electricity Board, *Long Term Generation Expansion Plan*, CEB, Colombo, 2011.

[4] RAM, *Sri Lankan Power sector-Firing the rain*, RAM Rating (Lanka), 2012. [E-book] Available: http://lra.com.lk/reports/power_sector_update.pdf.

[5] Abeygunawardana, A., "Feed - in - tariff in Sri Lanka", Energy Forum, Colombo, 2012.

[6] Streeter, A.L.E., and Jongh, D., "Factors influencing the implementation of clean energy interventions in low-incomeurban communities in South Africa," *Journal of Global Responsibility*, 4(1). 76-98. 2013.

[7] Ranasinghe, D.M.H.S.K., "Climate Change Mitigation Sri Lanka's Perspective," *International Forestry and Environment Symposium*, 15. 290-296.

[8] Ratnasiri, J., "Alternative energy - prospects for Sri Lanka", *Journal of the National Science Foundation of Sri Lanka*, 36(Special Issue). 89-114. 2008.

[9] Ming, W.X., and Xing, Z., "A new strategy of technology transfer to China", *International Journal of Operations & Production Management*, 19(5/6). 527-37. 1999.

[10] IPCC, *Special report on emission scenarios*, Intergovernmental panel of climate change, Cambridge University press, Cambridge, 2000.

[11] Shell, *Executive summary of the Shell Global Scenarios to 2025*, 2005, [E-book] Available: http://www.shell.com/content/dam/s hell/static/future-energy/downloads/shell-scenarios/shell-global-scenarios2025summary2005.pdf. [Accessed Jan. 5, 2015].

[12] Lizaso, F. and Reger, G. "Scenario-based roadmapping--a conceptual view", *EU-US Scientific Seminar on New Technology Foresight*, Forecasting & Assessment Methods, 2004.

[13] Saritas, O. and Aylen, J. "Using scenarios for roadmapping: the case of clean production", *Technological forecasting and social change*, 77(7). 1061-1075. 2010.

[14] Rikkonen, P. and Tapio, P. "Future prospects of alternative agro-based bioenergy use in Finland—constructing scenarios with quantitative and qualitative Delphi data", *Technological Forecasting and Social Change*, 76(7), 978-990, 2009.

[15] Organization For Economic Co-Operation And Development (OECD) and International Energy Agency(IEA), Energy to 2050: Scenarios for sustainable future, France, 2003.

[16] Hirsch, S., Burggraf, P., and Daheim, C. "Scenario planning with integrated quantification–managing uncertainty in corporate strategy building", *foresight*, 15(5), 363-374, 2013.

[17] UNIDO, *Technology Foresight Manual – Volume 01*, 2005. [E-book], Available: https://www.unido.org/foresight/registration/ dokums_raw/volume1_unido_tf_manual.pdf .

[18] Ricard, L. and Borch, K.., "From Future Scenarios to Roadmapping: A practical guide to explore innovation and strategy", *The 4th International Seville Conference on Future-Oriented Technology Analysis (FTA)*, May 12-13, 2011, 2012.

[19] Overland, E. "The Importance of Developing Long-Term Visions for Countries. The Norway2030-Vision: Lessons for Turkey", *Istanbul Forum*, 2003.

[20] Heinimö, J., Ojanen, V. and Kässi, T. "Views on the international market for energy biomass in 2020: results from a scenario study", *International Journal of Energy Sector Management*, 2(4), 547-569, 2008.

[21] dos Santos, M.F.R.F., Borschiver, S., and Couto, M.A.P.G. "The Application of Technology Roadmapping Method Related to Sustainable Energy: Case Study in the World and Brazil", *Chemical Engineering and Science*, 1(4), 67-74, 2013.

[22] Phaal, R., Farrukh, C.J.P., and Probert, D.R. "Technology roadmapping—a planning framework for evolution and revolution", *Technological forecasting and social change*, 71(1), 5-26, 2004.

[23] Kappel, T. "A. Perspectives on roadmaps: how organizations talk about the future", *The Journal of Product Innovation Management*, 18, 39-50, 2001.

[24] SLSEA, *A Guide To The Project Approval Process For On-Grid Renewable Energy Project Development*, Sustainable Energy Authority of Sri Lanka, Colombo, 2011.

[25] Ceylon Electricity Board, *Long Term Generation Expansion Plan 2013-2032*, CEB, Colombo, 2013.

[26] Loveridge, D. "The STEEPV acronym and process-a clarification", *Ideas in Progress, Paper,* 29, 2002.

[27] Fogg, B.J, *Persuasive technology: using computers to change what we think and do*, Morgan Kaufmann Publishers, Boston, 2003, 30-35.

[28] Hirsh, H., Coen, M.H., Mozer, M.C., Hasha, R. and Flanagan, J.L, "Room service, AI-style," *IEEE intelligent systems*, 14 (2). 8-19. Jul.2002.

[29] T. Eckes, *The Developmental Social Psychology of Gender*, Lawrence Erlbaum, 2000. [E-book] Available: netLibrary e-book.

27

Low-Power Air Conditioning Technology with Cold Thermal Energy Storage

Leila Dehghan[*], **Ahmad Fakhar**

Department of Mechanical Engineering, Faculty of Engineering, Azad University of Kashan, Iran
*Corresponding author: lleiladehghan1385@yahoo.com

Abstract Air conditioning of buildings is responsible for a large percentage of the greenhouse and ozone depletion effect, as refrigerant harmful gases are released into the atmosphere from conventional cooling systems. The vapor compression refrigeration is one of the many refrigeration cycles and is the most widely used method for air-conditioning of buildings. On the other hand, solar thermal energy can be used to efficiently cool in the summer. Single, double or triple iterative absorption cooling cycles are used in different solar thermal cooling system designs. Absorption chillers operate with less noise and vibration than compressor-based chillers, but their capital costs are relatively high. In this study, a system is proposed as a combination of the aforementioned systems and the power consumption is minimized using cold thermal energy storage (CTES).

Keywords: building cooling system, cold thermal energy storage

1. Introduction

Cooling of buildings can be achieved at very different energy consumption, ranging from zero energy for purely passive over low-energy consumption for earth heat exchange up to high electrical energy requirements for active compressor chillers. The application of different systems depends strongly on the cooling load, which has to be removed. If a building cannot be cooled by passive means such as night ventilation or earth heat exchange alone, active cooling technologies have to be employed. Today, the dominant cooling systems are electrically driven compression chillers, which have a world market share of about 90%. The average coefficient of performance (COP) of installed systems is about 3.0 or lower and only the best available equipment can reach a COP above 5.0. To reduce the primary energy consumption of chillers, thermal cooling systems offer interesting alternatives, especially if primary energy neutral heat from solar thermal collectors or waste heat from cogeneration units can be used. The main technologies for thermal cooling are closed-cycle absorption and adsorption machines, which use either liquids or solids for the sorption process of the refrigerant. Absorption chillers today are available in the range of 5 to 20,000 kW. In the last few years some new developments have been made in the medium-scale cooling range of 10 to 50 kW for water/lithium bromide and ammonia/water absorption chillers [1,2].

Kalogirou et al [3] presented modelling and simulation of an absorption solar cooling system. The system is modelled with the TRNSYS simulation program and the typical meteorological year file containing the weather parameters of Nicosia, Cyprus. Florides et al [4] presented the modelling, simulation and total equivalent warming impact of a domestic-size absorption solar cooling system. The system consists of a solar collector, storage tank, a boiler and a LiBr–water absorption refrigerator. Experimentally determined heat and mass transfer coefficients were employed in the design and costing of an 11 kW cooling capacity solar driven absorption cooling machine which, from simulations, was found to have sufficient capacity to satisfy the cooling needs of a well-insulated domestic dwelling. The system was modelled with the TRNSYS simulation program using appropriate equations predicting the performance of the unit. Assilzadeh et al presented a solar cooling system that has been designed for Malaysia and similar tropical regions using evacuated tube solar collectors and LiBr absorption unit. The modeling and simulation of the absorption solar cooling system is carried out with TRNSYS program. The typical meteorological year file containing the weather parameters for Malaysia is used to simulate the system. A comparative study was carried out by Fong et al [5] for the five types of solar cooling systems for a typical office in the subtropical Hong Kong. The results were worked out with the emphasis of suitable system control and operation in response to the year-round changing climatic and loading conditions. Based on the best year-round total of primary energy consumption, the order of the five types of solar cooling systems is: solar electric compression refrigeration, solar absorption refrigeration, solar adsorption refrigeration, solar solid desiccant cooling, and solar mechanical compression refrigeration. Tsoutsos et al [6] the performance and economic evaluation of a solar heating and cooling system of hospital in Crete, is studied using the transient simulation program (TRNSYS). The meteorological yearfile exploited the hourly weather data where produced by 30-year statistical process. The required data were obtained by Hellenic National Meteorological Service. Vidal et al [7] carried out an study on the hourly simulation of an ejector cooling cycle assisted by solar energy. The system is simulated using the TRNSYS program and the typical meteorological year (TMY) file that contains the weather data from Florianópolis, Brazil.

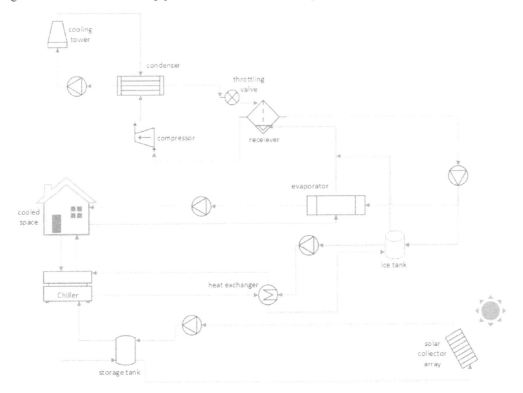

Figure 1. Schematics of the proposed cooling system by eliminating the adsorption chiller cooling tower

The ejector cycle uses R141b as the working fluid and a one-dimensional ejector is modelled in EES (Engineering Equation Solver). A full simulation model was developed by Eicker [8] for absorption cooling systems, combined with a stratified storage tank, steady-state or dynamic collector model and hourly resolved building loads. The model was validated with experimental data from various solar cooling plants. Sparber et al [9] reported that till 2007 there were 81 installed large scale solar cooling systems, eventually including systems which are currently not in operation. 73 installations are located in Europe, 7 in Asia, China in particular, and 1 in America (Mexico). 60% of these installations are dedicated to office buildings, 10% to factories, 15% to laboratories and education centers, 6% to hotels and the left percentage to buildings with different final use (hospitals, canteen, sport center, etc). They also cited that 56 installations are belong to absorption systems and the overall cooling capacity of the thermally driven chillers amounts to 9 MW 31% of it is installed in Spain, 18% in Germany and 12 % in Greece. Bong et al [10] designed and installed solar absorption chiller in Singapore. The system included 7 KW absorption chiller, heat pipe collectors with a total area of 32 m², a hot water storage tank, an auxiliary heater and a 17.5 KW cooling tower. They cited that the overall average cooling capacity provided was 4 KW, solar fraction of 39% and COP of 0.58. Balghouthi et al [11] accomplished a simulation using TRNSYS program in order to select and size different components of solar absorption chiller. They reported that solar absorption cooling systems were suitable for Tunisian's condition.

2. Description of the Proposed Systems

Figure 1 depicted two proposed cooling systems. In this scheme, the cooling tower for the adsorption chiller is replaced with a heat exchanger in which the water is cooled by recirculating the hot water in the ice storage tank.

In the design the ice is produced by pumping very cold liquid refrigerant HCFC-22 in commercial application through an array of pipes immersed in a tank of water. The tank is used as the heat sink for the adsorption chiller. The selected heat source for the chiller is the thermal energy from the sun. Technically, the design resembles process refrigeration. System components, that is, the compressor and condenser, pressure receivers, refrigerant pumps, evaporators, and ice tanks, are individually selected for the application and integrated to provide a reliable refrigeration system. Unlike direct-expansion systems, which rely on additional heat-transfer surface area to separate refrigerant vapor from liquid refrigerant, ice-on-pipe systems use a low-pressure receiver and a method called liquid overfeed to accomplish this. Chilled water or ice is produced by pumping cold liquid refrigerant to a chiller evaporator or an ice tank at a rate faster than that required to evaporate it. A two-phase solution of refrigerant liquid and vapor results, and is returned to the low-pressure receiver. This higher refrigerant flow rate is the reason for the system being referred to as liquid overfeed [12].

The refrigerant returned to the low-pressure receiver is nearly saturated. Refrigerant liquid that does not boil off in the evaporator is returned for a second pass. Note that the open or atmospheric design of the system dictates the use of a heat exchanger to separate ice water from the building cooling water loop. The cooling loop is normally a closed system [12].

2.1. Vapor Compression Cooling Cycle

The cycle used in the design called the ideal vapor-compression refrigeration cycle, and it is shown schematically in the above figure. The vapor-compression refrigeration cycle is the most widely used cycle for refrigerators, air-conditioning systems, and heat pumps. It consists of four processes: Isentropic compression in a compressor, Constant-pressure heat rejection in a condenser, Throttling in an expansion device, Constant-pressure heat absorption in an evaporator. In an ideal vapor-compression refrigeration cycle, the refrigerant enters the compressor at state 1 as saturated vapor and is compressed is entropically to the condenser pressure. The temperature of the refrigerant increases during this isentropic compression process to well above the temperature of the surrounding medium. The refrigerant then enters the condenser as superheated vapor at state 2 and leaves as saturated liquid at state 3 as a result of heat rejection to the surroundings. The temperature of the refrigerant at this state is still above the temperature of the surroundings.

The saturated liquid refrigerant at state 3 is throttled to the evaporator pressure by passing it through an expansion valve or capillary tube. The temperature of the refrigerant drops below the temperature of the refrigerated space during this process. The refrigerant enters the evaporator at state 4 as a low-quality saturated mixture, and it completely evaporates by absorbing heat from the refrigerated space. The refrigerant leaves the evaporator as saturated vapor and reenters the compressor, completing the cycle. In a household refrigerator, the tubes in the freezer compartment where heat is absorbed by the refrigerant serves as the evaporator. The coils behind the refrigerator, where heat is dissipated to the kitchen air, serve as the condenser.

2.1.1. High Pressure Side

On the high-pressure side of the system, the cold refrigerant vapor that collects at the top of the low-pressure receiver is drawn off by the compressor. After compression, the pressurized (and now hot) vapor passes to the condenser, where cooling tower water circulating through the shell causes the refrigerant to condense. The liquid refrigerant, still at high pressure, exits the condenser and enters a high-pressure receiver, where it is stored for later use. Refrigerant flow from the high-pressure receiver is regulated by a refrigerant metering device to ensure that a minimum liquid level is maintained in the low-pressure receiver.

This coil is actually a series of steel pipes immersed in a tank of water. Cold refrigerant (HCFC-22) is then pumped through these pipes to freeze the water that surrounds them. Bubbles flow around the steel pipes to agitate the water in the tank, sometimes by injecting air at the bottom. The rising air bubbles promote dense, even ice formation during the freezing cycle and uniform melting when the tank is discharged.

2.1.2. Low Pressure Side

The low-pressure receiver plays a critical role in the systems. It separates the two-phase refrigerant solution returning from the ice coil (or chiller evaporator) into liquid and vapor. Gravity induces this separation, causing the liquid refrigerant and oil to settle at the bottom of the receiver while pure refrigerant vapor collects at the top. As the compressor draws this vapor from the receiver, the liquid level falls. To ensure that there is always sufficient liquid in this vessel, a liquid level control adds refrigerant from the high-pressure receiver as needed.

Liquid overfeed systems require a separate oil return/recovery system. This is because the preferred compressor type (helical rotary/screw) expels significant amounts of oil into the discharge line. Entrained in the refrigerant, the oil makes its way through the condenser and high-pressure receiver, eventually ending up in the low-pressure receiver. There, the oil collects at the bottom of the tank (along with the liquid refrigerant) and cannot return to the compressor through the suction line. A separate oil-recovery system is needed to capture, distill, and return the oil to the compressor.

2.2. Solar Cooling System

When there is a source of inexpensive thermal energy at a temperature of 100°C to 200°C is absorption refrigeration. In this design, we used the most widely used absorption refrigeration system is the ammonia–water system, where ammonia (NH_3) serves as the refrigerant and water (H_2O) as the transport medium. Other absorption refrigeration systems include water–lithium bromide and water–lithium chloride systems, where water serves as the refrigerant. The latter two systems are limited to applications such as air-conditioning where the minimum temperature is above the freezing point of water.

It should be noted that this system looks very much like the vapor-compression system, except that the compressor has been replaced by a complex absorption mechanism consisting of an absorber, a pump, a generator, a regenerator, a valve, and a rectifier. Once the pressure of NH3 is raised by the components in the box (this is the only thing they are set up to do), it is cooled and condensed in the condenser by rejecting heat to the surroundings, is throttled to the evaporator pressure, and absorbs heat from the refrigerated space as it flows through the evaporator. So, there is nothing new there. Here is what happens in the box: Ammonia vapor leaves the evaporator and enters the absorber, where it dissolves and reacts with water to form NH3·H2O. This is an exothermic reaction; thus heat is released during this process. The amount of NH3 that can be dissolved in H2O is inversely proportional to the temperature. Therefore, it is necessary to cool the absorber to maintain its temperature as low as possible, hence to maximize the amount of NH3 dissolved in water. The liquid NH3-H2O solution, which is rich in NH3, is then pumped to the generator. Heat is transferred to the solution from a source to vaporize some of the solution. The vapor, which is rich in NH3, passes through a rectifier, which separates the water and returns it to the generator. The high-pressure pure NH3 vapor then continues its journey through the rest of the cycle. The hot NH3-H2O solution, which is weak in NH3, then passes through a regenerator, where it transfers some heat to the rich solution leaving the pump, and is throttled to the absorber pressure [13].

3. System Analysis

3.1. Vapor Compression Cooling

As described previously, a refrigeration system is considered which is operating between the pressure limits of 1.6 Mpa and 200 kpa with refrigerant-134a as the working fluid. The refrigerant leaves the condenser as a saturated liquid and is theflash chamber operates at 0.45 Mpa. Part of the refrigerant evaporates during this flashing

process, and this vapor is mixed with the refrigerant leaving the low-pressure compressor. The mixture is then compressed to the condenser pressure by the high-pressure compressor. The liquid in the flash chamber is throttled to the evaporator pressure and cools the refrigerated space as it vaporizes in the evaporator. The mass flow rate of the refrigerant through the low-pressure compressor is 0.11 kg. It is assumed that the refrigerant leaves the evaporator as a saturated vapor and the isentropic efficiency is 86% for the compressor. The results for this system and the one stage vapor compression system is shown in Table 1. It is clear that the choice of the overfeed system can increase the supply for the cooling load.

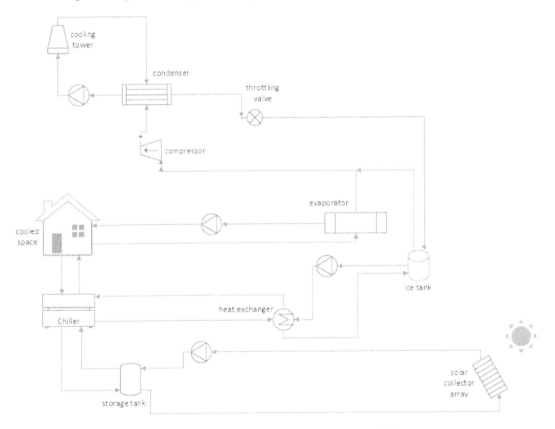

Figure 2. Single stage cooling system integrated absorption chiller

Table 1. Comparison of the One stage and overfeed systems

	Input power	Heat removal	COP
One stage	7.45 kW	19.33 kW	2.59
Overfeed	8.60 kW	18.54 kW	2.16
% changes	+15.4	-4.1	-16.6

3.2. Storage Analysis for Minimum Power

A MATLAB code is developed to simulate the overall system with the purpose that the power consumption of the system be minimized. It is achievable while the cold storage tank utilization is maximized. The thermal load can be found in Table 2. The simulation result is illustrated in Figure 3 where the energy delivery of the heat exchanger, evaporator and ice tank are compared.

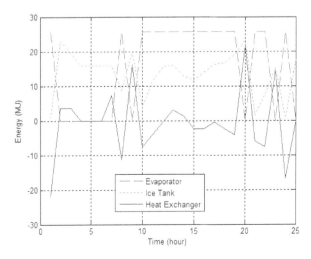

Figure 3. Simulation result for minimum system power consumption

Table 2. Thermal Load

Hour	Load
1	1
2	1
3	1
4	0
5	0
6	2
7	4
8	4.5
9	5
10	6
11	7
12	8
13	7.5
14	6.5
15	6.5
16	7
17	6.5
18	6
19	6
20	5.5
21	5
22	4
23	2.5
24	0

4. Conclusion

A MATLAB code is developed to simulate the daily performance of a vapor compression conditioning system integrated a cold thermal energy storage (ice storage). To minimize the overall power consumption of the conditioning system, cold storage side of the system should be used as much as possible because it just uses a small pump. Under certain load condition, the performance is simulated for the transferred energy through evaporator, ice tank and heat exchanger.

References

[1] Helm, M., Keil, C., Hiebler, S., Mehling, H., & Schweigler, C. (2009). Solar heating and cooling system with absorption chiller and low temperature latent heat storage: energetic performance and operational experience. *International journal of refrigeration, 32* (4), 596-606.

[2] Safarik, M., & Weidner, G. (2004). Neue 15 kW H2O-LiBr Absorptionskälteanlage im Feldtest für thermische Anwendungen. *Tagungsband, 3*, 159-171.

[3] Florides, G. A., Kalogirou, S. A., Tassou, S. A., & Wrobel, L. C. (2002). Modelling and simulation of an absorption solar cooling system for Cyprus. *Solar Energy, 72* (1), 43-51.

[4] Florides, G. A., Kalogirou, S. A., Tassou, S. A., & Wrobel, L. C. (2002). Modelling, simulation and warming impact assessment of a domestic-size absorption solar cooling system. *Applied Thermal Engineering, 22* (12), 1313-1325.

[5] Fong, K. F., Chow, T. T., Lee, C. K., Lin, Z., & Chan, L. S. (2010). Comparative study of different solar cooling systems for buildings in subtropical city. *Solar Energy, 84* (2), 227-244.

[6] Tsoutsos, T., Aloumpi, E., Gkouskos, Z., & Karagiorgas, M. (2010). Design of a solar absorption cooling system in a Greek hospital. *Energy and Buildings, 42* (2), 265-272.

[7] Vidal, H., Colle, S., & Pereira, G. D. S. (2006). Modelling and hourly simulation of a solar ejector cooling system. *Applied Thermal Engineering, 26* (7), 663-672.

[8] Eicker, U., & Pietruschka, D. (2009). Design and performance of solar powered absorption cooling systems in office buildings. *Energy and Buildings, 41* (1), 81-91.

[9] Sparber, W., Napolitano, A., & Melograno, P. (2007, October). Overview on worldwide installed solar cooling systems. In *2nd International conference on Solar Air Conditioning*.

[10] Bong, T. Y., Ng, K. C., & Tay, A. O. (1987). Performance study of a solar-powered air-conditioning system. *Solar Energy, 39* (3), 173-182.

[11] Balghouthi, M., Chahbani, M. H., & Guizani, A. (2005). Solar powered air conditioning as a solution to reduce environmental pollution in Tunisia. *Desalination, 185* (1), 105-110.

[12] Dincer, I., & Rosen, M. A. (2011). *Thermal Energy Storage: Systems and Applications.* John Wiley & Sons.

[13] Chinnappa, J. C. V., Crees, M. R., Srinivasa Murthy, S., & Srinivasan, K. (1993). Solar-assisted vapor compression/absorption cascaded air-conditioning systems. *Solar Energy, 50* (5), 453-458.

Permissions

List of Contributors

Sunday Olayinka Oyedepo
Mechanical Engineering Department, Covenant University, Ota3

Sana Ullah Khan
Sarhad University of Science & IT, PAKISTAN

Irfan Khan
Electrical Engineering, Sarhad University of Science & IT, PAKISTAN

Engr. Hashmat Khan and Engr. Qazi Waqar Ali
SARHAD University of Science & IT, PAKISTAN, MSc Electrical Power Engineering, BSc Electrical Engineering

Mohammad Sameti, Alibakhsh Kasaeian and Fatemeh Razi Astaraie
Department of Renewable Energies, Faculty of New Sciences and Technologies, University of Tehran, Tehran, Iran

Krishna Kumar, Omprakash Sahu
Department of Chemical Engineering, NIT Raipur, India

Mohammadreza Sedighi
Department of Mechanical Engineering, Islamic Azad University Nour Branch, Nour, Iran

Mostafa Zakariapour
Department of Mechanical Engineering, K.N.Toosi University of Technology, Tehran Iran

Pradipsaha, Md. FakhrulAlam, Ajit Chandra Baishnab and M. A. Islam
Department of Chemical Engineering and Polymer Science, Shahjalal University of Science and Technology (SUST), Sylhet, Bangladesh

MaksudurRahman Khan
Department of Chemical Engineering and Polymer Science, Shahjalal University of Science and Technology (SUST), Sylhet, Bangladesh
Faculty of Chemical and Natural Resources Engineering, University Malaysia Pahang, Gambang, Kuantan, Pahang, Malaysia

Seid Yimer and Omprakash Sahu
Department of Chemical Engineering, KIOT, Wollo University, Kombolcha, Ethiopia

Bezabih Yimer
Mersa Agricultural TVET College, Woldya, Ethiopia

Mohammad Sameti and Alibakhsh Kasaeian
Departmentof Renewable Energies, Faculty of New Sciences and Technologies, University of Tehran, Tehran, Iran

Seyedeh Sima Mohammadi
Department of Mechanical Engineering, South Tehran Branch, Islamic Azad University, Tehran, Iran

Nastaran Sharifi
Department of Mechanical Engineering, University of Zanjan, Zanjan, Iran

Mohammadjavad Mahdavinejad
Faculty of Art and Architecture, Tarbiat Modares University, Tehran, Iran

Sina Khazforoosh
Department of Architecture, University of Tehran-Kish International Campus, Iran

Danni Lei, Ting Yang, Baihua Qu, Jianmin Ma, Qiuhong Li, Libao Chen and Taihong Wang
Key Laboratory for Micro-Nano Optoelectronic Devices of Ministry of Education, Hunan University, Changsha, China
State Key Laboratory for Chemo/Biosensing and Chemometrics, Hunan University, Changsha, China

Askari. Mohammad Bagher
Department of Physics Azad University, North branch, Tehran, Tehran, Iran

Ajao K.R, Olabode O.F. and Sule O.
Department of Mechanical Engineering, University of Ilorin, Ilorin, Nigeria

Hossein Nasir Aghdam and Farzad Allahbakhsh
Department of Engineering, Ahar Branch, Islamic Azad University, Ahar, Iran

Seid Yimer, Omprakash Sahu
Department of Chemical Engineering, KIOT Wollo University, Kombolcha (SW), Ethiopia

M. Ghalamchi
Department of Energy Engineering, Science and Research Campus, Islamic Azad University, Tehran, Iran

M. Ghalamchi
Faculty of New Sciences &Technologies, University of Tehran, Tehran, Iran

T. Ahanj
Faculty of Nuclear Engineering, University of Shahid Beheshti, Tehran, Iran

Mohammad H. Ahmadi
Renewable Energies and Environmental Department, Faculty of New Science and Technologies, University of Tehran, Tehran, Iran

Hosyen Sayyaadi
Faculty of Mechanical Engineering-Energy Division, K.N. Toosi University of Technology, Tehran, Iran

Anuradha Tomar, Anushree Shrivastav, Saurav Vats, Manuja and Shrey Vishnoi
Department of Electrical & Electronics Engineering, Northern India Engineering College, New Delhi, India

Ibrahim Mabrouki, Zied Driss and Mohamed Salah Abid
Laboratory of Electro-Mechanic Systems (LASEM), National School of Engineers of Sfax (ENIS), University of Sfax, B.P. 1173

Ananda S. Amarasekara and Bernard Wiredu
Department of Chemistry, Prairie View A&M University, Prairie View, Texas, USA

Jun Xu
Department of Engineering Technology, Tarleton State University, Stephenville, USA

Abhik Milan Pal and Subhra Das
Renewable Energy Department, Amity School of Applied Science, Amity University, Gurgaon, India

N.B.Raju
National Institute of Solar Energy, Gurgaon, India

Mohammed Ben OUMAROU, Abdulrahim Abdulbaqi TOYIN and Fasiu Ajani OLUWOLE
Department of Mechanical Engineering, University of Maiduguri, PMB: 1069 Borno State, NIGERIA

Slah Driss, Zied Driss and Imen Kallel Kammoun
Laboratory of Electro-Mechanic Systems (LASEM), National School of Engineers of Sfax (ENIS), Univrsity of Sfax, Sfax, TUNISIA

Nnaemeka Vincent Emodi
College of Business, Law and Governance, James Cook University, P. O. Box 6811, Cairns Qld 4870, Australia

Nebedum Ekene Ebele
Department of Climate Change, Hallym University, Hallym University 1 Hallymdaehak-gil, Chuncheon, Gangwon-do, South Korea

Amila Withanaarachchi, Julian Nanayakkara and Chamli Pushpakumara
Department of Industrial management, University of Kelaniya, Sri Lanka

Leila Dehghan and Ahmad Fakhar
Department of Mechanical Engineering, Faculty of Engineering, Azad University of Kashan, Iran

Index